$$\aleph_0 = \infty$$

$$\aleph_0 < \aleph_1 = 2^{\aleph_0} < \aleph_2 = 2^{\aleph_1} < \aleph_3 = 2^{\aleph_2}$$

$$\aleph_i + \aleph_{i+1} = \aleph_{i+1}, \quad \aleph_i \cdot \aleph_{i+1} = \aleph_{i+1}$$

$$\mu(A) = \int_A f \, d\nu,$$

SETS MEASURES INTEGRALS

Dr. P. Todorovic

Library of Congress Control Number: 2011962854
ISBN: Hardcover   978-1-4691-3782-7
                 Softcover   978-1-4691-3781-0
                 Ebook   978-1-4691-3783-4

This book was printed in the United States of America.

To order additional copies of this book, contact:
Xlibris Corporation
1-888-795-4274
www.Xlibris.com
Orders@Xlibris.com
109504

SETS MEASURES INTEGRALS

Dr. P. Todorovic

$$\aleph_0 = \infty$$

$$\aleph_0 < \aleph_1 = 2^{\aleph_0} < \aleph_2 = 2^{\aleph_1} < \aleph_3 = 2^{\aleph_2}$$

$$\aleph_i + \aleph_{i+1} = \aleph_{i+1}, \quad \aleph_i \cdot \aleph_{i+1} = \aleph_{i+1}$$

$$\mu(A) = \int_A f \, d\nu$$

To my wife Živadinka

PREFACE

This book gives an account of a number of basic topics in set theory, measure, and integration. It is intended for graduate students in mathematics, probability and statistics, and computer sciences and engineering. It should provide readers with adequate preparations for further work in a broad variety of scientific disciplines.

The first chapter is an account of elements of set theory. It was written for readers having only modest background in this area of mathematics. Set theoretic methods have always constituted an important ingredient of mathematical reasoning. Our treatment of the subject is "informal," i.e., nonaxiomatic. That did not preclude extensive use of the Zermelo Axiom of Choice end its equivalents, the "Haüsdorff maximal principle" and the "Zorn lemma." A highly readable formal treatment can be found in "Axiomatic Set Theory" by Suppes.

The first three sections contain some basic notation and definitions. We also discuss briefly "paradoxes" brought out to the theory by the concept of infinity (the Russell paradox). The basic set theoretic operations and some properties of functions are also discussed in some detail (a formal definition of function, i.e., the definition in set theoretic terms is given in section 8).

The simplest infinite set is $N_+ = \{1, 2, \cdots\}$. To this set Cantor assigned the "transfinite" cardinal number $\aleph_0$. All sets equivalent to N_+ are said to be "countable" or "denumerable." Using ingenious arguments, Cantor established denumerability of many sets including the sets of rational and algebraic numbers, $N_+ \times N_+$ and many others. This is the subject of section 4.

One of the most remarkable achievements of G. Cantor was a proof that there are different "degrees" of infinity, one "larger" than another, something that was inconceivable at Cantor's time. With this discovery, the horizons of the subject were enormously expanded leading to the strange and remarkable world of the "Cardinal Arithmetic." This and related topics are discussed in sections 5, 6 and 7.

In section 8 we introduce the notion of "ordered pair" and prove it to be a legitimate set theoretic object. This enabled us to give a "proper" (set theoretic) definition of the Cartesian product and of a function. Various equivalent forms of the "Axiom of Choice" were outlined in section 9. In section 10, we investigate some ordering relations and their properties including the "well ordering" concept. The three maximal principles of set theory, the Haüsdorff maximal principles, the Zorn lemma, and the Kuratovski lemma are discussed in section 11. Their equivalence is proved in section 11 as well as the fact that that they imply the Axiom of Choice.

One of the crowning achievements of set theory is the theorem of Zermelo which states that every set can be well ordered, provided that the Axiom of Choice holds. A proof of this remarkable result is given in section 12. That the converse also holds is proved in section 13. A skeletal outline of the principle of "recursive definition" is given in section 14. In the rest of the chapter, we discuss some problems involving the concepts of "order type" (section 16) and of "ordinals" and "ordinal numbers" (section 17).

The turn of the last century saw the birth of "modern mathematics." Faced with the increasing use of the powerful methods of Cantorian set theory, Bair, Borel, Lebesgue, and other mathematicians of that time began to develop new approaches to mathematical analysis. Out of such efforts came the theory of measure (and integration) first to Euclidean spaces and later to abstract spaces by Radon, Young, Riesz, and Lebesgue himself. Other remarkable achievements of this period of time were the introduction of the concept of "outer measure" by Carathéodory and Hensel's discovery of p-adic numbers (circa 1900), which are essential for the development of set theory.

Chapter 2 contains reasonably comprehensive treatment of measure theory. Here we endeavor to emphasize more of those aspects of the theory which are most useful in applications, such as statistics or probability theory. For this reason our approach is based on σ-algebras rather than on σ-rings. Here is a brief sketch of its content.

Of the central importance here is the concept of "measurable space" $(X, \mathscr{A})$, where X is an arbitrary nonempty set and $\mathscr{A}$ is a σ-algebra of subsets of X. When X is finite or countably infinite $\mathscr{A}$ is usually $\mathscr{P}(X)$. Elements of $\mathscr{A}$ are called "measurable sets." Observe that there is no a preconceived idea of what "measurable" may mean, except that a subset $A \subset X$ is measurable if and only if $A \in \mathscr{A}$. For instance, θ and X are always measurable. Thus, measurability is purely a set theoretic concept which has nothing to do with any measure. The system $(X, \mathscr{A}, \mu)$, where $\mu: \mathscr{A} \to [0, \infty]$ is "axiomatically" σ-additive, is called a "measure space." The countably additive set function μ is a "measure" on $\mathscr{A}$.

Sections 17 and 18 are concerned with some same basic properties of measure. Since the concept of measure is an abstraction of the notions of length, volume, mass, etc., and these are defined on smaller classes, usually semialgebras, the problem here is that of extension. The problem of extending a measure defined on a semialgebra to the algebra generated by it is discussed in section 18.

Section 19 is of the central importance. Here we consider the problem of extending a measure μ defined on a semiring $\mathscr{S}$ of subsets of X (the length function λ, for instance) to a larger class (which is at least a σ-algebra) containing $\mathscr{S}$. The principal vehicle to achieve this is the Carathéodory concept of "outer measure" μ^* induced by μ, whose domain of definition is $\mathscr{P}(X)$. Then a subset $A \subset X$ is defined to be μ^*-measurable if

$$\mu^* (H) = \mu^* (H \cap A) + \mu^* (H \cap A^{\,c})$$

for all $H \in \mathscr{P}(X)$. It is an astonishing fact that the class $\mathscr{M}$ of all μ^*-measurable sets is a σ-algebra which contains $\mathscr{S}$. The restriction $\bar{\mu} = \mu^* \mid \mathscr{M}$ is a measure.

The concepts of "measurable cover" and "inner measure" are discussed in some detail in sections 20 and 21. The section 22 is concerned with various measures on the real line R, including the Lebesgue measure, Borel measure, and Lebesgue-Stieltjes measure. We also gave a short proof that there are subsets of R which are not Lebesgue measurable, provided that the Axiom of Choice holds.

Given measurable spaces $(X, \mathscr{A})$ and $(Y, \mathscr{B})$, a map $f: X \to Y$ is said to be "measurable" if

$$f^{-1} (\mathscr{B}) \subset \mathscr{A}.$$

Some basic properties of measurable functions are outlined in sections 23 and 24 for Y=R, the real line, and for $\mathscr{B}$ the Borel algebra of subsets of R. There we have established that the space $\mathbf{L}_0 = \mathbf{L}_0 (X, \mathscr{A})$ of measurable functions $f:X \to R$ is linear with the lattice structure. If $\{f_n\}_1^\infty$ is a sequence from $\mathbf{L}_0$ such that $f_n \to f$ pointwise on X then $f \in \mathbf{L}_0$. This result is the main thing we are after. The concept of "almost everywhere" (a.e) convergence of a sequence of measurable functions on $(X, \mathscr{A}, \mu)$ is discussed in section 25. Here we also proved Egorov's theorem and define the concept of "almost uniform" convergence.

The subject matter of section 26 is the concept of "convergence in measure" of a sequence $\{f_n\}_1^\infty$ of measurable functions. Here we study some features of this type of convergence and its relation to other modes of convergence. The following are two main results: First, if $\{f_n\}_1^\infty$ is a sequence of measurable functions from $\mathbf{L}_0 (X, \mathscr{A}, \mu)$ which is fundamental (Cauchy) in measure, then it converges in measure to a function $f \in \mathbf{L}_0$. Second, every sequence $\{f_n\}_1^\infty$ from $\mathbf{L}_0$ which converges in measure to a function $f \in \mathbf{L}_0$, contains a subsequence which converges (a.e) to f.

In chapter 3 we present elements of Integration theory. The Classical Integration theory as completed by Cauchy and Riemann by the end of eighteenth century has been, barring a few shortcomings, hugely successful in solving numerous problems in physics, theoretical mechanics, and engineering. One of its principal drawback is that the class of Riemann integrable functions defined on [a, b] (each such a function is bounded and (a.e) continuous) is incomplete. Most of these problems disappear in modern integration theory developed by many outstanding mathematicians of our times beginning with Henry Lebesgue.

The original work of Lebesgue, which was published in 1902, was concerned with integral on R with respect to the Lebesgue measure. As such it represents the single most important example, one which gave rise to the whole new theory.

An outline of the Lebesgue approach is presented in section 27 in the way that Lebesgue did it more than one hundred years ago. In this section we proved that every Riemann integrable function on $[a, b] \subset R$ is Lebesgue measurable and that its Riemann integral coincides with its Lebesgue integral.

The general theory of integral is discussed in section 28. Here we follow the oldest idea (sketched in the previous section) by defining the integral "axiomatically" for a class of simple functions. This construction is equivalent to the Lebesgue construction when we deal with step functions. Sections 29 and 30 deal with integration of positive functions. The results of central importance here are the all-powerful Lebesgue monotone convergence theorem and the Fatou lemma. The Banach space $\mathbf{L}_p$ was introduced in section 31.

A signed measure μ on a measurable space $(X, \mathscr{A})$ is a σ-additive set function such that $\mu:A \to (-\infty, \infty]$. In section 32, the Hahn decomposition theorem was proved which states that a signed measure splits the space X into two disjoint measurable parts one of which, say A, is μ-positive, meaning that $\mu(A_0) \geq 0$ for every $A_0 \in \mathscr{A} \cap A$, and another which is μ-negative. This decomposition is not unique. Using this result, Jordan proved that each signed measure μ is the difference of two measures μ^+ and μ^-, i.e. $\mu = \mu^+ - \mu^-$ such that μ^- is finite. Various features of these two measures and their relationships are also studied. Notice that this decomposition of μ is unique.

Let μ and ν be two measures on a measurable space $(X, \mathscr{A})$. If they are σ-finite, it was proved by Lebesgue that there exist two measures μ_1 and μ_2 with $\mu_1 << \nu$, $\mu_2 \perp \nu$ and $\mu = \mu_1 + \mu_2$. Here, for every $A \in \mathscr{A}$

$$\mu_2(A) = \mu(A \cap N) \qquad\qquad \mu_1(A) = \int_A f \, d\nu,$$

where $N \in \mathscr{A}$ is a fixed ν-null set and $f \in L_0(X, \mathscr{A}, \nu)$ is a non-negative uniquely defined up to equivalence. This is the main result of section 33. If $\mu << \nu$ it follows from this, (since $\mu_2(A) = 0$), that

$$\mu(A) = \int_A f \, d\nu.$$

This is the Radon-Nikodym theorem. In section 34 we also proved that the result holds if $(X, \mathscr{A}, \nu)$ is σ-finite measure space and μ an arbitrary measure on $\mathscr{A}$ such that $\mu << \nu$.

Recall that for $0 < p < \infty$, $L_p = L_p(X, \mathscr{A}, \mu)$ stands for the space of (numerical) measurable functions f for which $|f|^p$ is μ-integrable. Section 34 is concerned with some basic properties of L_p. Recall also that for $p \geq 1$, L_p is a normed linear space (see(6) of section 31) where the norm $\| \cdot \|_p$ of any $f \in L_p$, $1 \leq p < \infty$, is given by

$$\| f \|_p = (\int | f |^p \, d\mu)^{1/p}.$$

A sequence $\{f_n\}_1^\infty$ from L_p is said to converge in the mean of order p to a function $f \in L_p$ if $\| f_n - f \|_p \to 0$ as $n \to \infty$. According to the Riesz-Fisher theorem, L_p is a complete vector space. Observe that $\| f_n - f \|_p \to 0$ does not imply that $\| f_n \|_p \to \| f \|_p$. If, however, $f_n \to f$ (a.e) and $\| f_n - f \|_p \to 0$, then $\| f_n \|_p \to \| f \|_p$. Various inequalities involving $\| \cdot \|_p$ are proved. It appears that many of them are the convexity property. Of particular interest is the Jensen inequality. The principal result of section 34 is the Rietz theorem which states that every bounded linear functional Φ on L_p has the following representation: For any $f \in L_p$

$$\Phi(f) = \int f g \, d\mu$$

where $g \in L_q$ is fixed and $p + q = pq$.

Let $(X_1, \mathscr{A}_1, \mu_1)$ and $(X_2, \mathscr{A}_2, \mu_2)$ be measure spaces. Their product $(X, \mathscr{A}, \mu)$, where $X = X_1 \times X_2$, $\mathscr{A} = \mathscr{A}_1 \otimes \mathscr{A}_2$ and $\mu = \mu_1 \times \mu_2$ is also a measure space. This was proved in sections 36 and

37. The problem of integration of an $\mathscr{A}$-measurable function $h\colon X \to R$ was discussed in section 38. The principal result here is the Fubini theorem which states that if h is μ-integrable or $h \geq 0$, then its integral can be expressed in terms of the integrals with respect to μ_1 and μ_2. More specifically,

$$\int_X h\, d\mu = \int_{X_1} \left(\int_{X_2} h\, d\mu_2 \right) d\mu_1 = \int_{X_2} \left(\int_{X_1} h\, d\mu_1 \right) d\mu_2 .$$

Chapter 4 deals with infinite products of measurable and measure spaces. Given a family of measurable space $\{(X_t, \mathscr{K}_t); t\in T\}$, where the index set T is infinite, their product $(\Omega, \mathscr{B})$ is a measurable space, where $\Omega = \prod_T X_t$ and the σ-algebra $\mathscr{B}$ is constructed as follows: We call a measurable rectangle every subset $R \subset \Omega$ in the form

$$R = \prod_T A_t , \text{ where } A_t \in \mathscr{K}_t \text{ for all } t\in T,$$

and $A_t = X_t$ for all but finitely many $t\in T$. The class $\mathscr{S}^*$ of all measurable rectangles is a semialgebra and $\mathscr{B} = \sigma\{\mathscr{S}^*\}$. In essence, this is the content of section 39.

If each $(X_t, \mathscr{K}_t)$ is endowed with a measure μ_t the product

$$\mu = \prod_T \mu_t$$

is in general infinite on $\mathscr{S}^*$, unless each μ_t is a probability measure, i.e., $\mu_t(X_t)=1$ for all $t \in T$. In such a case $(X_t, \mathscr{K}_t, \mu_t)$ is referred to as probability space. The principal result of section 40 is proposition 3 which states that the set function μ is σ-additive on the algebra generated by $\mathscr{S}^*$. Thus, $(\Omega, \mathscr{B}, \mu)$ is a probability by space.

The rest of this chapter deals with elements of probability theory. Within the framework of mathematical disciplines the status of probability theory was somewhat uncertain until the publication of the Kolmogorov axiomatization in 1933, which is a milestone in the history of the subject. It was then recognized that the random variables are measurable functions on a probability space. The integral represented the "mathematical expectation" of the random variable. Thus, the measure theory provides a solid basis for determining what is deductive and what is inductive in a given probability situation. This formed the foundation for further development of the theory.

"Raison d'être" for probability theory is to provide mathematical tools for study of random phenomena. The nature of the problems that probability theory deals with gave rise to some new concepts regarding the structure of the probability measure μ, that do not exist in measure theory, such as independence, conditional probability, and conditional independence, the Markovian structure, and so on. In section 41, we present a mathematical analysis of a classical problem of gambling that played a central role in the evolution of probability theory.

Section 42 introduces more of probabilistic terminology and new concepts such as random variables and their distributions, joint and marginal. Finally, in section 43 we present the Kolmogorov construction.

CONTENTS

CHAPTER IV

Chapter I

SET THEORY

1. SET

What is a set? As George Cantor (1845–1918), the founder of modern set theory, famously explained,

"a set is a collection, regarded as an entity,

of separate objects of our perception or thought."

In other words, a set is a collection of objects, viewed as a single entity. This, of course, is not a formal definition of set but rather an attempt to convey an idea. The objects which constitute a set are called its members or elements. Some sets have special names. For instance, we say a "pride of lions", a "flock of sheep," or a "swarm of bees." Although a set is determined by its elements, much of set theory is not concerned with the nature of individual objects which constitute a set.

In this book sets are denoted by capital letters A, B, C, . . . while the elements of sets are denoted by the lower case letters a, b, c, ...Two sets A and B are said to be equal if and only if they consist of the same elements. We use the familiar symbol A=B to denote this fact. If two sets A and B are not equal, we write A≠B. Set is the first primitive notion of set theory.

The second primitive notion of set theory is the concept of "membership" or "belonging." To indicate that an object x is a member of a set K, we use the symbol (due to G. Peano) $x \in K$. We also write $x \notin K$ to indicate that x is not a member of the set K. The notion of set and membership in a set are fundamental in set theory in the sense that any other concept of the theory is defined in the terms of these two. For instance, the equality relation A=B is equivalent to

$$\forall x \, (x \in A) <=> (x \in B) => A = B$$

(which reads, for every x, x is a member of A if and only if it is a member of B).

If each member of a set A is also a member of a set B, we say that A is a "subset" of B and we write $A \subset B$. The "inclusion relation" $\subset$ so defined is obviously reflexive (i.e., $A \subset A$). In terms of the symbol $\in$, we have

$$A \subset B <=> \forall x \, (x \in A) => x \in B.$$

If we have that $A \subset B$ and at the same time that $B \subset A$, then necessarily $A = B$. In other words, we prove that $A = B$ by showing that every element of A is an element of B, and conversely, that every element of B is also an element of A.

Practical considerations compel us to introduce the notion of "empty" or "void" set, i.e., the set which has no elements. We use the symbol θ to denote this set. The empty set is a subset of every set, i.e., $\theta \subset A$ for every set A. For if it is not so, there exists an element of θ which does not belong to A. But this is obviously false since θ does not have elements. The phrase used in such a case is $\theta \subset A$ "vacuously." It is clear that θ is unique since if there are two empty sets, θ_1 and θ_2 then $\theta_1 \subset \theta_2$ and $\theta_2 \subset \theta_1$, implying that $\theta_1 = \theta_2$.

Any set which consists of a single element, say ω, is called "singleton" and is denoted by the symbol $\{\omega\}$. Notice that $\omega \in \{\omega\}$, $\theta \subset \{\omega\}$ and if $x \in A$ then $\{x\} \subset A$. If a set K has only finitely many elements,

$\alpha_1, \alpha_2, \ldots, \alpha_n$, we say that K is finite and write $K=\{\alpha_1, \alpha_2, \ldots, \alpha_n\}$. Due to their frequent occurrence, special symbols are reserved for certain sets. The symbol Z is used to denote the set of all integers (positive, negative, and zero):

$Z = \{\ldots -3, -2, -1, 0, 1, 2, 3, \ldots\}$

$N_+ = \{1, 2, 3, \ldots\}$

$\omega = \{0, 1, 2, \ldots\}$

Q = the set of all rational numbers.

R = the set of all real numbers, $(-\infty, \infty)$.

Since a set is determined by its elements, it seems only reasonable to define a set by specifying its elements when it is possible. This is the "constructive" method for set formation. Another method is by means of the "Axiom of Choice" which will be discussed later. The constructive method is based on the following principle: If Ω is a set and $\pi(\cdot)$ is a property, then the set A which consists of the elements of Ω having this property is defined by

$$A = \{\omega \in \Omega \; ; \; \pi(\omega) \text{ is true}\}$$

or abbreviated, $A=\{\omega \in \Omega; \pi(\omega)\}$. The set A can, of course, be empty if no element of Ω has the property $\pi(\cdot)$. For instance,

$$\{\omega \in \Omega; \omega \neq \omega\} = \theta, \quad \{\omega \in \Omega; \omega = \omega\} = \Omega \quad (\text{every } \omega \in \Omega \text{ is equal to itself}).$$

If we are given a property $p(\cdot)$ and it is desired to form the set of all objects having this property, i.e.,

$$\{x; p(x) \text{ is true}\},$$

the assumption that this is always possible leads to the well-known **"antinomy"** termed the **"Russell paradox."** Antinomies, in the technical sense, are those catastrophes of reason whereby the mind is compelled by logic itself to draw absurd conclusions (Rebecca Goldstein).

Russell's paradox

Denote by $\mathscr{L}$ the set of all sets, i.e.,

$$\mathscr{L} = \{A; A \text{ is a set}\}$$

Since by assumption $\mathscr{L}$ is a set, it follows that $\mathscr{L} \in \mathscr{L}$. From this it seems reasonable to conclude that some elements of $\mathscr{L}$ are members of themselves. Denote by

$$\mathscr{A} = \{A \in \mathscr{L}, A \notin A\}$$

Can it be that $\mathscr{A} \in \mathscr{L}$? If this is true, then either $\mathscr{A} \in \mathscr{A}$ or $\mathscr{A} \notin \mathscr{A}$. If $\mathscr{A} \in \mathscr{A}$, it clearly follows from its definition that $\mathscr{A} \notin \mathscr{A}$, which is absurd. If $\mathscr{A} \notin \mathscr{A}$, then it follows that $\mathscr{A} \in \mathscr{A}$, which is also absurd. Consequently, $\mathscr{A} \in \mathscr{L}$ is impossible, so that $\mathscr{A} \notin \mathscr{L}$.

The most important consequence of this conclusion is that there exists a set namely $\mathscr{A}$, which is not an element of $\mathscr{S}$, which is supposed to be set of all sets. Consequently, the "universal" set does not exist, or in other words, nothing contains everything.

To avoid paradoxes (brought out by the concept of infinity) in the process of sets formation, it is necessary to start with a set and then use the constructive principle which has been outlined in the previous paragraph.

Let Ω be a nonempty set. The collection of all subsets of Ω, including the empty set θ, is called **the power** of the sets Ω and is denoted by the symbol $\mathscr{P}(\Omega)$. Thus, $\mathscr{P}(\Omega)$ is a set whose elements are sets. In the following, the collections whose members are sets will be called "classes" and will be denoted by the capital script letters $\mathscr{A}$, $\mathscr{B}$, $\mathscr{S}$, etc. Of course, there are collections whose members are classes. If a set Ω consists of n elements, it is not difficult to see that its power set $\mathscr{P}(\Omega)$ consists of exactly 2^n subsets.

Remark 1

In modern set theory, the notion of set and membership in a set are fundamental in the sense that any other mathematical concept is defined in terms of these two. In other words, the set theory sees each mathematical object as a set. The empty set θ is postulated as an axiom.

Circa 1880, Giuseppe Peano devised a method to express the natural numbers 0, 1, 2, . . . in terms of the empty set θ. Peano defined 0 as the empty set θ. One was then defined as $\{\theta\}$. Two was defined as the set $\{\theta, \{\theta\}\}$, etc. In other words,

$$0 = \theta,\; 1 = \{\theta\},\; 2 = \{\theta, \{\theta\}\},\; 3 = \{\theta, \{\theta\}, \{\theta, \{\theta\}\}\}, \ldots.$$

Notice that each term in this sequence consists of the sets preceding it in the sequence. Notice also that

$$\mathscr{P}(\theta) = \{\theta\} = 1,\; \mathscr{P}(\mathscr{P}(\theta)) = \mathscr{P}(\{\theta\}) = \{\theta, \{\theta\}\} = 2, \ldots.$$

From this we have that

$$0 = \theta;\; 1 = \{\theta\} = \{0\};\; 2 = \{\theta, \{\theta\}\} = \{0, 1\};$$

$$3 = \{\theta, \{\theta\}, \{\theta, \{\theta\}\}\} = \{0, 1, 2\}, \ldots,$$

or in general

$$n = \{0, 1, 2, \ldots, n-1\}.$$

Note that each member of n is also a subset of n.

2. OPERATIONS ON SETS

To avoid paradoxes and to simplify our presentation, we shall, from now on, assume that all sets under considerations are subsets of a fixed set Ω, which we shall call "space." There are certain set theoretic operations that can be performed on the elements of $\mathscr{P}(\Omega)$.

Let $\{A, B\} \subset \mathscr{P}(\Omega)$ be arbitrary. The "difference" $A{-}B$ is the set which consists of those elements of A which do not belong to B. In symbols

$$A{-}B = \{\omega \in \Omega;\ \omega \in A \text{ and } \omega \notin B\}.$$

If $A = \Omega$, the difference is called the "complement" of B and is denoted by B^C, i.e.,

$$\Omega{-}B = B^C.$$

The "intersection" $A \cap B$ is the set of those elements ω which belong to both A and B. In symbols

$$A \cap B = \{\omega \in \Omega;\ \omega \in A \text{ and } \omega \in B\}$$

The sets A and B are said to be "disjoint" if they do not have common elements, i.e., if

$$A \cap B = \theta$$

The union $A \cup B$ is the set of elements each of which belongs to at least one of the sets A or B. In other words,

$$A \cup B = \{\omega \in \Omega;\ \omega \in A \text{ or } \omega \in B\}.$$

Each of the operations $\cap$ and $\cup$ is

i. Associative: $\qquad A \cup (B \cup C) = (A \cup B) \cup C,$

$\qquad\qquad\qquad\quad A \cap (B \cap C) = (A \cap B) \cap C.$

ii. Commutative: $\qquad A \cup B = B \cup A$ and $A \cap B = B \cap A.$

In addition, $\cap$ is distributive with respect to $\cup$, and $\cup$ is distributive with respect to $\cap$. In other words,

iii. Distributive: $\qquad A \cap (B \cup C) = (A \cap B) \cup (A \cap C),$

$\qquad\qquad\qquad\quad A \cup (B \cap C) = (A \cup B) \cap (A \cup C).$

The operation complement has the following properties which are easy to verify:

$$(A^C)^C = A;\ \theta^C = \Omega;\ \Omega^C = \theta;\ A \cup A^C = \Omega;\ A \cap A^C = \theta;$$

$$A{-}B = A \cap B^C;\ (A \cup B)^C = A^C \cap B^C;\ (A \cap B)^C = A^C \cup B^C.$$

The "symmetric difference" of two sets A and B, denoted by $A \triangle B$, is the set defined by

$$A \triangle B = (A{-}B) \cup (B - A).$$

The concepts of union and intersection of two sets can be extended to arbitrary many sets. For instance, if $\{A_t\} \subset \mathscr{P}(\Omega)$ is an arbitrary class of sets, the intersection $\bigcap_t A_t$ is the set which consists of all those elements $\omega \in \Omega$ which belong to every A_t. The union $\bigcup_t A_t$ is the set which consists of those $\omega \in \Omega$ that belong to at least one of the sets A_t. It is not difficult to establish that

$$\left(\bigcup_t A_t\right)^C = \bigcap_t A_t^c \qquad \left(\bigcap_t A_t\right)^C = \bigcup_t A_t^c,$$

which is known as the "De Morgan rule."

A sequence of nonempty disjoint sets $\{A_i\} \subset \mathcal{P}(\Omega)$ is said to form a partition of Ω if $\bigcup_i A_i = \Omega$. The union of any a sequence of sets can be written as the union of pairwise disjoint sets. For instance, if $\{A_i\}_1^\infty$ is a sequence from $\mathcal{P}(\Omega)$

$$\bigcup_{i=1}^\infty A_i = A_1 \cup \bigcup_{k=2}^\infty \left(A_k - \bigcup_{i=1}^{k-1} A_i\right).$$

Let $\{A_t; t \in I\} \subset \mathcal{P}(\Omega)$ be an arbitrary family of subset of Ω, where $I \subset R = (-\infty, \infty)$ is an index set. We shall often use notations

$$\bigcup_{t \in I} A_t = \sup_{t \in I} A_t, \qquad \bigcap_{t \in I} A_t = \inf_{t \in I} A_t.$$

If we have two index sets, I_1 and I_2, it seems reasonable to conclude that

$$\bigcap_{t \in I_1 \cup I_2} A_t = \left(\bigcap_{t \in I_1} A_t\right) \cap \left(\bigcap_{s \in I_2} A_s\right).$$

Now, if we suppose that $I_2 = \theta$, then since $I_1 \cup \theta = I_1$, to preserve the last identity we are compelled to assume that

$$\bigcap_{s \in \theta} A_s = \Omega.$$

In a similar fashion, we infer that

$$\bigcup_{s \in \theta} A_s = \theta.$$

With an arbitrary infinite sequence of sets $\{A_k\}_1^\infty$ from $\mathcal{P}(\Omega)$ we associate two limits, the set A_* and the set A^* defined by

$$A_* = \liminf A_n, \qquad A^* = \limsup A_n.$$

The set A_* consists of those elements ω which belong to almost all A_n (all but a finite number of them).

This set is called the limit inferior of $\{A_k\}_1^\infty$. It seems clear from this definition that

$$A_* = \bigcup_{k=1}^\infty \bigcap_{i=k}^\infty A_i$$

The second limit A^* is the set of all those elements ω which belong to infinitely many A_n. It is called the limit superior of $\{A_k\}_1^\infty$. It seems clear from its definition that

$$A^* = \bigcap_{k=1}^\infty \bigcup_{i=k}^\infty A_i$$

Since for every $k=1, 2, \ldots$

$$\bigcap_{i=k}^{\infty} A_k \subset \bigcup_{i=k}^{\infty} A_k \, ,$$

it is apparent that $A_* \subset A^*$. When $A^* = A_*$ we say that the sequence $\{A_k\}_1^{\infty}$ converges and the set A defined by

$$A = A_* = A^*$$

is its limit. In this case we write

$$A = \lim_{k \to \infty} A_k .$$

A sequence of sets $\{M_k\}_1^{\infty}$ from $\mathscr{P}(\Omega)$ is called "increasing" if

$$M_1 \subset M_2 \subset \dots .$$

On the other hand, if

$$M_1 \supset M_2 \supset \dots$$

the sequence $\{M_k\}_1^{\infty}$ is called "decreasing." Sequences of sets which are either increasing or decreasing are called "monotone."

Every monotone sequence converges. For instance, if $\{M_k\}_1^{\infty}$ is decreasing we have that

$$M_* = \bigcup_{k=1}^{\infty} \bigcap_{i=k}^{\infty} M_i = \bigcup_{k=1}^{\infty} \bigcap_{i=1}^{\infty} M_i = \bigcap_{i=1}^{\infty} M_i$$

$$M^* = \bigcap_{k=1}^{\infty} \bigcup_{i=k}^{\infty} A_i = \bigcap_{k=1}^{\infty} M_k .$$

Thus,

$$\lim_{k \to \infty} M_k = \bigcap_{i=1}^{\infty} M_i$$

If, on the other hand, $\{M_k\}_1^{\infty}$ is increasing, its limit is equal to

$$\lim_{k \to \infty} M_k = \bigcup_{k=1}^{\infty} M_k$$

We often use notation

$$M_k \downarrow \bigcap_{i=1}^{\infty} M_i$$

if the sequence of sets is decreasing and

$$M_k \uparrow \bigcup_{i=1}^{\infty} M_i$$

if the sequence of sets is increasing .

Problems and complements

1. Let $\{A, B, C, D\} \subset \mathscr{P}(\Omega)$; show that the following holds:

i. $(A - B) - (B - C) = (A - B)$;   ii. $(A - C) \cup (B - C) = (A \cup B) - C$;

iii. $(A-B) - (A- C) = (A \cap C) - B$;   IV. $A \cap (B \, \Delta \, C) = (A \cap B) \, \Delta \, (A \cap C)$;

V. $(\bigcup\limits_{k=1}^{n} A_k) \cap B = \bigcup\limits_{k=1}^{n} (A_k \cap B)$;   VI. $A \, \Delta \, (B \, \Delta \, C) = (A \, \Delta \, B) \, \Delta \, C$;

2. If $\{A_i\}_1^n$ and $\{B_i\}_1^n$ are subsets of Ω show that

     i. $\bigcup\limits_{i=1}^{n} A_i - \bigcup\limits_{i=1}^{n} B_i \subset \bigcup\limits_{i=1}^{n} (A_i - B_i)$;

     ii. $\bigcap\limits_{i=1}^{n} A_i - \bigcap\limits_{i=1}^{n} B_i \supset \bigcap\limits_{i=1}^{n} (A_i - B_i)$;

3. Show that

     i. if $A \subset B$, $\mathscr{P}(A) \subset \mathscr{P}(B)$.

     ii. if C, D are subsets of Ω, then

$$\mathscr{P}(C \cap D) = \mathscr{P}(C) \cap P(D);$$

$$\mathscr{P}(C \cup D) \supset \mathscr{P}(C) \cup P(D);$$

     Hint:

     i. if $G \in \mathscr{P}(A)$ then $G \subset A \Rightarrow G \subset B \Rightarrow G \in \mathscr{P}(B)$.

     ii. if $G \in \mathscr{P}(C) \cap \mathscr{P}(D) \Rightarrow G \subset C$ and $G \subset D \Rightarrow G \subset C \cap D \Rightarrow G \in \mathscr{P}(C \cap D)$.

On the other hand, if $G \in \mathscr{P}(C \cap D) \Rightarrow G \subset C \cap D \Rightarrow G \subset C$ and $G \subset D \Rightarrow G \in \mathscr{P}(C)$ and

$G \in \mathscr{P}(C) \Rightarrow G \in \mathscr{P}(C) \cap \mathscr{P}(D)$.

4. Let A_n be the set of all rational numbers whose absolute value is less then $\frac{1}{n}$. Show that

$$\bigcap\limits_{1}^{\infty} A_n = \{0\}.$$

     Hint:

     If Q is the set of all rational numbers then $A_n = (-\frac{1}{n}, \frac{1}{n}) \cap Q$.

5. Let B_n be the set of all positive rational numbers less then $\frac{1}{n}$. Show that $\bigcap\limits_{n=1}^{\infty} B_n = \theta$.

6. Prove De Morgan's rule.

     Hint:

     Let us show that

$$(\bigcap\limits_{j} A_j)^c = \bigcup\limits_{j} A_j^c.$$

Clearly, for each k

$$\bigcap_j A_j \subset A_k \;=>\; (\bigcap_j A_j)^c \supset A_k^c \;=>\; (\bigcap_j A_j)^c \supset \bigcup_k A_k^c.$$

On the other hand,

$$\omega \in (\bigcap_j A_j)^c \;=>\; \omega \notin \bigcap_j A_j \;=>\; \omega \notin A_j,$$

$$=>\; \omega \in A_j^c \;=>\; \omega \in \bigcup_j A_j^c.$$

Consequently,

$$(\bigcap_j A_j)^c \subset \bigcup_j A_j^c.$$

This combined with the first inclusion proves the claim .

7. Let $\{A_j\}_1^\infty$ and $\{B_k\}_1^\infty$ are subsets of Ω. Show that

i. $\displaystyle\bigcap_{j=1}^\infty (A_j \cap B_j) = (\bigcap_{j=1}^\infty A_j) \cap (\bigcap_{j=1}^\infty B_j)$;

ii. $\displaystyle\bigcup_{j=1}^\infty (A_j \cup B_j) = (\bigcup_{j=1}^\infty A_j) \cup (\bigcup_{j=1}^\infty B_j)$;

iii. $\displaystyle(\bigcap_{j=1}^\infty A_j) \cap (\bigcup_{k=1}^\infty B_k) = \bigcap_{j=1}^\infty \bigcup_{k=1}^\infty (A_j \cap B_k)$;

iv. $\displaystyle(\bigcap_{j=1}^\infty A_j) \cup (\bigcap_{k=1}^\infty B_k) = \bigcap_{j=1}^\infty \bigcap_{k=1}^\infty (A_j \cup B_k)$;

8. Show that

i. $\displaystyle\bigcap_{j=1}^\infty A_j \subset \liminf A_n \subset \limsup A_n \subset \bigcup_{j=1}^\infty A_j$

ii. $\limsup A_n^c = (\liminf A_n)^c$;

iii. if $A_n \subset B_n$ for all $n = 1, 2, \ldots$ then

$$\liminf A_n \subset \liminf B_n \text{ and } \limsup A_n \subset \limsup B_n.$$

iv. $\liminf (A_n \cap B_n) = (\liminf A_n) \cap (\liminf B_n)$;

v. $\limsup (A_n \cup B_n) = (\limsup A_n) \cup (\limsup B_n)$;

Hint:

iii. $\displaystyle A_j \subset \bigcup_{k=n}^\infty B_k$ for all $j = n, n+1, \ldots$: Thus $\displaystyle\bigcup_{j=n}^\infty A_j \subset \bigcup_{k=n}^\infty B_k.$

vi. Since $A_n \cap B_n \subset A_n$ and $A_n \cap B_n \subset B_n$ clearly then

$$\bigcap_{n=k}^\infty (A_n \cap B_n) \subset \bigcap_{n=k}^\infty A_n \text{ and } \bigcap_{n=k}^\infty (A_n \cap B_n) \subset \bigcap_{n=k}^\infty B_n. \text{ Therefore,}$$

$$\liminf (A_n \cap B_n) \subset \liminf A_n \text{ and } \liminf (A_n \cap B_n) \subset \liminf B_n.$$

From this we have that $\lim \inf (A_n \cap B_n) \subset (\lim \inf A_n) \cap (\lim \inf B_n)$.

On the other hand, if $\omega \in (\lim \inf A_n) \cap (\lim \inf B_n)$, ω belongs to almost all A_n and to almost all B_k. Consequently, it belongs to almost all $A_n \cap B_k$ and therefore to almost all $(A_n \cap B_n)$. Therefore $\omega \in \lim \inf (A_n \cap B_n)$.

9. Show that

    i. $(\lim \inf A_n) \cup (\lim \inf B_n) \subset \lim \inf (A_n \cup B_n)$;

    ii. $\lim \sup (A_n \cap B_n) \subset (\lim \sup A_n) \cup (\lim \sup B_n)$;

    Hint:

    i. $\lim \inf A_n \subset \lim \inf (A_n \cup B_n)$ and $\lim \inf B_n \subset \lim \inf (A_n \cup B_n)$.

10. Prove the following identities

    i. $B - \lim \sup A_n = \lim \inf (B - A_n)$;

    ii. $B - \lim \inf A_n = \lim \sup (B - A_n)$;

    iii. $A \,\Delta\, \lim \sup A_n \subset \lim \sup (A \,\Delta\, A_n)$;

11. Let $\{G_n\}_1^\infty$ be a disjoint sequence. Show that

$$\lim G_n = \theta.$$

    Hint

    $\lim \inf G_n = \theta$; on the other hand if $\omega \in \lim \sup G_n$ then $\omega \in \bigcup_{k=n}^{\infty} G_k$ for every $n = 1, 2, \ldots$. This is not possible because if $\omega \in \bigcup_{k=n}^{\infty} G_k$ it must belong to a G_N where $N \geq n$. Consequently, $\omega \notin \bigcup_{k=N+1}^{\infty} G_k$, and $\lim \sup G_n = \theta$.

3. FUNCTIONS

Informally, a "function" (the terms map, mapping, operator, and function are synonymous) f on a set X to a set Y is a rule which assigns to each $x \in X$ a unique member $y \in Y$, denoted by $y = f(x)$, and called the "value" of the f at x or the "image" of x in Y under f. We use symbol $f: X \to Y$ to express this fact. Often, instead of "f" we shall use the symbol $f(\cdot)$. The set X is called the "domain of definition" of $f(\cdot)$ and the set of values $f(x)$ taken by $f(\cdot)$ while x runs through X, i.e.,

$$\{y \in Y; \text{ there is } x \in X \text{ with } y = f(x)\}$$

is called the "range" of $f(\cdot)$. For any $A \subset X$, we define

$$f(A) = \{f(x); x \in A\} \subset Y \tag{1}$$

the "image" of A in Y under f. Obviously, $f(X)$ is the range of $f(\cdot)$

A map $f: X \to Y$ is said to be "injective" if for any $\{x_1, x_2\} \subset X$, $(x_1 \neq x_2)$, we have that $f(x_1) \neq f(x_2)$. We say that f is "surjective" if $f(X) = Y$. If $f(\cdot)$ is both surjective and injective, it is called "bijective." For an injective map f, we say that it is one-to-one; if $f(X) \subset Y$ we say that f maps X into Y, and if $f(X) = Y$, we often say that f maps X onto Y. A bijective map f is often called the "one-to-one correspondence."

Let $f: X \to Y$ be a map and $G \subset Y$. The "inverse image" $f^{-1}(G)$ in X is the set of all $x \in X$ such that $f(x) \in G$, i.e.,

$$f^{-1}(G) = \{x \in X; f(x) \in G\} \tag{2}$$

If $f(X) \subset Y$ and $G \subset Y - f(X)$ then $f^{-1}(G) = \theta$. Thus, a map f is surjective if and only if $f^{-1}(G) \neq \theta$ for any nonempty $G \subset Y$. In such a case we clearly have that

$$f(f^{-1}(G)) = G .$$

A map $f: X \to Y$ is bijective if and only if for each $y \in Y$, $f^{-1}(\{y\})$ is a singleton in X. Thus, every bijection f induces a map $h: Y \to X$, where $h(y)$ is the point $x \in X$ defined by $\{x\} = f^{-1}(\{y\})$. The map $h: Y \to X$ is called the "inverse" of $f(\cdot)$ and is denoted also by f^{-1}. Notice that f^{-1} is bijective since $(f^{-1})^{-1} = f$.

Let $f: X \to Y$ be an arbitrary map. As we have seen, it induces a map (denoted also by f)

$$f: \mathscr{P}(X) \to \mathscr{P}(Y) \quad \text{defined by (1).}$$

Some properties of this map are given in the following:

Proposition 1

For any $\{A, B\} \subset \mathscr{P}(X)$ we have:

i. If $A \subset B$, then $f(A) \subset f(B)$.

ii. $f(A \cup B) = f(A) \cup f(B)$.

iii. $f(A \cap B) \subset f(A) \cap f(B)$.

Proof:

i. If $x \in A$, $f(x) \in f(A)$; but since $x \in B$, $f(x) \in f(B)$.

ii. Since $A\subset A\cup B$ and $B\subset A\cup B$ clearly $f(A)\cup f(B)\subset f(A\cup B)$. On the other hand, if $y\in f(A\cup B)$ there exists $x\in A\cup B$ such that $y=f(x)$. Since $x\in A$ or B, $f(x)\in f(A)$ or $f(B)$. Thus, $f(A\cup B)\subset f(A)\cup f(B)$, which proves (ii).

iii. $A\cap B\subset A$ and $A\cap B\subset B$ so $f(A\cap B)\subset f(A)\cap f(B)$. ♥

For an arbitrary family $\{A_t\}\subset\mathscr{P}(X)$ we have, which is easy to verify, that

$$f(\bigcup_t A_t)=\bigcup_t f(A_t)\ \text{ and }\ f(\bigcap_t A_t)\subset\bigcap_t f(A_t).$$

It follows from (2) that the map f induces the inverse map

$$f^{-1}:\mathscr{P}(Y)\to\mathscr{P}(X),$$

which preserves the basic set theoretic operations. Specifically, $f^{-1}(\theta)=\theta$ and $f^{-1}(G^c)=(f^{-1}(G))^c$, for every $G\in\mathscr{P}(Y)$. If $\{G_t\}\subset\mathscr{P}(Y)$ we have

$$f^{-1}(\bigcap_t G_t)=\bigcap_t f^{-1}(G_t)\ \text{ and }\ f^{-1}(\bigcup_t G_t)=\bigcup_t f^{-1}(G_t)$$

For the combination of $f(\cdot)$ and $f^{-1}(\cdot)$ the following obviously holds: if $G\subset Y$ and $A\subset X$

$$f(f^{-1}(G))=G\cap f(X)\quad\text{and}\quad f^{-1}(f(A))\supset A. \tag{3}$$

Indeed (see (1)),

$$f(X)\cap G=\{f(x);\ x\in X\}\cap G$$
$$=\{f(x);\ x\in X\}\cap\{f(x);\ x\in f^{-1}(G)\}$$
$$=\{f(x);\ x\in f^{-1}(G)\}=f(f^{-1}(G))$$

The inclusion in (3) is obvious.

Let $A\subset X$; a map $f:X\to Y$ induces a new map $\hat{f}:A\to Y$ such that $\hat{f}(x)=f(x)$ for each $x\in A$. The new map $\hat{f}$ is called the "restriction" of $f(\cdot)$ to A and it is often designated by the symbol $f|A$. It is clear that $f|\theta=\theta$. Let $h:A\to Y$ be a map which coincides with $f(\cdot)$ on the set $A\subset X$, i.e., $h=f|A$, then the map $f(\cdot)$ is called an "extension" of h over X. The following result is often useful:

Proposition 2

Let X be a set and $\{A_t;\ t\in I\}\subset\mathcal{P}(X)$ such that $\bigcup_I A_t=X$. For each $t\in I$ let $f_t:A_t\to Y$ be given and assume that for any $s,t\in I$

$$f_t|A_s\cap A_t=f_s|A_s\cap A_t. \tag{4}$$

Then there exists one and only one map $f:X\to Y$ such that for every $t\in I$, $f|A_t=f_t$.

Proof:

For every $x\in X$ define $f(x)=f_t(x)$ if $x\in A_t$. If, in addition, $x\in A_s$ then the condition (4) yields $f(x)=f_t(x)=f_s(x)$. Thus, the map $f(\cdot)$ so defined on X is clearly an extension of each f_t over X. It is also unique because each $x \in X$ belongs to some A_t, so any map that satisfies the requirement (4) will have to assume the value $f_t(x)$ at x, that is the same value as f. ♥

Remark 1

If $\{A_t ; t \in I\}$ is a partition of X, i.e., $A_t \neq \theta$ for each $t \in I$ and $A_t \cap A_s = \theta$ for each $s \neq t$ from I, and if for each $t \in I$ a map $f_t : A_t \to Y$ is given, then there is a unique $f : X \to Y$ which is an extension of each f_t. Indeed, for each s, $t \in I$, $f_t | A_s \cap A_t = f_s | A_s \cap A_t = \theta$. Thus, the requirement (4) is satisfied.

Next, we shall introduce some special functions which are needed later. Let X be a set; with every $A \subset X$ we associate a function $I_A(\cdot)$ on X defined by

$$I_A(x) = \begin{cases} 1 & \text{if} \quad x \in A \\ 0 & \text{if} \quad x \notin A \end{cases}$$

This map is called the "indicator function" of the set A. The set of all maps $h : X \to \{0, 1\}$ is usually denoted by 2^X. It is clear that

$$2^X = \{I_A ; A \in \mathcal{P}(X)\} = \{0, 1\}^X \tag{5}$$

Let $f : X \to Y$ and $h : Y \to Z$ be two maps. The composition $\psi = h \circ f$ is the map $\psi : X \to Z$ defined by

$$\psi(x) = h(f(x)).$$

This map induces a map $\psi : \mathcal{P}(X) \to \mathcal{P}(Z)$ whose inverse ψ^{-1} is given by

$$\psi^{-1} = (h \circ f)^{-1} = f^{-1} \circ h^{-1}. \tag{6}$$

Here we have two compositions $h \circ f$ and $f^{-1} \circ h^{-1}$, and we claim that (6) holds. Indeed, given $C \subset Z$

$$x \in (h \circ f)^{-1}(C) <=> h(f(x)) \in C <=> f(x) \in h^{-1}(C) <=>$$
$$x \in f^{-1}(h^{-1}(C)) <=> x \in (f^{-1} \circ h^{-1})(C),$$

which proves (6).

The mapping $i_X : X \to X$ defined by $i_X(x) = x$ for every $x \in X$ is called the "identity" map on X. For a bijective map $f : X \to Y$ we have that $f^{-1}(f(x)) = x$. In other words, $f^{-1} \circ f = i_X$. If, on the other hand, we have two maps $f : X \to Y$ and $h : Y \to X$ such that

$$h \circ f = i_X \tag{7}$$

then necessarily f is injective and h is surjective. Indeed, if for any $x \neq x'$ from X we have that $f(x) = f(x')$, then

$$i_X(x) = h(f(x)) = h(f(x')) = i_X(x') => x = x'.$$

Surjectivity of h is obvious from (7).

Two sets X and Y are said to be "equivalent" if there exists a bijective map $f : X \to Y$. In this case, for every $A \subset X$

$$f(A^c) = (f(A))^c. \tag{8}$$

Indeed, $f(X) = Y$, $f(A) \cap f(A^c) = f(A \cap A^c) = \theta$ and $f(A) \cup f(A^c) = f(A \cup A^c) = f(X) = Y$.

Problems and complements

1. Let $f{:}X{\to}Y$ be arbitrary. The obvious inclusion $f(A{\cap}B){\subseteq}f(A){\cap}f(B)$ holds for any A, B$\subset$X. Give a counter example to disprove equality $f(A{\cap}B){=}f(A){\cap}f(B)$.

Hint:

Take X=[0,1], $f(\cdot)$=c (constant) A=[0, 0.5), B=[0.5, 1]; clearly then $f(A{\cap}B){=}\theta$, $f(A){\cap}f(B){=}\{c\}$.

2. For a mapping $f{:}X{\to}Y$ the following statements are equivalent:

(i) $f(\cdot)$ is an injection on X;

(ii) $f(A{\cap}B) = f(A){\cap}f(B)$;

(iii) $A{\cap}B = \theta$ implies $f(A){\cap}f(B) = \theta$.

Hint:

(i) => (ii); If $y{\in}f(A){\cap}f(B)$ then $y{\in}f(A)$ and $y{\in}f(B)$. Thus, there is $x_0 {\in}A$ with $f(x_0){=}y$ and $x_1 {\in}B$

with $f(x_1){=}$ y. Since $f(\cdot)$ is injective $x_0{=}x_1 {\in}(A{\cap}B)$, which implies that $y{=}f(x_0){\in}f(A{\cap}B)$. Consequently,

$f(A){\cap}f(B){\subset}f(A{\cap}B)$. Since the converse always holds, (i) => (ii) is true.

(ii) => (iii); it is obvious.

(iii) => (i); if $f(x_0) = f(x_1)$ for some $x_0 {\neq} x_1$ then by setting $A{=}\{x_0\}$ and $B{=} \{x_1\}$ we have $A{\cap}B{=}\theta$

while $\{f(x_0)\}{=}f(A){\cap}f(B)$ which is a contradiction. Thus , $f(x_0) {\neq} f(x_1)$.

3. Let $f{:}X{\to}Y$ be arbitrary. Show that for any $G_1, G_2{\subset}Y$ we have

(i) $f^{-1}(G_1^c) = (f^{-1}(G_1))^c$; (ii) $f^{-1}(G_1){\subset}f^{-1}(G_2)$ if $G_1{\subset}G_2$

(iii) $f^{-1}(G_1{\cap}G_2) = f^{-1}(G_1){\cap}f^{-1}(G_2)$; $f^{-1}(G_1{\cup}G_2) = f^{-1}(G_1){\cup}f^{-1}(G_2)$.

4. Prove that for any $f{:}X{\to}Y$

$f^{-1}(f(A)) \supset A$ for every $A{\subset}X$

Hint:

Since $f^{-1}(f\{x\}) \ni x$ for all $x{\in}X$,

$$f^{-1}(f(A)) = f^{-1}(\bigcup_{x{\in}A} \{f\{x\}\}) = \bigcup_{x{\in}A} f^{-1}\{f\{x\}\} \supset \bigcup_{x{\in}A} \{x\} = A$$

5. Show that for any $A{\subset}X$ and $G{\subset}Y$

$$f(f^{-1}(G){\cap}A) = G{\cap}f(A)$$

Hint:

Set $A{=}f^{-1}(G_0)$, then from (3) we have: $f(A){=}G_0{\cap}f(X)$. Thus,

$$f(f^{-1}(G) \cap A) = f(f^{-1}(G{\cap}G_0)) = G{\cap}G_0{\cap} f(X) = G{\cap} f(A)$$

6. If $f{:}X{\to}Y$ is surjective, show that there exists $h{:}Y{\to}X$ satisfying $f \circ h = i_Y$

Hint:

If f is surjection, then according to the axiom of choice, for every $y{\in}Y$ there is at least one $x{\in}X$ such that $f(x)=y$. Now define $h(y)=x$. Clearly then

$$f(h(y)) = f(x) = y.$$

7. Let A_1 and A_2 be two sets. The "Cartesian product" $A_1 \times A_2$ is by definition the set of all maps $f:\{1,2\}\to A_1\cup A_2$ such that $f(1)\in A_1$ and $f(2)\in A_2$. Show that $A_1 \times A_2 \neq \theta \Leftrightarrow A_1\neq\theta$ and $A_2\neq\theta$.

    Hint:

    In other words, $A_1 \times A_2$ is the set of all (ordered) pairs (x_1, x_2) where $x_1\in A_1$ and $x_2\in A_2$ (each $f\in A_1 \times A_2$ is a "vector" $f=(f(1), f(2))$). If $A_1 \times A_2 \neq\theta$, there exists a member $(x_1, x_2)\in A_1 \times A_2$ where $x_1\in A_1$ and $x_2\in A_2$ implying that $A_1\neq\theta$ and $A_2\neq\theta$. Conversely, if $A_1\neq\theta$ and $A_2\neq\theta$, there are $x_1\in A_1$ and $x_2\in A_2$; clearly then $(x_1, x_2)\in A_1 \times A_2\neq\theta$. It is clear that $\theta \times \theta = \theta$, $\theta \times A = A \times \theta = \theta$.

8. Extend the definition of Cartesian product to a finite collection of sets, say $\{A_i\}_1^n$. Is the product associative?

    Hint:

    The Cartesian product of the family $\{A_i\}_1^n$, denoted by $A_1 \times A_2 \times \ldots \times A_n$ is the set of all maps

$$f: \{1,2, \ldots, n\} \to \bigcup_{i=1}^{n} A_i$$

such that $f(i)\in A_i$ for each $i=1, 2, \ldots, n$. Clearly, each $f\in A_1 \times A_2 \times \ldots \times A_n$ is in the form

$$f = (f(1), f(2), \ldots, f(n)).$$

Strictly speaking, $((x_1, x_2), x_3)$ is different from the element (a set, as a matter of fact) $(x_1, (x_2, x_3))$. However, we shall always assume that

$$A_1 \times A_2 \times A_3 = (A_1 \times A_2) \times A_3 = A_1 \times (A_2 \times A_3)$$

4. COUNTABLE SETS

The notion of infinity has troubled philosophers and mathematicians since antiquity. In 1831, Gauss, in a letter to Schumacher, wrote : "I protest against an infinite quantity as an actual entity; this is never allowed in mathematics. The infinity is merely a 'façon de parler' (a manner of speaking)."

Approximately forty-five years later, Dedekind, the last student of Gauss, formalized the concept of infinity by defining a set to be infinite if it is equivalent to a proper subset of itself. In 1882, he sent a letter to Cantor informing him of his discovery. Dedekind's work in general and on analysis of continuum in particular, was one of Cantor's major sources of inspiration, which helped him to set the foundations of a solid mathematical theory of infinity. Cantor's groundbreaking discovery that there are different "degrees" of infinity revolutionized and expanded our understanding of this concept and lead to the modern set theory.

Let us recall that two sets A and B are **"equivalent"** if there exists a bijection $f:A \to B$. We use the symbol $A \sim B$ to denote this fact. If A and B are finite sets, equivalence means that they have the same number of elements. Since f^{-1} is also a bijection, $B \sim A$ so that the relation $\sim$ is symmetric. The relation $\sim$ is also reflexive since $A \sim A$. Finally, it is easy to verify that $A \sim B$ and $B \sim C$ imply that $A \sim C$, which is the "transitivity property" of $\sim$. To summarize, the relation $\sim$ is reflexive, symmetric, and transitive.

Remark 1

It is the common sense to call a set F finite if it contains exactly n elements, where $n \in N_+$. If F is not finite, it is infinite. This approach is reasonable and technically sound, and will be used subsequently. There are several non-numerical definitions of finitude. One of such definitions is due to Dedekind. **Let us recall Dedekind's definition of infinity: a set K is infinite if it is equivalent to a proper part of itself. If this is not the case, the set K is finite. It is interesting to mention that the numerical and Dedekind's definition are equivalent if we assume the Axiom of Choice (corollary 1 of section 5).** The set N_+ is clearly infinite in both senses. **Any set A which is equivalent to N_+ is said to be "countable."** In this case, there exists a bijection $f:N_+ \to A$ such that for each $i \in N_+$, $f(i)=a_i \in A$. Consequently, all the elements of the set A can be written as a sequence

$$A = \{a_1, a_2, a_3, \ldots\}.$$

Thus, all what is needed to show that an infinite set is countable is to find a way to write all its elements as a sequence.

Proposition 1

Every subset of a countable set is either finite or countable.

Proof:

An easy argument shows that it suffices to consider the set N_+. Let $A \subset N_+$ and denote by n_1 the least element of A. Let n_2 be the least element of $A - \{n_1\}$, and so on, inductively. If $n_1, n_2, \ldots, n_{k-1}$ are elements of A then n_k is defined as the least member of the set $A - \{n_1, n_2, \ldots n_{k-1}\}$. If for some $m \in N_+, A - \{n_1, n_2, \ldots, n_m\} = \theta$, it is clear that the set A is finite and that $\{n_1, n_2, \ldots, n_m\}$ are its elements. Otherwise, $A = \{n_1, n_2, n_3, \ldots\}$ which is obviously a countable subset of N_+. ♥

Corollary 1.

If C is an infinite set and $f:C \to N_+$ is an injection, then since $C \sim f(C)$ and $f(C) \subset N_+$, $f(C)$ is infinite and countable; hence, $C \sim N_+$.

Proposition 2

Let Λ be an infinite set. If $f:N_+ \to \Lambda$ is surjection then $\Lambda \sim N_+$.

Proof:

Given $\lambda \in \Lambda$, denote by $K_\lambda = f^{-1}(\{\lambda\})$, then $\{K_\lambda ; \lambda \in \Lambda\}$ is obviously a partition of N_+. Let D be a set which consists of exactly one element from each K_λ (here we invoke the Axiom of Choice). In other words, $D \cap K_\lambda = \{i\} \subset K_\lambda$. The restriction of $f(\cdot)$ to D, i.e., $f|D$ is clearly a bijection $f|D:D \to \Lambda$, so that $D \sim \Lambda$. Thus, $D \subset N_+$ is an infinite set and according to proposition 1, countable. This proves the proposition. ♥

Proposition 3

The set $N_+ \times N_+$ is countable.

Proof:

The map $\varphi:N_+ \times N_+ \to N_+$, defined by $\varphi(i, k)=2^i \cdot 3^k$ is an injection. For, if there exists $(j, n) \in N_+ \times N_+$ such that $2^i \cdot 3^k = 2^j \cdot 3^n$, then for i<j, we have by dividing both sides with 2^i, that $3^k = 2^{j-i} \cdot 3^n$. This is obviously impossible since on the left-hand side is an odd number and on the right-hand side is an even number. Thus, i<j is not possible. In a similar fashion we can show that j<i is not possible. Thus, i=j, and as a consequence it follows that $3^k = 3^n$, implying that k=n. This proves that the map φ is injective. The proof of our proposition now follows from corollary 1. ♥

Corollary 2

The Cartesian product of a finite number of copies of N_+ is also countable, i.e.,

$$\underbrace{N_+ \times N_+ \times ... \times N_+}_{n\text{ - tmes}} \sim N_+$$

for any $n \in N_+$. Consequently, so is the set of all sequences $(k_1, k_2, \ldots, k_n)$, where each $k_i \in N_+$, i=1, 2, . . . , n.

Corollary 3

Let $\{B_i\}_1^\infty$ be a sequence of countably infinite sets, where $B_k = \{b_{1k}, b_{2k}, b_{3k}, \ldots\}$ and consider the union

$$D = \bigcup_{k=1}^{\infty} B_k$$

Since the map $h: N_+ \times N_+ \to D$ defined by $h(i, j)=b_{ij}$, is obviously a surjection, the set D is countable.

Corollary 4

The set Q_+ of positive rational numbers is countable. Indeed, since each element of Q_+ is the ratio of two positive integers, the map $f:N_+ \times N_+ \to Q_+$ defined by $f(m, n) = m/n$ is clearly a surjection. Thus, according to proposition 2, Q_+ is countable. It is not difficult to see that the set Q of all rational numbers is also countable.

Let us recall that a root of a (nonzero) polynomial with integer coefficients is called an "algebraic" number. Denote by $\mathcal{A}$ the set of algebraic numbers. It is clear that $Q \subset \mathcal{A}$. The next proposition shows that $Q \sim \mathcal{A}$.

Proposition 4.

The set $\mathcal{A}$ of all algebraic numbers is countable.

Proof:

Since every polynomial

$$P_n(x) = a_n x^n + a_{n-1} x^{n-1} + \ldots a_1 x + a_0$$

is uniquely determined by its (integer) coefficients, according to corollary 2 the collection of all polynomials $P_n(\cdot)$ for n fixed, is countable. According to corollary 3, the collection of all polynomials is also countable. Finally, by the fundamental theorem of algebra $P_n(x)=0$ has n roots. Invoking again corollary 3, the assertion follows. ♥

Remark 2

The nonalgebraic numbers are called **"transcendental."** The elements of the set Q^c, often called **"irrational numbers,"** can be obtained in various ways from the set Q. Frequently, they are limits of the convergent sequences of rational number. For instance, it is clear that $\{(1+1/n)^n\}_1^\infty$ is a sequence from Q, whose limit is $e \approx 2.7182818284590$. To show that $e \notin Q$ we start with the well-known formula

$$e = \sum_{k=0}^{n} \frac{1}{k!} + \frac{t}{n \cdot n!}, \qquad \text{where } 0 < t < 1.$$

If $e \in Q$ we would have for some m, n from N_+ that

$$\frac{m}{n} = \sum_{k=0}^{n} \frac{1}{k!} + \frac{t}{n \cdot n!}.$$

Multiplying both sides with n!, we obtain on the left-hand side of this equation a positive integer and on the right-hand side the sum of an integer and of the fraction t/n, which is contradiction and which proves that $e \notin Q$. As a matter of fact, in 1873, Hermite proved that $e \notin \mathcal{A}$. This result provided an example to settle the long controversy whether transcendental numbers exist or not.

Problems and complements

1. Show that R ~ (0, 1)

Hint:

The mapping $f:R \to (0, 1)$ defined by

$$f(x) = \frac{1}{\pi} \arctan x + \frac{1}{2}$$

is a bijection.

2. Prove that A~B and B~C => A~C .

Hint:

If $f:A \to B$ and $h:B \to C$ are bijections then

$$h \circ f:A \to C$$

is also bijection.

3. Show that a set A is countable if and only if all its elements can be written in a form of a sequence.

Hint:

Assume that $A = \{a_1, a_2, \ldots\}$ then the mapping $f:A \to N_+$ defined by $f(a_n)=n$ is a bijection. Thus, $A \sim N_+$. Conversely, if $A \sim N_+$, there exists a bijection $h:N_+ \to A$. Denote by $a_1 = f(1)$, $a_2 = f(2)$, ... to obtain $\{a_1, a_2, \ldots\}=A$.

4. Show that the set G_n of all ordered n-tuples of the form $(i_1, i_2, \ldots, i_n)$ where each $i_j \in N_+$, $j=1, \ldots, n$, is countable.

Hint:

Clearly,

$$G_n = \{(i_1, \ldots, i_n); (i_1, \ldots, i_n) \in N_+ \times \ldots \times N_+\}.$$

5. Show that the class of all finite subsets of N_+ is countable.

Hint:

Denote by F_n the class of all subsets of N_+ consisting of n elements. Clearly, $F_1 \sim N_+$, and for $n \geq 2$ $F_n \sim N_+ \times \ldots \times N_+ \sim N_+$. Now the set

$$F = \bigcup_{n=1}^{\infty} F_n$$

is countable according to corollary 3.

6. Show that the set of irrational numbers is not countable.

Hint:

The set Q of rational numbers is countable. Denote by $I=R-Q$ the set of irrational numbers, then $R=I \cup Q$. If I is countable then $I \cup Q$ is countable, which is contradiction. Thus, the set I or irrational numbers is not countable.

7. Any real number which is not algebraic is called "transcendental." Show that the set of transcendental numbers is not countable.

Hint:

Let T⊂R be the set of transcendental numbers. Clearly T=R−$\mathcal{A}$ and since R=T∪$\mathcal{A}$, T obviously cannot be countable. It is of some interest to point out that in the year 1874, when Cantor proved the existence of transcendental numbers, he was not able to exhibit a single transcendental number.

8. Let $\mathcal{J}$ be the class of all disjoint nondegenerate intervals of the real line. Show that the class $\mathcal{J}$ is countable.

Hint:

Pick one rational number from each interval.

9. Show that $(0, 1] \sim (0, 1)$.

Hint:

The map h: $(0, 1] \to (0, 1)$ defined by

$$h(x) = \sum_{i=1}^{\infty} (3 \cdot 2^{-i} - x) \, I_{(2^{-i}, \, 2^{-i+1}]}(x)$$

is a bijection.

5. CARDINALS

After establishing that so many sets are countable, Cantor conjectured that all infinite sets are countable. Eventually, he recognized that his assumption was false when he tried and failed to prove that the set $(0, 1)$ is countable. The following result is due to Cantor.

Proposition 1

The set $(0, 1)$ is not countable.

Proof:

Assume to the contrary that $(0, 1)$ is countable, then all its elements can be written as a sequence $\{c_n\}_1^\infty$, where

$$c_1 = 0 \, . \, \alpha_{11} \, \alpha_{12} \ldots \alpha_{1n} \cdots$$

$$c_2 = 0 \, . \, \alpha_{21} \alpha_{22} \ldots \alpha_{2n} \cdots$$

$$\ldots\ldots\ldots\ldots\ldots\ldots\ldots\ldots\ldots\ldots$$

$$c_n = 0 \, . \, \alpha_{n1} \, \alpha_{n2} \ldots \alpha_{nn} \cdots$$

$$\ldots\ldots\ldots\ldots\ldots\ldots\ldots\ldots\ldots$$

$$\ldots\ldots\ldots\ldots\ldots\ldots\ldots\ldots\ldots$$

and α_{ik} is the k-th digit in the decimal expansion of c_i. Consider the decimal

$$0.\alpha_{11} \, \alpha_{22} \ldots \alpha_{nn} \cdots,$$

and replace α_{11} with $d_1 \neq \alpha_{11}$, where d_1 can be any of the digits $0, 1, \ldots, 9$. Next, replace α_{22} with $d_2 \neq \alpha_{22}$ where d_2 can be any of the digits $0, 1, \ldots, 9$, and so on. Write

$$b = 0, d_1 d_2 \ldots d_n \ldots.$$

It is clear that $b \neq c_i$ for any $i=1, 2, \ldots$ and since $b \in (0, 1)$ this clearly implies that the sequence $\{c_i\}_1^\infty$ does not contain all the numbers of the interval $(0, 1)$. Thus, $(0, 1)$ is not a countable set. ♥

As a consequence of this proposition it appears that there are different "degrees" of infinity one "larger" than another, something that at Cantor's time was inconceivable. Another fascinating discovery made by Cantor is formulated in the next proposition:

Proposition 2

$X \nsim \mathcal{P}(X)$ for any set X.

Proof:

If $X = \theta$, $\mathcal{P}(\theta) = \{\theta\}$; if $X \neq \theta$ is a finite set consisting of n elements, $\mathcal{P}(X)$ consists of 2^n subsets. To prove that the proposition holds for an infinite X, assume to the contrary that $X \sim \mathcal{P}(X)$. Then there exists a bijection $h: X \to \mathcal{P}(X)$. Now, consider the set

$$B = \{x \in X; \, x \notin h(x)\}.$$

Since $B \in \mathcal{P}(X)$ there exists an $y \in X$ such that $h(y) = B$. Either $y \in B$ or $y \notin B$. If $y \in B$, then $y \notin h(y) = B$, which is absurd. If, on the other hand, $y \notin B$, then $y \in h(y) = B$, absurd again. These contradictions show that the set B does not belong to the range $h(X)$ of the map h. Consequently, the map h cannot be a bijection. ♥

With recognition that there are different infinities, it became necessary to compare the "sizes" of infinite sets. But what is the size of an infinite set? The size of a finite set is the number of its elements. For an infinite set the answer is not quite clear. However, for comparison of two infinite sets this is not essential.

Let us recall that a set A is said to be countable if it is equivalent to the set N_+, i.e., if $A \sim N_+$. Clearly, any two countable sets are equivalent. We then say that they have the same "cardinality" and write

$$\text{Card } A = \text{Card } N_+.$$

The notion of "cardinality" of a set is one among many concepts introduced by Cantor. It represents (loosely speaking) an extension to infinite sets the notion of size of a finite set.

Definition 1

Let F and G be two arbitrary sets. If $F \sim G$ we say they have the same cardinality and we write Card F=Card G. It is clear from this definition that two finite sets are equivalent if and only if they have the same number of elements. Thus, if a set F consists of n elements, Card F=n. If F_1 and F_2 are finite sets with sizes n_1 and n_2 respectively, then Card F_1<Card F_2 if and only if n_1<n_2.

Cardinality of the set N_+ (or of any other countable set) is always denoted by the symbol $\aleph_0$ (read: aleph-zero). Thus, Card $N_+ = \aleph_0$. According to proposition 3 and corollary 2 of section 4

$$\text{Card } (N_+ \times N_+) = \aleph_0 \quad \text{and} \quad \text{Card } (N_+ \times N_+ \times \ldots \times N_+) = \aleph_0$$

for any finite number of factors N_+ in the Cartesian product. The following proposition is fundamental:

Proposition 3

Every infinite set S contains a countable subset.

Proof:

The proof in the framework of the naïve set theory is trivial. Select an element $x_1 \in S$; then from the set $S - \{x_1\}$ we select an element x_2, and so on. This procedure yields a sequence of elements of S, $x_1, x_2, \ldots$, which represents a countable subset of the set S.

The problem here is with the selection procedure of elements $x_1, x_2, \ldots$. The process of choosing an element from S is repeated infinitely many times, never to be completed, is somewhat vague. A proper proof requires the Axiom of Choice.

Since S is infinite, it contains for every $n \in N_+$ a subset consisting of n! elements. Denote by S_n the class of all subsets of S consisting of exactly n! elements. Clearly, $S_n \cap S_m = \theta$ for any $n \neq m$ from N_+. Now, according to the Axiom of Choice, there exists a class $\mathcal{K}$ consisting of exactly one member (set) from each S_n. In other words,

$$\mathcal{K} = \{M_1, M_2, \ldots\},$$

where each $M_n \in S_n$ consists of exactly n! elements . Obviously, the number of elements in $M_1 \cup M_2 \cup \ldots$

$\cup M_{n-1}$ is less than $(n-1) \cdot (n-1)! < n!$. Now consider

$$K_n = M_n - \bigcup_1^{n-1} M_i \, ;$$

clearly, $\{K_n\}_1^\infty$ is a sequence of finite pairwise disjoint subsets of S. Applying once more the Axiom of Choice, we obtain a set C which consists of exactly one element from each K_n. Thus $C \subset S$ is the required countable subset.   ♥

      If A and B are two sets and A is equivalent to a subset of B, the only logical conclusion from this is that

$$\text{Card } A \le \text{Card } B.$$

If, in addition, $A \not\sim B$, it seems only reasonable to conclude that Card A<Card B.

      Definition 2

**   If A and B are two sets such that (i) $A \not\sim B$; (ii) there is $B_0 \subset B$ such that $A \sim B_0$, then we postulate that**

Card A < Card B.

      For example, we have established that (0, 1) is not countable (proposition 1). On the other hand, $N_+ \sim (0, 1) \cap Q$ (Q being the set of rational numbers). Consequently,

$$\aleph_0 < \text{Card } (0, 1).$$

Cantor used the symbol **c** (for continuum) to denote the cardinality of the set (0, 1), i.e.,

$$\mathbf{c} = \text{Card } (0, 1).$$

Actually, every interval (a, b) has cardinality of continuum, for the map $f{:}(0, 1) \to (a, b)$ defined by $f(x)=(b-a)x+a$ is bijection. In fact,

$$\text{Card } R = \mathbf{c}$$

because the map $h{:}(0, 1) \to R$ defined by

$$h(x) = (2x-1)/x(1-x)$$

is clearly a bijection.

      According to proposition 2, for any set X we have

$$\textbf{Card X} < \textbf{Card } \mathcal{P}\textbf{(X)} \tag{1}$$

For, $X \not\sim \mathcal{P}(X)$ and X is equivalent to the subclass of $\mathcal{P}(X)$ consisting of all singletons of X.

      So far we have identified two cardinal numbers $\aleph_0$ and **c**. From (1) it seems apparent that there are sets which are neither finite or countable nor equivalent to R. For instance, the collection of all real functions defined on the interval [0, 1]. To show this denote by

$$R^{[0,\,1]} = \text{the collection of all maps } f{:}[0, 1] \to R.$$

Clearly, $2^{[0,\,1]} \subset R^{[0,\,1]}$, so that Card $2^{[0,\,1]} \le$ Card $R^{[0,\,1]}$. But

$$\text{Card } 2^{[0,\,1]} = \text{Card } \mathscr{P}([0, 1]),$$

so that according to (1)

$$c = \text{Card}\,([0, 1]) < \text{Card}\,\mathscr{P}([0, 1]) = \text{Card}\,2^{[0, 1]} \leq \text{Card}\,R^{[0, 1]}$$

Consequently,

$$c < \text{Card}\,R^{[0, 1]} = f$$

Later we shall show that the $f = \text{Card}\,2^{[0, 1]} = 2^{c}$

As far as the infinite sets are concerned, we have identified so far several classes of infinite sets, namely the class of countable sets, the class of sets having the cardinality of continuum, and the class of sets equivalent to the set $R^{[0, 1]}$. If we set Card $\theta = 0$, we are familiar thus far with cardinals

$$0\ 1, 2, \ldots, n, \ldots ; \aleph_0, c, f,$$

listed in increasing order.

Proposition 4

The set $S_0 = (N_+ \times N_+ \times \ldots) = N_+^{N_+}$ is not countable (actually, Card $S_0 = c$).

Proof:

By contradiction, if the set S_0 is countable, then all its elements can be written as a sequence, say

$\{z_k\}_1^{\infty}$, where for each $k=1, 2, \ldots$

$$z_k = (n_{1k}, n_{2k}, \ldots) \quad \text{and} \quad n_{ij} \in N_+.$$

But the sequence

$$z = (n_{11}+1, n_{22}+1, \ldots)$$

is obviously not a member of $\{z_k\}_1^{\infty}$, which contradicts assumption that $\{z_n\}_1^{\infty}$ contains all the sequences of natural numbers. This proves the proposition. ♥

Proposition 5

Let S be an infinite non countable set. If a finite or countable set C is removed from S, the remaining set is not countable.

Proof:

Let $R = S - C$, then $S = R \cup C$. If the set R is finite or countable, then $R \cup C$ is countable, which is a contradiction. ♥

Proposition 6

Let $\{A_k\}_1^{\infty}$ and $\{B_k\}_1^{\infty}$ be a sequence of disjoint sets, i.e., $A_i \cap A_j = \theta$ and $B_i \cap B_j = \theta$ if $i \neq j$. If $A_k \sim B_k$ for each $k=1, 2, \ldots$ then

$$\bigcup_k A_k \sim \bigcup_k B_k$$

Proof:

Since $A_k \sim B_k$ there exists a bijection $f_k : A_k \to B_k$, $k=1, 2, \ldots$. According to remark 1 of section 3 there exists a map

$$f: \bigcup_1^\infty A_k \;\to\; \bigcup_1^\infty B_k$$

which is an extension of each f_k over $\bigcup_1^\infty A_k$. Clearly, f is also bijection, proving the proposition.   ♥

Corollary 1

The sets R and S in proposition 5 are equivalent (proving that the numerical and Dedekind's definition of infinity are equivalent). In other words, every infinite set contains an equivalent proper subset. Indeed, let $A \subset R$ be countable and denote by K=R−A, then

$$R = K \cup A \quad \text{and} \quad S = K \cup (A \cup C)$$

Here K~K and A~A∪C; from this and proposition 6, it follows that R~ S.

Corollary 2

If M is an infinite set and C countable and such that C∩M=θ, then C∪M~M. Indeed, let $A \subset M$ be countable and consider K=M−A. Then M=K∪A and obviously M∪C=K∪(A∪C). Here K~K, A~(A∪C). The assertion follows from proposition 6.

What can we say about two sets X and Y if all we know is that X is equivalent to a subset of Y and that Y is equivalent to a subset of X? Cantor's conjecture was that X~Y, but he was not able to prove it. A proof was provided by F. Bernstein. The following is a precise formulation and a proof of Bernstein's equivalence theorem.

Proposition 7

Let A and B be two sets; if there is an injection f:A→B and also an injection h:B→A, then A~B.

Proof:

Without any loss of generality we may assume that A∩B=θ. Write A_1=A−h(B), then h(B)=A−A_1.

Set:

$$f(A_1) = B_1 \qquad\qquad h(B_1) = A_2$$

$$f(A_2) = B_2 \qquad\qquad h(B_2) = A_3$$

$$\cdots\cdots\cdots\cdots\cdots\cdots\cdots\cdots\cdots\cdots\cdots\cdots$$

$$\cdots\cdots\cdots\cdots\cdots\cdots\cdots\cdots\cdots\cdots\cdots\cdots$$

$$f(A_n) = B_n \qquad\qquad h(B_n) = A_{n+1}$$

$$\cdots\cdots\cdots\cdots\cdots\cdots\cdots\cdots\cdots\cdots\cdots\cdots\cdots .$$

Continuing this procedure, we obtain two sequences of disjoint sets $\{A_k\}_1^\infty$ and $\{B_k\}_1^\infty$. Since f is an injection and $f(A_n)=B_n$, it follows that $A_n \sim B_n$ for every n=1, 2, ... Thus, according to proposition 6

$$\bigcup_1^\infty A_k \;\sim\; \bigcup_1^\infty B_k$$

The sets $A^* = A - \bigcup_1^\infty A_k$ and $B^* = B - \bigcup_1^\infty B_k$ are equivalent. Indeed, since h is an injection

$$h(B^*) = h(B) - \bigcup_1^\infty h(B_k) = A - A_1 - \bigcup_2^\infty A_k = A - \bigcup_1^\infty A_k = A^*$$

Consequently, $A^* \sim B^*$. The proof now follows from

$$A = A^* \cup \left(\bigcup_1^\infty A_k\right) \quad \text{and} \quad B = B^* \cup \left(\bigcup_1^\infty B_k\right)$$

and from proposition 6. ♥

Corollary 3

According to the last proposition, if X and Y are two sets such that $\text{Card } X \leq \text{Card } Y$ and $\text{Card } Y \leq \text{Card } X$, then necessarily $\text{Card } X = \text{Card } Y$.

Corollary 4

No two of the following three relations can occur simultaneously: $\text{Card } A < \text{Card } B$, $\text{Card } A > \text{Card } B$ and $\text{Card } A = \text{Card } B$. Indeed, assume, for instance, that simultaneously $\text{Card } A < \text{Card } B$ and that $\text{Card } A > \text{Card } B$. The first relation means that $A \not\sim B$ and that $A \sim B_1$, where $B_1 \subset B$. The second one means $A \not\sim B$ and that $B \sim A_1$, where $A_1 \subset A$. But then according to the Bernstein theorem $A \sim B$, which is a contradiction.

Proposition 8

Let A_1, A_2, A_3 be sets such that

$$\text{Card } A_1 < \text{Card } A_2 \quad \text{and} \quad \text{Card } A_2 < \text{Card } A_3,$$

then $\text{Card } A_1 < \text{Card } A_3$.

Proof:

We must show that (i) $A_1 \not\sim A_3$ and that (ii) $A_1 \sim C$ where $C \subset A_3$.

(i) If $A_1 \sim A_3$ then since $A_1 \sim B \subset A_2$ it follows that $B \sim A_3$, which is impossible.

(ii) Clearly $A_1 \sim B \subset A_2$ while $A_2 \sim C \subset A_3$; thus, $B \sim C_1 \subset C$ so that $A_1 \sim C_1 \subset A_3$, which proves the assertion. ♥

The cardinals of infinite sets were called by Cantor the "transfinite" cardinals. If A is an infinite set, then for any $n \in N_+$, $n < \aleph_0 \leq \text{Card } A$ (every infinite set contains a countable subset). Let us recall that for every set X, $\mathcal{P}(X) \sim 2^X$ (see (5) of section 3). In an analogy with the finite sets, it is custom to write

$$\text{Card } \mathcal{P}(X) = 2^{\text{Card } X}. \tag{2}$$

Then, according to (1),

$$\aleph_0 < 2^{\aleph_0} < 2^{2^{\aleph_0}} < \dots. \tag{3}$$

An outstanding open problem concerning the transfinite cardinals is the following: Is there any cardinal between $\aleph_0$ and $2^{\aleph_0}$? The "continuum hypothesis" conjectures that the answer is no. Given an arbitrary infinite set X the generalized continuum hypothesis conjectures that there is no cardinal between Card X and Card $\mathcal{P}(X)$.

Remark 1

We have established that $\aleph_0 < \mathbf{c}$; since by the continuum hypotheses (which is taken for granted), there are no cardinals between $\aleph_0$ and $2^{\aleph_0}$, it is clear that $\mathbf{c} \geq 2^{\aleph_0}$. On the other hand, since $\mathbf{c} < \mathbf{f}$ we have that $2^{\aleph_0} \leq \mathbf{c} < \mathbf{f}$. From this it seems reasonable to conclude that

$$\mathbf{c} = 2^{\aleph_0}.$$

Later, in section 7, we shall give a proof of this conjecture. Until then we shall simply assume that it holds.

Remark 2

Bernstein's theorem is a simple consequence of the "comparability theorem" for ordinals (see corollary 4 of section 16)

Problems and complements

1. If X and Y are infinite sets and $f:X\to Y$ is an injection such that $Y-f(X)$ is at most countable, show that $X\sim Y$.

   Hint:

   Since $X\sim f(X)$

   $$Y = f(X)\cup(Y-f(X))\sim X\cup(Y-f(X))\sim X$$

2. Show that the following sets are equivalent:

   $$(0, 1), [0, 1], (0, 1], (-\infty, \infty), (a, b).$$

   Hint:

   Use corollary 2.

3. Let $f:X\to Y$ be a surjection; if X is countable so is Y.

   Hint:

   For any $y\in Y$, $f^{-1}(\{y\})\subset X$. Since for any $y_1\neq y_2$ from Y, $f^{-1}(\{y_1\})\cap f^{-1}(\{y_2\})=\theta$,

$\{f^{-1}(\{y\}); y\in Y\}$ is a partition of X, i.e.,

$$\bigcup_{y\in Y}f^{-1}(\{y\}) = X$$

It is now clear that Y must be countable.

4. If $A\not\sim B$ but $A\sim A^*$ and $B\sim B^*$, then $A^*\not\sim B^*$.

   Hint:

   If $A^*\sim B^*$ then $A\sim A^*\sim B^*\sim B$, so that $A\sim B$ which is a contradiction.

5. A cardinal is a common characteristic of the collection of mutually equivalent sets. Thus, in definition 2, it is implicitly assumed that Card $A <$ Card B implies Card $A^* <$ Card B^* for any $A^*\sim A$ and any $B^*\sim B$. Show this.

   Hint:

   There is $B_1\subset B$ such that $A\sim B_1$. Let $B_1^*\subset B^*$ satisfies $B_1\sim B_1^*$, then

   $$B_1^*\sim B_1\sim A\sim A^*$$

Thus, A^* is equivalent to a subset of B^*. To show that $A^*\not\sim B^*$, we proceed as follows. Assume that $A^*\sim B^*$. Then since $B\sim B^*$, we would have that

$$B\sim B^*\sim A^*\sim A,$$

which is a contradiction. Thus, $A^*\not\sim B^*$.

6. If A_0, A_1, and A_2 are sets such that $A_2\subset A_1\subset A_0$, and $A_2\sim A_0$, show that $A_1\sim A_0$.

   Hint:

   There exists a bijection $h:A_0\to A_2$, so that $h(A_0)=A_2$. Denote by $h(A_1)=A_3$, then clearly $A_1\sim A_3$. In general, we have

$$h(A_{2n}) = A_{2n+2} \Rightarrow A_{2n}\sim A_{2n+2} \qquad n = 0, 1, \ldots$$

$$h(A_{2n+1}) = A_{2n+3} \Rightarrow A_{2n+1} \sim A_{2n+3} \qquad\qquad n = 0, 1, \ldots .$$

Since $\{A_k\}_0^\infty$ is decreasing sequence we have that

$$h(A_{2n} - A_{2n+1}) = h(A_{2n}) - h(A_{2n+1}) = A_{2n+1} - A_{2n+3}$$

Consequently, $(A_{2n} - A_{2n+1}) \sim (A_{2n+1} - A_{2n+3})$ for all $n=0, 1, 2, \ldots .$ Set

$$D = \bigcap_{k=1}^{\infty} A_k : \text{ then } A_0 = D \cup (A_0 - A_1) \cup (A_1 - A_2) \cup (A_2 - A_3) \cup \ldots$$

$$= D \cup \left(\bigcup_{k=1}^{\infty} (A_{2k-1} - A_{2k}) \right) \cup \left(\bigcup_{k=0}^{\infty} (A_{2k} - A_{2k+1}) \right)$$

On the other hand,

$$A_1 = D \cup (A_1 - A_2) \cup (A_2 - A_3) \cup (A_3 - A_4) \cup \ldots$$

$$= D \cup \left(\bigcup_{k=1}^{\infty} (A_{2k-1} - A_{2k}) \right) \cup \left(\bigcup_{k=1}^{\infty} (A_{2k} - A_{2k+1}) \right).$$

This proves that $A_1 \sim A_0$.

7. Prove the proposition 7 by means of the previous exercise.

Hint:

Let A and B be sets such that $A \sim B_1 \subset B$ and $B \sim A_1 \subset A$. There is a bijection $h: B \to A_1$; denote by

$h(B_1) = A_2$. Clearly, $A \supset A_1 \supset A_2$ and $A \sim A_2 \sim B_1$. Next, since $A \sim A_1$ and since $A_1 \sim B$, we have that $A \sim B$.

8. If $f: X \to Y$ is a surjection show that

$$\text{Card } Y \leq \text{Card } X$$

Hint:

For any $y \in Y$ set $A_y = f^{-1}(\{y\})$. The collection $\{A_y ; y \in Y\}$ forms a partition of X such that

$f(A_y) = \{y\}$. There exists a set A_0 which consists of one element from each A_y (the Axiom of Choice).

Clearly, $A_0 \subset X$ and $A_0 \sim Y$, which proves the claim.

9. Show that the collection 2^{N_+} of all sequences whose elements are 0 or 1 is not countable.

Hint:

Let $f: \mathcal{P}(N_+) \to 2^{N_+}$ be the map defined as follows: With each $K \in \mathcal{P}(N_+)$ we associate a unique

sequence $f(K) = (\varepsilon_1, \varepsilon_2, \ldots \varepsilon_n, \ldots)$ where each $\varepsilon_i = 1$ if $i \in K$ and $\varepsilon_i = 0$ if $i \notin K$. For instance, if $K = \{1, 3, 5\}$

$$f(K) = (1, 0, 1, 0, 1, 0, \ldots).$$

This clearly indicates that $\mathcal{P}(N_+) \sim 2^{N_+}$.

10 If a set $A \sim A_0$ and a set $B \sim B_0$, show that $A \times B \sim A_0 \times B_0$.

Hint:

Let $f_1: A \to A_0$ and $f_2: B \to B_0$ be bijections, then the map $\Psi: A \times B \to A_0 \times B_0$ defined by

$$\Psi(a, b) = (f_1(a), f_2(b))$$

is clearly a bijection.

6. ELEMENTS OF THE CARDINAL ARITHMETIC

In this section we discuss in an informal fashion elements of the "**cardinal arithmetic.**" According to definition 2 of section 5, if A and B are two sets such that $A \not\sim B$ but $A \sim B_0$ where $B_0 \subset B$, then we postulate that Card A<Card B. This is one of the basic statements of the cardinal arithmetic. We now introduce the operations "addition", "product" and "exponentiation" of cardinals.

Let A and B be two disjoint sets: we define the sum of their respective cardinals as the cardinal of their union, i.e.,

$$\textbf{Card A + Card B = Card } (A \cup B). \tag{1}$$

For instance, if $A=(0, 1)$ and $B=[1, 2)$, then as we know Card $(0, 1)=\mathbf{c}$, Card $[1, 2)=\mathbf{c}$ and Card $(0, 2)=\mathbf{c}$, yielding $\mathbf{c}+\mathbf{c} = \mathbf{c}$.

Given any two sets A_1 and A_2 (not necessarily disjoint) we define the product of their cardinals as follows:

$$\textbf{(Card } A_1) \cdot (\textbf{Card } A_2) = \textbf{Card } (A_1 \times A_2). \tag{2}$$

Let us recall that a cardinal is a characteristic of a family of equivalent sets. Thus, in order that (1) makes sense we have to show that it holds for any other two disjoint sets A_0 and B_0 such that $A_0 \sim A$ and $B_0 \sim B$. But this is trivial since, according to the proposition 6 of section 5, $A \cup B \sim A_0 \cup B_0$. Thus,

$$\text{Card } (A_0 \cup B_0) = \text{Card } (A \cup B) = \text{Card A} + \text{Card B}$$

$$= \text{Card } A_0 + \text{Card } B_0,$$

and similarly for the product. It follows from (1) that the operation "addition" of cardinals is commutative. If we have three disjoint sets A_1, A_2 and A_3 then from $(A_1 \cup A_2) \cup A_3 = A_1 \cup (A_2 \cup A_3)$ we obtain that

$$\text{Card } (A_1 \cup A_2) + \text{Card } A_3 = \text{Card } A_1 + \text{Card } (A_2 \cup A_3)$$

(the operation "addition" is associative). Finally, since Card $(A_1 \times A_2)=$Card $(A_2 \times A_1)$ we clearly have that the product is commutative i.e.,

$$(\text{Card } A_1) \cdot (\text{Card } A_2) = (\text{Card } A_2) \cdot (\text{Card } A_1)$$

There is no difficulty about extending (1) to infinitely many summands. If $\{A_t ; t \in T\}$ is a family of pairwise disjoint sets, we define

$$\sum_T \text{Card } A_t = \text{Card } (\bigcup_T A_t) \tag{3}$$

and similarly for the product

$$\prod_T \text{Card } A_t = \text{Card } (\prod_T A_t). \tag{4}$$

Here we use the same symbol Π for the ordinary product and for the "Cartesian product." When factors under the product sign are sets, this is the Cartesian product. For instance, for a class $\{M_k ; k \in N_+\}$ we have:

$$\prod_k M_k = M_1 \times M_2 \times \dots$$

Consider again a family of pairwise disjoint sets $\{A_t ; t \in T\}$.

Proposition 1

If $A_s \sim A_t$ for every s, $t \in T$, then

$$\sum_T \text{Card } A_t = (\text{Card } T)(\text{Card } A),$$

where $A \sim A_t$.

Proof:

For every $t \in T$ we have a bijection $f_t : A \to A_t$. Let Ψ be the map $\Psi : T \times A \to \bigcup_T A_t$ defined by

$\Psi(t, a) = f_t(a)$. It is clear that Ψ is a bijection. Consequently, $T \times A \sim \bigcup_T A_t$; form this, (2) and (3) the

assertion follows. ♥

Let us consider some examples. From corollary 3 of section 4 we deduce that

$$\aleph_0 = \aleph_0 + \aleph_0 + \ldots$$

On the other hand, if we write $\{B_k\}_1^\infty = \{B_k ; k \in N_+\}$, where Card $B_k = \aleph_0$ for all k=1, 2, . . . , we obtain

from the last proposition that

$$\aleph_0 \cdot \aleph_0 = \aleph_0 + \aleph_0 + \ldots,$$

which implies that

$$\aleph_0 \cdot \aleph_0 = \aleph_0 .$$

From $R_+ = \bigcup_{k=1}^\infty [k-1, k)$ it follows that

$$c = c + c + \ldots .$$

On the other hand, by writing $R_+ = \{[k-1, k); k \in N_+\}$ we obtain from proposition 1 that

$$\aleph_0 \cdot c = c.$$

Let X and Y be two sets; recall that the collection of all maps $f : X \to Y$ is denoted by Y^X. We define

$$\text{Card } (Y^X) = (\text{Card } Y)^{\text{Card } X}. \tag{5}$$

Proposition. 2

Let X_1 and X_2 be two disjoint sets, then for any set Y

$$Y^{X_1 \cup X_2} \sim Y^{X_1} \times Y^{X_2} .$$

Proof

Let $f_i \in Y^{X_i}$, i=1, 2, then according to proposition 2 of section 3, there exists one and only one map

$f : X_1 \cup X_2 \to Y$ such that $f_i = f | X_i$. Thus, the map $\Psi : Y^{X_1 \cup X_2} \to Y^{X_1} \times Y^{X_2}$, defined by

$\Psi(f) = (f_1, f_2)$ is clearly a bijection , which proves the proposition. ♥

It is now clear that the following holds

$$\mathrm{Card}\,(Y^{X_1 \cup X_2}) = \mathrm{Card}\,(Y^{X_1} \times Y^{X_2}).$$

From this, (1), (2) and (5) we obtain that

$$(\mathrm{Card}\,Y)^{\mathrm{Card}\,X_1 + \mathrm{Card}\,X_2} \tag{6}$$
$$= (\mathrm{Card}\,Y)^{\mathrm{Card}\,X_1} \cdot (\mathrm{Card}\,Y)^{\mathrm{Card}\,X_2},$$

As we have seen, $c = 2^{\aleph_0}$; according to (6)

$$c \cdot c = 2^{\aleph_0} \cdot 2^{\aleph_0} = 2^{\aleph_0 + \aleph_0} = 2^{\aleph_0} = c.$$

In other words, $c \cdot c = c$.

Remark 1

A simple argument shows that equation (6) can be generalized as follows: Let $\{X_t\,;\, t \in T\}$ be a family of disjoint sets. Then

$$(\mathrm{Card}\,Y)^{\sum_T \mathrm{Card}\,X_t} = \prod_T (\mathrm{Card}\,Y)^{\mathrm{Card}\,X_t}. \tag{7}$$

If for every $s, t \in T$, $X_s \sim X_t$ we obtain from (7) that

$$(\mathrm{Card}\,Y)^{(\mathrm{Card}\,T)(\mathrm{Card}\,X)} = \{(\mathrm{Card}\,Y)^{(\mathrm{Card}\,X)}\}^{(\mathrm{Card}\,T)}$$

where $X \sim X_t$.

Here are some special cases: The set of all infinite sequences whose elements are 0 or 1 is $\{0, 1\}^{N_+}$ or simply 2^{N_+}. The cardinal number of this set is obviously $2^{\aleph_0}$. Let us show that $c^{\aleph_0} = c$; indeed,

$$c^{\aleph_0} = (2^{\aleph_0})^{\aleph_0} = 2^{\aleph_0} = c.$$

The set of all sequences whose elements are positive integers is, obviously $N_+^{N_+}$. We have that

$$\mathrm{Card}\,(N_+)^{N_+} = \aleph_0^{\aleph_0} = c.$$

Indeed, $c = 2^{\aleph_0} \le \aleph_0^{\aleph_0} \le c^{\aleph_0} = c$. The set R^{N_+} of all sequences of real numbers has also cardinality of continuum, for $\mathrm{Card}\,(R^{N_+}) = c^{\aleph_0} = c$. Finally, $\mathrm{Card}\,(R^R) = c^c = f$. Thus, the cardinality of all maps $f{:}R \to R$ is $f = c^c$.

Remark 2

We have established so far that

$$\text{(i) } \aleph_0 \cdot \aleph_0 = \aleph_0; \qquad \text{(ii) } \aleph_0 \cdot c = c; \qquad \text{(iii) } c \cdot c = c.$$

On the other hand, it follows from proposition 2 of the section 5 that for any set A

$$\mathrm{Card}\,A < 2^{\mathrm{Card}\,A}.$$

Consequently, the only way to get higher cardinals is through exponentiation. Since the set of all maps $f{:}R \to R$ is R^R, we have that

$$f = \mathrm{Card}\,(R^R) = c^c = (2^{\aleph_0})^c = 2^c. \tag{8}$$

Let A be a set; we have established so far that

$$\text{Card } (A \times A) = \text{Card } A,$$

if Card $A = \aleph_0$ or c. The fundamental theorem of the cardinal arithmetic asserts tat this is true for any infinite set.

Proposition 3

For any infinite set A

$$\text{Card } (A \times A) = \text{Card } A.$$

Since the proof of this proposition requires concepts of set theory which will be introduced later, it will be postponed until section 11. Until then it will be taken for granted.

Proposition 4

Given two cardinals α_1 and α_2, at least one of which is transfinite (i.e., a cardinal of an infinite set) then we have:

$$\text{(i) } \alpha_1 + \alpha_2 = \alpha_1 \cdot \alpha_2 = \max(\alpha_1, \alpha_2)$$

$$\text{(ii) If } \alpha_1 < \alpha_2 \text{ then } \alpha_2 - \alpha_1 = \alpha_2$$

Proof:

Assume that $\alpha_1 \leq \alpha_2$ then according to proposition 3, we have:

$$\alpha_2 \leq \alpha_1 + \alpha_2 \leq \alpha_2 + \alpha_2 = 2\alpha_2 \leq \alpha_2 \cdot \alpha_2 = \alpha_2,$$

as claimed. Next, let $X \subset Y$ and $X \nsim Y$, then according to (i)

$$\text{Card } Y = \text{Card } X + \text{Card } (Y\text{-}X) = \text{Card } (Y\text{-}X). \qquad \qquad ♥$$

Remark 3

The second part of the proposition generalizes proposition 5 of section 5. That is, removal from a set Y a set of a "smaller" cardinality does not reduce the cardinality of Y. Notice also that $A \times A \sim A$ implies

$$\text{Card } A \leq \text{Card } A + n \leq \text{Card } A + \text{Card } A = 2 \text{ Card } A$$

$$\leq n \text{ Card } A \leq \text{Card } A \cdot \text{Card } A = \text{Card } A.$$

Problems and complements

1. Show that the class of all intervals of R with rational end points is countable.

Hint:

Each such interval is determined by two rational numbers, say r and s, $r<s$. Thus the class is equivalent to the set

$$\{(r, s)\in Q\times Q; r<s\} \subset Q\times Q,$$

where $Q\subset R$ is the set of all rationals.

2. Let $\{E_k\}_1^\infty$ be a sequence of disjoint sets such that $\text{Card } E_n = c$ for all $n=1, 2, \ldots$. Show that

$$\text{Card}\left(\bigcup_{k=1}^\infty E_k\right) = c.$$

Hint:

Each $E_n \sim [n-1, n)$.

3. If $A\sim A_0$ and $B\sim B_0$, where $A\cap B=A_0\cap B_0=\theta$, show that $A\cup B\sim A_0\cup B_0$.

Hint:

There exist bijections $f_1:A\to A_0$ and $f_2:B\to B_0$, then the map $f:A\cup A_0 \to B\cup B_0$ defined by $f|A=f_1$ and $f|B=f_2$ is a bijection (proposition 2 of section 3). In other words, f is the unique extension of each f_i to $A\cup B$.

4. Show that $A\times B \sim A_0\times B_0$.

Hint:

By proposition 4

$\text{Card } A \cdot \text{Card } B = \text{Card } A + \text{Card } B = \text{Card } A_0 + \text{Card } B_0$

$= \text{Card } A_0 \cdot \text{Card } B_0$.

5. (Continuation). If $\aleph<\aleph'$ show that $\aleph^{\aleph'}=2^{\aleph'}$.

Hint:

Clearly, $\aleph^{\aleph'}\geq 2^{\aleph'}$; on the other hand, since $\aleph<2^{\aleph}$

$$(\aleph)^{\aleph'} \leq (2^{\aleph})^{\aleph'} = 2^{\aleph\cdot\aleph'} = 2^{\aleph'}.$$

6. Let X be an infinite set with $\text{Card } X = \aleph'$. Show that for any set $B\subset X$ with $\text{Card } B= \aleph<\aleph'$, we have that

$$\text{Card }(X-B) = \aleph'.$$

Hint:

$X = B\cup(X-B) \Rightarrow \text{Card } X = \text{Card } B + \text{Card }(B-X)$.

7. Let X be an infinite set and denote by $\mathscr{F}$ the collection of all finite subsets of X. Show that $\text{Card } X= \text{Card } \mathscr{F}$.

Hint:

For each $n\in N_+$, let $\mathscr{F}_n$ be the collection of all n-element subsets of X. Clearly, $\mathscr{F}_m\cap \mathscr{F}_n =\theta$ for all $m \neq n$. In addition Card $\mathscr{F}_n$ = Card X for all n=1, 2, ... Since $\mathscr{F} = \cup\{\mathscr{F}_n ; n \in N_+\}$, according to proposition 1

$$\text{Card } \mathscr{F} = \aleph_0 \cdot \text{Card } X = \text{Card } X.$$

8. Show that the class of all open intervals of R has the cardinality of continuum.

Hint:

Each interval of R is determined by two numbers s, $t \in R$ with s<t. Since for each s fixed, the set of ordered pairs $\Gamma_s = \{(s, t); t>s\} \sim (s, \infty)$, Card $\Gamma_s = c$. Taking into account that $\Gamma_u \cap \Gamma_v = \theta$ and that $\Gamma_u \sim \Gamma_v$, $\{\Gamma_s ; s \in R\}$ is a disjoint class. According to proposition 1 section 3

$$\text{Card } \bigcup_s \Gamma_s = \text{Card } R \cdot \text{Card } \Gamma_t = c \cdot c = c.$$

9. Show that the set C[0, 1] of all real continuous functions on [0, 1] has the cardinality of continuum.

Hint:

The set of constant functions on [0, 1] has the cardinality of continuum. Thus, Card C[0, 1] $\geq c$. Now write the set of all rational numbers from [0, 1] as a sequence $z_1, z_2, \ldots$; since for any $f \in C[0, 1]$,

$$(f(z_1), f(z_2), \ldots) \in R^{N_+},$$

and every continuous function is completely determined by its values on the set of rationales $\{z_i\}_1^\infty$, it follows that Card C[0, 1] $\leq c^{\aleph_0} = c$. Thus Card C[0, 1] $= c$.

7. CANTOR'S TERNARY SET AND FUNCTION

Recall that each $t \in [0, 1]$ has a unique **"binary expansion"** (see problem 1)

$$t = \sum_{i=1}^{\infty} 2^{-i} \varepsilon_i = . \varepsilon_1 \varepsilon_2 \varepsilon_3 \ldots \tag{1}$$

where $\varepsilon_i = 0, 1$ for all $i = 1, 2, \ldots$, except when t is in the form $m/2^n$, $m = 1, 2, \ldots, 2^n - 1$ (the "binary rationals"). Each such a t has two binary expansions

$$t = . \varepsilon_1 \varepsilon_2 \ldots \varepsilon_{n-1} \, 0 \, 1 \, 1 \ldots$$

$$t = . \varepsilon_1 \varepsilon_2 \ldots \varepsilon_{n-1} \, 1 \, 0 \, 0 \ldots .$$

In the following we shall adopt one with infinitely many 0's. With this convention, the representation (1) is unique.

Consider the set $\{0, 1\}^{N_+}$ of all sequences $(\varepsilon_1, \varepsilon_2 \ldots)$ and denote by $S_0 \subset \{0, 1\}^{N_+}$ the subset which consists of those members having all $\varepsilon_i = 1$ for $i \geq k$, where $k = 1, 2, \ldots$. For instance,

$$(\varepsilon_1, \varepsilon_2, \ldots, \varepsilon_{k-1}, 1, 1, \ldots) \in S_0.$$

It seems clear that the set S_0 is countable; thus (see corollary 3 of section 5),

$$\{0, 1\}^{N_+} - S_0 \sim \{0, 1\}^{N_+}.$$

Consequently, the map

$$\Psi : \{0, 1\}^{N_+} - S_0 \to [0, 1]$$

defined by

$$\Psi((\varepsilon_1, \varepsilon_2, \ldots)) = \sum_{i=1}^{\infty} 2^{-i} \varepsilon_i$$

is obviously a bijection, implying that

$$\text{Card } \{0, 1\}^{N_+} = 2^{\aleph_0} = \mathbf{c}, \tag{2}$$

as we conjectured (see remark 1 of section 5) earlier. In addition, since for any set X, $2^X \sim \mathcal{P}(X)$ we have for $X = N_+$ that

$$\text{Card } \mathcal{P}(N_+) = \mathbf{c}.$$

The Cantor set C is a subset of the interval [0, 1] which consists of those $t \in [0, 1]$ which have a "ternary expansion" in which the digit 1 does not occur. More precisely, the map

$$h : \{0, 2\}^{N_+} \to [0, 1] \tag{3}$$

defined as follows:

$$h((d_1, d_2, \ldots)) = \sum_{i=1}^{\infty} d_i 3^{-i}, \qquad \text{where } d_i = 0, 2$$

for every $i = 1, 2, \ldots$, is an injection. The image

$$h(\{0, 2\}^{N_+}) = C. \tag{4}$$

The Cantor set is endowed with many interesting properties which are not immediately obvious from (4). To elucidate somewhat its basic structure, we shall give a geometric construction of this set.

From the closed interval $[0, 1]$ remove the "open middle third", i.e., the interval $(1/3, 2/3)$. What is left is the closed set

$$F_1 = [0, 1/3] \cup [2/3, 1].$$

From each of the closed intervals of F_1 we remove the open middle third to obtain the closed set

$$F_2 = [0, 1/3^2] \cup [2/3^2, 3/3^2] \cup [6/3^2, 7/3^2] \cup [8/3^2, 1],$$

and so on. We claim that

$$C = \bigcap_1^\infty F_n . \tag{5}$$

To characterize the points of this set, we express each $t \in [0, 1]$ in the "ternary" system

$$t = \sum_{i=1}^\infty d_i 3^{-i} = .d_1 d_2 d_3 \ldots ,$$

each $d_i = 0, 1, 2$. For some $t \in [0, 1]$ there are two possible ternary expansions, for instance,

$$1/3 = .1\,0\,0 \ldots = .0\,2\,2 \ldots ; \quad 2/3 = .2\,0\,0 \ldots .$$

However, each $t \in (1/3, 2/3)$ is in the form

$$t = .1\,d_2\,d_3 \ldots .$$

Thus, if we remove $(1/3, 2/3)$, each $t \in F_1$ has ternary expansion with $d_1 \neq 1$. In a similar fashion we deduce that the ternary expansion of each $t \in F_2$ has $d_1 \neq 1$ and $d_2 \neq 1$, and so on. Consequently, the intersection $\bigcap_1^\infty F_n$ consists exactly of those $t \in [1, 0]$ whose ternary expansion does not contain the digit 1. This proves (5).

From (4) we have that Card $C = c$ and from (5) we conclude that C is compact. The set C is also dense in itself. Indeed, if

$$t = .d_1\,d_2\,d_3 \ldots , \quad d_i = 0, 2$$

is an arbitrary point of C, and $t_n \in C$ is the element whose ternary expansion differs from that of t only in the digit d_n, then if it was 0 for t, let it be 2 for t_n, and conversely. So we obtain a sequence $\{t_n\}_1^\infty$ of points from C such that $t_n \to t$. Consequently, each point of C is an accumulation point of C. Thus, C is a **"perfect"** set.

To obtain the closed set F_n we removed from F_{n-1} exactly 2^{n-1} open intervals each one with the same length 3^{-n}. Thus, the total length of all open intervals removed from the interval $[0, 1]$ to obtain F_n is

$$\frac{1}{3} + \frac{2}{3^2} + \frac{2^2}{3^3} + \ldots + \frac{2^{n-1}}{3^n} = 1 - (\frac{2}{3})^n .$$

This implies that the total length of the removed open intervals from $[0, 1]$ to obtain the set C is 1. Thus, the

Cantor set contains no intervals; in other words, its interior $\overset{0}{C}=\theta$, and consequently it is a **"nowhere dense"** subset of R.

Remark 1

Given any $t_1, t_2 \in C$ such that $t_1 < t_2$, then t_1 and t_2 are in different intervals of some closed set F_n. Consequently, the interval (t_1, t_2) contains an open interval which is a subset of the open set $[0, 1]-C$.

We are now in a position to define the Cantor function which, similarly to Cantor's set, is a rich source of counterexamples. Let η be a map

$$\eta : [0, 1] \to N_+ \cup \{\infty\}$$

defined by

$$\eta(t) = \inf \{k;\, d_k = 1\} \quad \text{and} \quad \eta(t) = \infty \quad \text{if } t \in C.$$

The function $\Phi: [0, 1] \to [0, 1]$ defined by

$$\Phi(t) = \sum_{i=1}^{\eta(t)-1} d_i 2^{-i-1} + 2^{-\eta(t)} \tag{6}$$

is called the Cantor function.

The function Φ is a nondecreasing continuous surjection. For instance, since $0, 1 \in C$, $\eta(0) = \eta(1) = \infty$ and consequently $\Phi(0) = 0$ and $\Phi(1) = 1$. It is easy to see that $\eta(t) = 1$ for every $t \in [1/3, 2/3]$. Consequently, $\Phi(t) = 1/2$ on this interval. As a matter of fact, it is constant on every "open middle third" interval removed from $[0, 1]$ in constructing the Cantor set **C** (see Fig. 1).

Remark 2

The p-adic representation of real numbers in $(0, 1)$ was discovered by Hensel (circa 1900). Given any $x \in (0, 1)$ and an integer $p \geq 2$ we can write

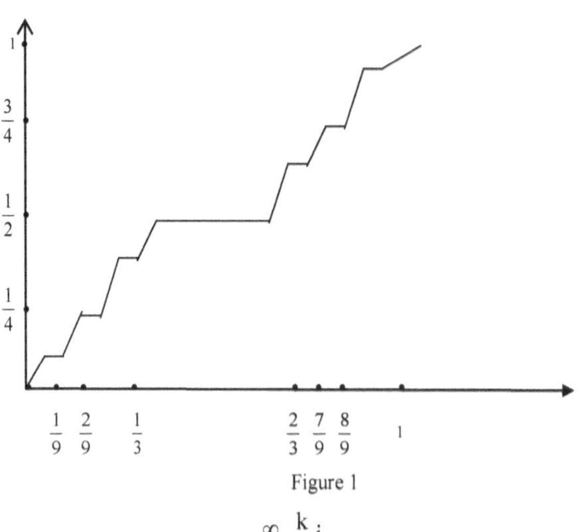

Figure 1

$$x = \sum_{i=1}^{\infty} \frac{k_i}{p^i}$$

where for each i=1, 2, . . . , k_i is an integer such that $0 \leq k_i < p$. For p=2 we have the "binary expansion" and for p=3 it is the "ternary expansion."

Problems and complements

1. Show that each $t \in [0, 1]$ has the binary expansion

$$t = \sum_{i=1}^{\infty} 2^{-i} \varepsilon_i, \qquad \varepsilon_i = 0, 1.$$

Hint:

Split $[0, 1]$ into $\Delta_0 = [0, 2^{-1})$ and $\Delta_1 = [2^{-1}, 1]$. Then split Δ_0 into two parts $\Delta_{00} = [0, 2^{-2})$ and $\Delta_{01} = [2^{-2}, 2^{-1})$; then do the same with Δ_1 to obtain $\Delta_{10} = [2^{-1}, 3 \cdot 2^{-2})$ and $\Delta_{11} = [3 \cdot 2^{-2}, 1]$. At the n-step of this procedure we split $[0, 1]$ into 2^n disjoint subintervals

$$\Delta_{\varepsilon_1 \varepsilon_2 \cdots \varepsilon_n} = [\sum_{i=1}^{n} 2^{-i} \varepsilon_i, \sum_{i=1}^{n} 2^{-i} \varepsilon_i + 2^{-n})$$

where $\varepsilon_i = 0, 1$ for all $i = 1, 2, \ldots$. Clearly,

$$\Delta_{\varepsilon_1} \supset \Delta_{\varepsilon_1 \varepsilon_2} \supset \ldots \supset \Delta_{\varepsilon_1 \varepsilon_2 \cdots \varepsilon_n} \supset \ldots;$$

thus for any $t \in [0, 1]$

$$|t - \sum_{1}^{n} 2^{-i} \varepsilon_i| \le 2^{-n}$$

for every $n = 1, 2, \ldots$.

2. Let $t = . \varepsilon_1 \varepsilon_2 \varepsilon_3 \ldots$ be the binary expansion of a number $t \in [0, 1]$. Each ε_i is obviously a function of t, i.e., $\varepsilon_i = \varepsilon_i(t)$. Show that

$$\int_0^1 \varepsilon_i(t)\, dt = 1/2 \text{ for each } i = 1, 2, \ldots.$$

Hint:

A simple argument yields that

$$\varepsilon_n(t) = \sum_{i=1}^{2^{n-1}} I_{[\frac{2i-1}{2^n}, \frac{2i}{2^n})}(t), \qquad n = 1, 2, \ldots.$$

For instance,

$$\varepsilon_1(t) = I_{[\frac{1}{2}, 1)}(t).$$

3. Let $\{r_k(t)\}_1^{\infty}$ be a sequence of functions on $[0, 1]$ defined by

$$r_0(t) = 1; \quad r_n(t) = 1 - 2\varepsilon_n(t), \qquad n = 1, 2, \ldots,$$

(the "Rademacher functions"). Show that

(a) $\int_0^1 r_n(t) = 0$ and $\int_0^1 r_n^2(t) = 1 \qquad$ for all $n = 1, 2, \ldots$.

(b) $\int\limits_0^1 r_i(t)\, r_j(t) = 0$ if $i \neq j$

Hint

Divide the interval $[0, 1]$ into 2^n equal parts then $r_n(\cdot)$ alternatively takes on these segments values $+1$ and -1 (see the previous problem). Thus, $r_n(\cdot)$ is a periodic function with period $p=2^{-n}$. In other words, $\{r_n(\cdot)\}_1^\infty$ is a normalized orthogonal system. The part (a) is trivial. To prove (b) assume $i<j$, and let $I_1, I_2, \ldots, I_{2^i}$ be the partition of $[0, 1]$ into 2^i equivalent parts. Now, partition each I_k into 2^{j-i} equal parts on which $r_j(\cdot)$ is constant and takes values -1 and $+1$, alternatively. Thus,

$$\int\limits_{I_k} r_j(t)\, dt = 0$$

and $(k=1, 2, \ldots, 2^n)$

$$\int\limits_{I_k} r_i(t)\, r_j(t)\, dt = (-1)^{k+1} \int\limits_{I_k} r_j(t)\, dt = 0.$$

Consequently,

$$\int\limits_0^1 r_i(t)\, r_j(t)\, dt = \sum_{k=1}^{2^n} \int\limits_{I_k} r_i(t)\, r_j(t)\, dt = 0.$$

4. Find the binary expansion for $1/3$ and ternary for $20/81$.

Hint:

$1/3 = .010101 \ldots$; $20/81 = .2122 \ldots$.

5. Write explicitly the interval Δ_{00110} .

Hint:

Clearly $\Delta_{00110} = [p, p+2^{-5})$ where

$$p = 0 \cdot 2^{-1} + 0 \cdot 2^{-2} + 2^{-3} + 2^{-4} + 0 \cdot 2^{-5} = 3 \cdot 2^{-4}$$

6. Show that the map

$$f: \{0, 2\}^{N+} \to [0, 1]$$

defined by

$$f((.d_1\, d_2\, d_3 \ldots .)) = \sum_{i=1}^\infty d_i 3^{-i}; \quad d_i = 0, 2$$

is injective.

Hint:

If

$$\sum_{i=1}^\infty d_i 3^{-i} = \sum_{i=1}^\infty d_i' 3^{-i}, \qquad \text{where } d_i' = 0, 2$$

we must show that $d_i = d'_i$ for all $i=1, 2, \ldots$. Assume that $d_i \neq d'_i$ for some n; let

$$k = \inf\{p; d_p \neq d'_p\}$$

then clearly $|d_k - d'_k| = 2$. Since $|d_n - d'_n| \leq 2$ for each $n = 1, 2, \ldots$, we have:

$$0 = \left| \sum_{n=k}^{\infty} (d_n - d'_n)/3^n \right| \geq 3^{-k} \left(|d_k - d'_k| - \sum_{n=k+1}^{\infty} |d_n - d'_n|/3^n \right)$$

$$\geq 3^{-k} \left(2 - \sum_{n=1}^{\infty} 2/3^n \right) = 3^{-k}$$

which is absurd. Thus, $d_i = d'_i$ for every $i = 1, 2, \ldots$, proving that the map f is an injection.

7. Show that for any $t_1, t_2 \in C$ such that $t_1 < t_2$

$$\Phi(t_1) < \Phi(t_2).$$

    Hint:

    Clearly $\eta(t_i) = \infty$ for $i=1,2$. Set

$$t_1 = .d_1\, d_2\, d_3 \ldots, \qquad t_2 = .d'_1\, d'_2\, d'_3 \ldots$$

and denote by

$$j = \inf\{i; d_i < d'_i\}$$

then

$$\Phi(t_2) - \Phi(t_1) = \sum_{i=j}^{\infty} (d'_i - d_i)/2^{i+1} > 1/2^j$$

$$- \sum_{i=j+1}^{\infty} 1/2^i = 0.$$

In a similar fashion it can be shown that for any $t_1, t_2 \in [0, 1]$ with $t_1 < t_2$ we have that

$$\Phi(t_1) \leq \Phi(t_2) \text{ and that } 0 \leq \Phi(t_2) - \Phi(t_1) < 1/2^{j-1}.$$

8. Let $\{\Phi_n(\cdot)\}_1^{\infty}$ be a sequence of nondecreasing continuous functions $\Phi_n : [0, 1] \to [0, 1]$ defined as follows:

$$\Phi_1(t) = 1/2 \text{ if } t \in (1/3, 2/3)$$

and linear on each closed interval of the set F_1.

$$\Phi_2(t) = \begin{cases} 1/4 & \text{if } t \in (1/3^2, 2/3^2) \\ 1/2 & \text{if } t \in (1/3, 2/3) \\ 3/4 & \text{if } t \in (7/3^2, 8/3) \end{cases} \text{ and linear on every closed interval of the set}$$

F_2. In general, define $\Phi_n(t) = k\, 2^{-n}$ on the k-th open interval among 2^n disjoint open intervals

$[0, 1]-F_n$ and linear on each closed interval of F_n.

(i) Show that the sequence $\{\Phi_n\}_1^\infty$ converges uniformly to a limit Φ.

(ii) Show that the map Φ is the Cantor function.

Hint:

(i) Clearly, $|\Phi_{n+1}(\cdot)-\Phi_n(\cdot)| < 2^{-k}$; consequently, $\Sigma(\Phi_{n+1}-\Phi_n)$ converges uniformly on $[0, 1]$ and therefore $\{\Phi_n\}$ converges uniformly on $[0, 1]$.

(ii) The limit $\Phi(\cdot)$ is clearly continuous and nondecreasing map $\Phi:[0, 1] \to [0, 1]$, constant on any open middle third removed from $[0, 1]$ to obtain the closed sets $F_1, F_2, \ldots$.

9. Show that

(i) $1/4 \in C$; (ii) $1/8 \notin C$; (iii) $1/9 \in C$; (iv) $2/9 \notin C$.

Hint:

$1/4 = .02020 \ldots$; $1/8 = .010101 \ldots$

$1/9 = .00222 \ldots$; $2/9 = .01222 \ldots$

8. ORDERED PAIRS

Most of set theory introduced up to this point has not been concerned either with the nature of the elements of sets or with their mutual relationship. In this section we deal, to a limited extent, with some of these questions. We shall begin with the notion of "ordered pair."

Consider a set $\{x, y\}$; if we desire to order the elements of this set so that x come in the first place and y in the second, we may use the notation (x, y) and call x the "first coordinate" and y the "second coordinate." The system (x, y) is called an **"ordered pair."**

In order to be a "legitimate object" of set theory, this new concept must be either defined in terms of sets, if it is possible, or if not, we must consider (x, y) as another "primitive notion" of the theory of sets and endow it with properties that we desire; for instance that $(x, y)=(u, v)$ if and only if $x=u$ and $y=v$.

It is remarkable fact that it can be done, i.e., that (x, y) can be defined (as was shown by Kuratowski) in terms of sets as follows:

$$(x, y) = \{\{x\}, \{x, y\}\}. \tag{1}$$

In other words, (x, y) is a class (and therefore an object of set theory) which consists of two sets; a singleton $\{x\}$ and the set $\{x, y\}$. **With this definition the basic principle of set theory is preserved: to make new sets out of old ones**.

The newly defined object has some curious properties. In addition to the fact that $\{x, y\} \in (x, y)$, we also have that $(x, y) \cap (y, x) = \{\{x, y\}\}$. To demonstrate the "reasonableness" of the definition (1), we must show that this new structure has the properties which are expected of ordered pairs, namely, if (x, y) and (u, v) are ordered pairs, then $(x, y)=(u, v)$ if and only if $x=u$ and $y=v$. To this end we need the following:

Lemma 1

If $(u, a) = (u, b)$ then $a = b$.

Proof:

Clearly, if $(u, a) = (u, b)$ then $\{u, a\}=\{u, b\}$. Consequently, $a \in \{u, b\}$ and if $a \neq u$, then clearly $a=b$. If $u=a$, then $(u, a)=(a, a)=\{\{a\}\}$. Hence, $\{a\}=\{a, b\}$ and since $b \in \{a, b\}$ necessarily $b \in \{a\}$. Thus, in both cases $a=b$, which proves the lemma. ♥

Proposition 1

$$(x, y) = (u, v) <=> x = u, \ y=v$$

Proof:

If $x=u$ and $y=v$ then $\{x\}=\{y\}$ and $\{x, y\}=\{u, v\}$ from which follows that $(x, y)=(u, v)$. Conversely, assume that $(x, y)=(u, v)$, then since $\{x\} \in (x, y)=(u, v)$, either $\{x\}=\{u\}$ or $\{x\}=\{u, v\}$. In the first case $x=u$; in the second case, since $u \in \{u, v\}=\{x\}$, it implies that $x=u$. Thus, in either case $x=u$, implying that $(x, y)=(x, v)$. This proves the proposition. ♥

We are now in a position to give a proper definition of the Cartesian product of two sets, say X and Y. **The Cartesian product of X and Y, denoted by X×Y, is the set of all ordered pairs (x, y) where $x \in X$ and $y \in Y$. In other words,**

$$X \times Y = \{(x, y); x \in X, y \in Y \}$$

The concept of function was introduced in section 3. Here we shall give it a proper definition in terms of ordered pairs.

A function, or a map, from a set X to a set Y is a subset $f \subset X \times Y$ satisfying the following conditions:

(i) If (x, y) and (x, y') are members of f then necessarily $y=y'$.

(ii) For any $x \in X$, $(x, y) \in f$ for some $y \in Y$.

The set of ordered pairs that constitutes a function is sometimes called the "graph" of the function.

By means of the notion of function the concept of Cartesian product can be extended to any family of sets, say $\{A_t ; t \in T\}$, which is usually denoted by the symbol

$$\prod_{t \in T} A_t \quad \text{or simply by} \quad \prod_T A_t ,$$

and defined as the family of maps

$$h : T \to \bigcup_T A_t$$

such that $h(t) \in A_t$ for every $t \in T$. If $A_t = A$ for every $t \in T$, we write

$$\prod_t A_t = A^T .$$

Here, of course, A^T is the set of all maps $h : T \to A$.

From this definition it seems rather obvious that

$$\prod_t A_t \neq \theta \quad \text{implies that } A_t \neq \theta \text{ for all } t \in T.$$

The converse is not obvious to hold. As a matter of fact, if $A_t \neq \theta$ for every $t \in T$ there is no way to prove that $\prod_t A_t \neq \theta$ using the standard axioms of set theory.

One of the principles of set theory is the postulate that the Cartesian product of any (nonempty) family of nonempty sets is a nonempty set. This postulate is often called the "Axiom of Choice." In the next section we shall give another formulation (due to Zermelo) of the axiom of choice, and then prove that

$$\prod_t A_t \neq \theta \iff A_t \neq \theta \text{ for every } t \in T.$$

Next, we shall define a new concept which is called **"relation."** Let X and Y be two sets. **Any** subset $\mathcal{R} \subset X \times Y$ is called a "relation." For instance, any function $f : X \to Y$ is a relation. An element $x \in X$ is said to be $\mathcal{R}$-related to an element $y \in Y$ if $(x, y) \in \mathcal{R}$. Thus a relation is a set of ordered pairs. If $\mathcal{R}$ is a relation it is often convenient to express the fact that $(x, y) \in \mathcal{R}$ by a more suggestive $x \, \mathcal{R} \, y$. Notice that $x \, \mathcal{R} \, y$ does not imply necessarily that $y \, \mathcal{R} \, x$.

If $\mathcal{R} \subset X \times X$ then we say that $\mathcal{R}$ is a relation in X. Thus, any relation in X is a subset of $X \times X$. The empty set θ is a relation since it is a subset of any set. There are all kinds of relations having various properties. **A relation $\mathcal{R}$ in X is called a "preorder" if it is "reflexive" and "transitive."** In other words, if $x\mathcal{R}x$ for all $x \in X$, and if from $x\mathcal{R}y$ and $y\mathcal{R}z$, it follows that $x\mathcal{R}z$. If a preorder $\mathcal{R}$ in X is also "antisymmetric", i.e., from $x\mathcal{R}y$ and $y\mathcal{R}x$, it follows that $y=x$, then $\mathcal{R}$ is called a "partial ordering."

A relation $\mathcal{R}$ in a set X is said to be an "equivalence relation" if:

(i) $x\,\mathcal{R}\,x$ for all $x \in X$ (reflexive)

(ii) $x\,\mathcal{R}\,y \Rightarrow y\,\mathcal{R}\,x$ (symmetric)

(iii) $x\,\mathcal{R}\,y$ and $y\,\mathcal{R}\,z \Rightarrow x\,\mathcal{R}\,z$ (transitive)

The fundamental significance of equivalence relations is that they generate "equivalence classes." For instance, in measure theory we encounter measurable functions which are equal almost everywhere. Almost everywhere equality a.e is obviously an equivalence relation. Analysis of equivalence classes of objects rather than objects themselves is often much simpler.

Let $\mathcal{R}$ be an equivalence relation in a set X, and x an element of X. The "equivalence class" of x (with respect to $\mathcal{R}$) is the set

$$C_X = \{y \in X; x\,\mathcal{R}\,y\}.$$

It is clear from this definition that $x \in C_X$ (reflexivity). Also, any element $y \in X$ such that $x\,\mathcal{R}\,y$ belongs to C_X. The following result concerning the equivalence classes is fundamental.

Proposition 2

Any two equivalence classes C_X and C_y are either disjoint or identical.

Proof:

If $C_X \cap C_y \neq \theta$ for some $x,\, y \in X$, then $x\mathcal{R}y$. Indeed, for any $t \in C_X \cap C_y$ we have that $x\mathcal{R}t$ and $y\mathcal{R}t$; but due to symmetry we have that $x\mathcal{R}t$ and $t\mathcal{R}y$, yielding that $x\mathcal{R}y$. Next, if $x_0 \in C_X$ then $x\mathcal{R}x_0$; hence $x_0\mathcal{R}x$ (due to symmetry) and since $x\mathcal{R}y \Rightarrow x_0\mathcal{R}y$ (transitivity) and obviously $y\mathcal{R}x_0$. Thus $x_0 \in C_y$ implying that $C_X \subset C_y$. Similarly, if $y_0 \in C_y$, then $y\mathcal{R}y_0$ and $y_0\mathcal{R}y$; since $x\mathcal{R}y$, $y\mathcal{R}x$, implying that $y_0\mathcal{R}x$ and $x\mathcal{R}y_0$. Thus, $y_0 \in C_X \Rightarrow C_y \subset C_X$. Hence $C_X = C_y$.   ♥

Remark 1

Given an equivalence relation $\mathcal{R}$ in a set X we have just proved that

$$C_X = C_y \iff x\,\mathcal{R}\,y;$$

hence $C_X \cap C_y = \theta$ if and only if x and y are not related. Consequently, the distinct equivalence classes (with respect to $\mathcal{R}$) form a partition of the set X.

Problems and complements

1. Let x, y∈X; show that $(x, y)∈\mathcal{P}(\mathcal{P}(X))$.

    Hint:

    Since {x}⊂X and {y}⊂X, we have that {x}∪{y}={x, y}⊂X. Thus,

$$(x, y) = \{\{x\}, \{x, y\}\} ⊂ \mathcal{P}(X) => (x, y) ∈ \mathcal{P}(\mathcal{P}(X)).$$

2. Show that: (i) ∪(x, y) = {x, y}; (ii) ∩(x, y)={x}.

    Hint:

    (i) ∪(x, y) = {x}∪{x, y} = {x, y}

    (ii) ∩(x, y) = {x}∩{x, y} = {x}

3. Determine (x, y)∪(y, x) and (x, y)∩(y, x).

    Hint:

    (x, y) ∪{y, x} = {{x}, {y}, {x, y}}

    (x, y) ∩{y, x} = {{x , y}}.

4. If x∈X and y∈Y, show that $(x, y)∈\mathcal{P}(\mathcal{P}(X∪Y))$.

    Hint:

    Clearly, {x}⊂X, {y}⊂Y so that {x}∪{y}={x, y}⊂X∪Y. Hence, {{x}, {x, y}}⊂$\mathcal{P}(X∪Y)$ => (x, y)

∈$\mathcal{P}(\mathcal{P}(X∪Y))$.

5. Show that A×B = θ <=> A = θ or B = θ.

    Hint

    (By contradiction). Assume that A×B = θ; If A≠θ and B≠θ, then there is an a∈A and b∈B; hence,

(a, b)∈A×B = θ, which is contradiction. Thus, either A=θ or B=θ. Now assume A=θ or B=θ, then no a∈A or

b∈B=> there is no (a, b)∈A×B =>A×B=θ.

6. Show that

$$A×B = B×A <=> A = θ \text{ or } B = θ \text{ or } A = B.$$

    Hint:

    Assume that A≠θ, B≠θ and A≠B. There exists an x such that x∈A∆B. For definiteness assume that

x∈A-B and let y∈B, then (x, y)∈A×B and consequently (x, y)∈B×A. But then x∈B which is contradiction.

Sufficiency of the condition is obvious.

7. If X and Y are sets and A⊂X, B⊂Y. Show that (i) A×B⊂X×Y; (ii) A×B = (A×Y)∩(X×B).

    Hint:

    (i) Obvious;

    (ii) If (x, y)∈A×B, x∈A and y∈B; hence x∈A and y∈Y and x∈X and y∈Y. Consequently, (x,y) ∈

(A×Y)∩(X×B). Conversely, if (x, y)∈(A×Y)∩(X×B) then x∈A and x∈X => x∈A; y∈B and y∈Y => y∈B.

Thus, (x, y)∈A×B.

8. Show that $(A_1∩A_2)×B=(A_1× B) ∩ (A_2 × B)$.

Hint:

Obviously $(A_1 \cap A_2) \times B \subset (A_i \times B)$, $i=1,2$, which implies that $(A_1 \cap A_2) \times B \subset (A_1 \times B) \cap (A_2 \times B)$. Conversely, if $(x, y) \in (A_1 \times B) \cap (A_2 \times B)$ then $x \in A_1$, $y \in B$ and $x \in A_2$ and $y \in B$. Hence, $x \in A_1 \cap A_2$ and $y \in B \Rightarrow x \in (A_1 \cap A_2) \times B$.

9. Show that $A \times (B_1 \cap B_2) = (A \times B_1) \cap (A \times B_2)$

Hint:

See the previous problem.

10. Show that $(A_1 \times B_1) \cap (A_2 \times B_2) = (A_1 \cap A_2) \times (B_1 \cap A_2)$.

Hint:

Since $A_1 \cap A_2 \subset A_i$ and $B_1 \cap B_2 \subset B_i$, $i=1, 2$, it follows that $(A_1 \cap A_2) \times (B_1 \cap B_2) \subset (A_i \times B_i)$, $i=1,2$. For the converse consider $(x, y) \in (A_1 \times B_1) \cap (A_2 \times B_2)$ and proceed as in the problem 8.

11. Let $\mathcal{R}$ be a relation defined in the set $Z = \{ \ldots, -2, -1, 0, 1, 2, \ldots \}$ as follows:

$$x \, \mathcal{R} \, y \text{ if and only if } x = y \pmod 2$$

(i.e., x-y is divisible by 2). Show that $\mathcal{R}$ is an equivalence relation. Find C_X and $\mathcal{R} \subset Z \times Z$.

Hint:

$\mathcal{R}$ is trivially an equivalence relation. If we write $Z = Z_e \cup Z_0$ where $Z_e = \{ \ldots, -4, -2, 0, 2, 4, \ldots \}$ and $Z_0 = \{ \ldots, -3, -1, 1, 3, \ldots \}$, $C_{2k} = Z_e$ and $C_{2k+1} = Z_0$ and

$$\mathcal{R} = \{(x, y) \in Z \times Z; \, x, y \in Z_e \cup Z_0 \}.$$

12. Let $\mathcal{R}$ be a relation in the set $A = [0, 1]$ defined by

$$\mathcal{R} = \{(x, y) \in A \times A; \, |x-y| \le 1/2 \}.$$

Is $\mathcal{R}$ an equivalence relation?

Hint:

The transitivity property does not hold.

13. Recall that a relation $\mathcal{R}$ in X is called a "partial ordering" if it is reflexive, transitive and "antisymmetric" (the last one meaning $x\mathcal{R}y$ and $y\mathcal{R}x \Rightarrow x=y$). Show that the inclusion $\subset$ is a partial ordering.

Hint:

Given a nonempty class of sets $\mathcal{K} \subset \mathcal{P}(X)$, it can be partially ordered by inclusion. Hence two members A and B of $\mathcal{K}$ are related if either $A \subset B$ or $B \subset A$. It is easy to see that $\subset$ is a partial ordering.

9. THE AXIOM OF CHOICE

The **Axiom of Choice is** an "existential postulate" independent of any other axiom of set theory. It is also one of the most controversial. The example listed below illustrates why that is so. At present there are many versions of this axiom. The following one is due to Zermelo.

Let $\mathcal{T}$ be a class of nonempty disjoint sets. There exists a set S which consists of exactly one element from each member of $\mathcal{T}$.

We have already used this axiom to prove that every infinite set contains a proper countable subset (proposition 3 of section 5). In section 8 we stated that one of the principles of set theory is the postulate that the Cartesian product of an arbitrary family of nonempty sets is a nonempty set. We shall show that this postulate is equivalent to the Axiom of Choice.

The Axiom of Choice is an existential axiom, which is not to be understood in the sense that one can construct the **set S** which, as the next example shows, is often impracticable.

Example 1

Let $\Re$ be a relation in the set $[0, 1]$ defined as follows:

$$x \,\Re\, y \quad \text{if } x\text{-}y \in Q \text{ (the set of rationals).}$$

It is not difficult to verify that the relation $\Re$ is an equivalence relation. For any $x \in [0, 1]$ the equivalence class C_x of x, with respect to $\Re$, is obviously

$$C_x = \{y \in [0, 1]; x\text{-}y \in Q\},$$

and for any $r \in [0, 1] \cap Q$, $C_r = [0, 1] \cap Q$. Clearly,

$$C_{x_1} \cap C_{x_2} = \theta \qquad \text{if } x_1 - x_2 \notin Q.$$

By the Axiom of Choice, there exists a set S which consists of one and only one element from each set C_x .

By applying the Axiom of Choice to the family of equivalence classes $\{C_x\}$ we proved the existence of a set (subset of $[0, 1]$) which is "unspecifiable" (i.e., there is no a rule $\pi(\cdot)$ which would define S as the collection of all x's such that $\pi(x)$ is true) and the set by itself can hardly be imagined.

The non- constructive nature of the Axiom of Choice is one of the principal reasons for many controversies associated with this axiom. However, in spite of all its shortcomings, applications of the axiom have lead to some stunning results. One of the most spectacular (and an affront to intuition) is the "Banach-Tarski paradox." By an ingenious train of reasoning, they proved, using the axiom, that a sphere of a fixed radius, can be decomposed into a finite number of parts and put together again in such a way as to form two spheres with the given radius.

At present the Axiom of Choice enjoys rather wide acceptance for compelling reasons, both extrinsic and intrinsic. It established itself as a powerful and pervading tool in almost all branches of mathematical sciences. In addition, as it was proved by Sierpinski, the general continuum hypothesis implies the Axiom of Choice. However, it should be pointed out that, from the stand point of set theory, there are still deep and not well understood problems associated with this axiom.

As we have mentioned at the outset of this section there are many equivalent forms of the axiom of choice. For our purposes the next one is essential.

Let $\mathcal{K}$ be a nonempty family of nonempty sets. There exists a map, called "choice function"

$$h : \mathcal{K} \to \cup \mathcal{K}$$

of the family $\mathcal{K}$, such that $h(A) \in A$ for every $A \in \mathcal{K}$ (here, as usual, $\cup \mathcal{K} = \cup \{A;\ A \in \mathcal{K}\}$).

Next we shall prove that these two postulates are equivalent.

Proposition 1

There exists a choice function if and only if the axiom of choice holds.

Proof:

Let $\mathcal{T}$ be a class of nonempty disjoint sets. If $h: \mathcal{T} \to \cup \mathcal{T}$ is a choice function, then

$$S = \{h(H);\ H \in \mathcal{T}\}$$

is the desired set (S consists of exactly one element from each member of $\mathcal{T}$).

Next, assume that the axiom of choice holds and let $\mathcal{K}$ be a nonempty class of nonempty sets. For every $H \in \mathcal{K}$ define

$$L_H = \{(H, u);\ u \in H\}.$$

Clearly $L_H \cap L_{H'} = \theta$ if $H \neq H'$; consequently,

$$\mathcal{H} = \{L_H;\ H \in \mathcal{K}\}$$

is a collection of nonempty disjoint sets. Then, according to the axiom of choice there exists a set h consisting of exactly one element from each member of $\mathcal{H}$, i.e.,

$$h = \{(H, u);\ H \in \mathcal{K}\}.$$

Clearly, h is a set of ordered pairs (H, u) with $u \in H$, which is obviously a map

$$h : \mathcal{K} \to \cup \mathcal{K}$$

satisfying $h(H) \in H$ for every $H \in \mathcal{K}$. Thus, h is a choice function. ♥

Proposition 2

Let $\{A_t;\ t \in T\}$ be a family of sets. If the axiom of choice holds then

$$\prod_t A_t \neq \theta \iff A_t \neq \theta \text{ for all } t \in T.$$

Proof:

Assume that $\prod_t A_t \neq \theta$ then there exists a map $h \in \prod_t A_t$ such that $h(t) \in A_t$ for all $t \in T$, implying that $A_t \neq \theta$ for all $t \in T$. Conversely, if $A_t \neq \theta$ for all $t \in T$, the axiom of choice stipulates the existence of a map

$$h_0 : \{A_t;\ t \in T\} \to \bigcup_t A_t$$

such that $h_0(A_t) \in A_t$. Set

$$h(t) = h_0(A_t)$$

then clearly $h: T \to \bigcup_t A_t$ is an element of $\prod_t A_t$, proving that $\prod_t A_t \neq \theta$.

     Let X be a nonempty set; denote by

$$\mathscr{P}_X = \mathscr{P}(X) - \{\theta\},$$

then according to the Axiom of Choice there exists a choice function

$$h : \mathscr{P}_X \to X,$$

which is called the "choice function of the set X.

Problems and complements

1. Let $h:X\to Y$ be a surjection. Prove that there exists a subset $X_0\subset X$ such that

$$h_0 = h \mid X_0:$$

is a bijection, if the axiom of choice holds.

Hint:

Consider the class $\mathfrak{I}\subset P(X)$ defined by

$$\mathfrak{I} = \{h^{-1}(\{y\}); y\in Y\}.$$

Clearly, $\mathfrak{I}$ is a collection of nonempty disjoint sets. By the axiom of choice there exists a set $X_0\subset X$ which

consists of exactly one element from each $h^{-1}(\{y\})$. Clearly then

$$h_0 : X_0 \to Y$$

is a bijection.

2. Find a choice function of the sets (i) N_+ ; (ii) $R=(-\infty, \infty)$.

Hint:

(i) On the collection $P_{N_+} = P(N_+)-\{\theta\}$ define a map $h: P_{N_+}\to N_+$ as follows:

$$h(M) = \min(M) \text{ for every } M\in P_{N_+}.$$

Clearly, $h(M)\in M$.

(ii) This is an open problem; at present there is no choice function of the set R.

3. Show that a choice function of any nonempty set X is a surjection.

Hint:

Let $h: P_X \to X$ be a choice function. We must show that for any $x\in X$ there exists at least one set

$A\in P_X$ such that $h(A)=x$. But this clearly holds for $A=\{x\}$ since from $h(\{x\})\in\{x\}$ we have that $h(\{x\})=x$.

4. Show that a set is Dedekind infinite if and only if it is infinite.

Hint:

Recall that a set X is Dedekind infinite if it is equivalent to a proper part of itself (see remark 1 of section 4). Let X be an infinite set; according to proposition 3 of section 5, X contains a countable subset, say $C\subset X$. Next, from corollary 2 of section 5 we have that $X\sim(X-C)$, proving that the set X is Dedekind infinite. Conversely, if a set is Dedekind infinite it must be infinite because no finite set is equivalent to a proper part of itself.

5. Let $Q\subset R$ be the set of rationals and denote by

$$S + r = \{x + r; x\in S\}$$

where $r\in [0, 1]\cap Q$ and S is the set defined in example 1. Show that $(r, r'\in [0, 1]\cap Q)$.

$$(S + r) \cap (S + r') = \theta \text{ if } r \neq r'.$$

Hint:

If $(S+r)\cap(S+r')\neq\theta$ and $y\in(S+r)\cap(S+r')$, then $y=x_0+r$ (since $y\in(S+r)$) and $y=x_1+r'$ (since $y\in S+r'$).

Hence, $x_1 - x_0 = r' - r$, which is not possible since x_0 and x_1 belong to different equivalence classes.

6. Denote by

$$H = \bigcup_{r \in [-1, 1] \cap Q} (S + r)$$

Show that $[0, 1] \subset H \subset [-1, 2]$.

    Hint:

    It is clear that each $S + r \in [-1, 2]$ for every $r \in [-1, 1] \cap Q$ so that $H \subset [-1, 2]$. On the other hand, if $x \in [0, 1]$ there exists an $y \in S$ such that $x - y = r$ for some $r \in [-1, 1] \cap Q$. Thus, $x = y + r \in S + r$, which implies that $x \in H$ and consequently that

$$[0, 1] \subset H.$$

10. ORDERING RELATIONS

Recall that a relation $\Re$ in a set X is a subset of X×X. When $(x, y) \in \Re$ we say that x is related to y and write x$\Re$y. If "related" means some sort of "order" between x and y, we say that $\Re$ is an **"ordering relation"** and replace the symbol $\Re$ with (a more suggestive) symbol $<$. Thus, $x < y$ is read "x precedes y" or "y follows x." We also say that x is a "predecessor" for y and that y is a "successor" for x. Notice that not all elements of the set X are related. If for two elements $x, y \in X$ either $x<y$ or $y<x$ we say that x and y are "comparable." Otherwise they are not comparable.

Recall (see section 8) an ordering relation $<$ in a set X is said to be a **"partial ordering"** if it is antisymmetric, reflexive and transitive. Then, the system $(X, <)$ is called a "partially ordered set." If every two elements in X are comparable we say that $(X, <)$ is **"totally ordered set."** Notice that a total ordering $<$ is not necessarily reflexive. However, the total ordering $\leq$ defined as follows: $x \leq y$ if and only if $x<y$ or $x=y$, is reflexive. Ordering by inclusion $\subset$ in a family of sets $\mathscr{K}$ is always a partial ordering. Here two members A, B $\in \mathscr{K}$ are related if either A$\subset$B or B$\subset$A.

Let $(X, <)$ be a partially ordered set. An element $m \in X$ is said to be **"minimal"** if and only if $x<m$ implies that $x=m$, **and maximal** if and only if $m<x$ implies that $x=m$. In other words, an element $m \in X$ is minimal if it does not have a predecessor and maximal if it does not have a successor. Clearly, a partially ordered set may have many minimal and maximal elements.

An element $s \in X$ is said to be the "first" or **"smallest"** in $(X, <)$ if and only if $s<x$ for every $x \in X$. If, on the other hand, $x<s$ for every $x \in X$, the element $s \in X$ is called the "**greatest**." Clearly, a partially ordered set may have at most one first element and at most one greatest element (although they may not exist). It is also clear that a smallest element, if it exists, it is also minimal but not conversely and similarly for a greatest element. Notice that smallest and largest members are related to any other member of the partially ordered set $(X, <)$.

Remark 1

The statement "$m \in X$ is minimal" means that any other member x of $(X, <)$ is either not related to m or if it is related, $m<x$. On the other hand, $s \in X$ is the smallest if and only if $s<x$ for every $x \in X$. Thus, the smallest element is related to every element $x \in X$, and every element of $(X, <)$ is related to the largest element.

Example 1

Let X be a nonempty set, then $(\mathscr{P}(X), \subset)$ is partially ordered set. It is easy to see that every singleton $\{x\} \in \mathscr{P}(X)$ is a minimal element. The empty set θ is the smallest and X is the largest member of the partially ordered set. However, the only maximal element is the set X.

Let $(X, <)$ be a partially ordered set: a subset C$\subset$X is said to be a "chain" in $(X, <)$ if any two elements of C are related. In other words, $(C, <)$ is a totally ordered subset. A chain C*$\subset$X is said to be "maximal" if no other chain in $(X, <)$ contains C* as a proper subset.

Let B be a subset of a partially ordered set $(X, <)$. An element $w \in X$ (if it exists) is called an "**upper bound**" of the set B if for every $x \in B$, $x < w$. If w^* is an upper bound of B and $w^* < w$, where w is any other upper bound of B, then w^* is called the "**least upper bound**" of B and is often denoted by $w^* = \sup B$. It seems rather obvious that $\sup B$ is uniquely determined if it exists.

If every chain in an partially ordered set $(X, <)$ has a "least upper bound" in X, the set $(X, <)$ is said to be "**strictly inductive**"; otherwise, i.e., if every chain has just an upper bound in X, the set $(X, <)$ is called "**inductive**."

Example 2

Consider the set $R_+^2 = \{(x_1, x_2); x_1 \geq 0, x_2 \geq 0\}$, where $\geq$ is the "usual" ordering on the real line.

Define an ordering relation $<$ in R_+^2 as follows: if $y, z \in R_+^2$ where $y = (y_1, y_2)$ and $z = (z_1, z_2)$ then $y < z$ if and only if $y_1 \leq z_1$, $y_2 \leq z_2$. For instance, the element $(7, 5)$ precedes $(9, 7)$, since $(7, 5) < (9, 7)$. However, the elements $(4, 5)$ and

$(5, 3)$ are not related. It is easy to verify that $(R_+^2, <)$ is a partially ordered set. Since any $x \in R_+^2$ has a

successor, the set $(R_+^2, <)$ has no maximal element. The only minimal (and the smallest) element is $(0, 0)$. Consider the subset

$$B = \{(x_1, x_2) \in R_+^2 ; x_1^2 + x_2^2 \leq 1\}.$$

Any $y = (y_1, y_2)$ with $y_1 \geq 1$, $y_2 \geq 1$ is an upper bound of B and $\sup B = (1, 1)$. The subset $C \subset R_+^2$ defined by

$C = \{(x, x); x > 0\}$ is a (maximal) chain. The subset $C_{N_+} = \{(n, n)\}_1^\infty$ is a chain such that $C_{N_+} \subset C$.

If $x, y \in R_+^2$ are related and $x < y$, the set of points on the line segment

$$(1 - t)\, x + t\, y, \ 0 \leq t \leq 1$$

forms a chain in $(R_+^2, <)$ which can be written as

$$C(x, y) = \{(1 - t)x + t\, y; \ 0 \leq t \leq 1\}.$$

This chain is bounded and $\max C(x, y) = y$. The partially ordered set $(R_+^2, <)$ is not inductive.

Totally ordered sets $(X, <)$ defined below are of considerable interest in set theory.

An ordered set $(X, <)$ is called "well ordered" (or an "ordinal") if X and every nonempty subset $B \subset X$ has a first element; that is, there exists a $x_0 \in B$ such that $x_0 < x$ for every $x \in B$.

Clearly, every well ordered set is totally ordered since each pair $\{x, y\} \subset X$ has a first element. The empty set θ is vacuously well ordered and so is every singleton. The standard example of a well ordered set is $(N_+, <)$ where $<$ is the usual ordering relation on the real line.

Example 3

The set $(Q_+, <)$, where Q_+ is the set of positive rational numbers and $<$ the usual ordering on the real line, is totally ordered. A well ordering of the set Q_+ is given below

$$(Q_+, \precsim) = \{1, 2, 3, \ldots; \frac{1}{2}, \frac{3}{2}, \frac{5}{2}, \ldots; \frac{2}{3}, \frac{5}{3}, \frac{7}{3}, \ldots; \ldots\}.$$

Clearly, the set on the right-hand side contains every positive rational number. Every subset of this set contains fractions with the smallest denominator. Among the fractions with this denominator there is one with the smallest numerator. This fraction is the first element of this subset. Obviously 1 is the first element in $(Q_+, \precsim)$. Notice also that $(p/q) \precsim (p'/q')$ if $q<q'$.

Remark 2

Every element, if it is not the largest one, of a well ordered set has an immediate successor and, if it is not the first element, it has an immediate predecessor.

Proposition 1

Let $(\mathcal{U}, \subset)$ be a partially ordered set, where $\mathcal{U}$ is a class of nonempty sets. If we assume that the Axiom of Choice holds, there exists a mapping $f: \mathcal{U} \to \mathcal{U}$ such that $A \subset f(A)$ for all $A \in \mathcal{U}$ and $f(A)=A$ if A is a maximal member.

Proof:

With each $H \in \mathcal{U}$ we associate a nonempty class

$$\mathcal{A}_H = \{B \in \mathcal{U}; B \supset H\}.$$

Since $\subset$ is reflexive $H \in \mathcal{A}_H$ and if H is a maximal element, then clearly $\mathcal{A}_H = \{H\}$. Define

$$\overset{o}{\mathcal{M}} = \{\mathcal{A}_H; H \in \mathcal{U}\};$$

According to the Axiom of Choice there exists a choice function

$$h: \overset{o}{\mathcal{M}} \to \mathcal{U} \text{ where } h(\mathcal{A}_H) \in \mathcal{A}_H$$

for every $\mathcal{A}_H \in \overset{o}{\mathcal{M}}$. Then the map $f(H) = h(\mathcal{A}_H)$ is the required function. $\quad \blacktriangledown$

Corollary 1

Since for every $A \in \mathcal{A}_H$ we have that $H \subset A \subset f(A)$, it follows that

$$f(A) \in \mathcal{A}_H \Rightarrow f(\mathcal{A}_H) \subset \mathcal{A}_H.$$

The partially ordered set $(\mathcal{A}_H, \subset)$ is clearly inductive if $(\mathcal{U}, \subset)$ is inductive, for if $\mathcal{C} \subset \mathcal{A}_H$ is a chain, $\cup\mathcal{C} \in \mathcal{U}$ and since $H \subset \cup\mathcal{C}$ it follows that $\cup\mathcal{C} \in \mathcal{A}_H$.

Problems and complements

1. Let C[0, 1] be the set of continuous functions on the set [0, 1]. Are the following relations in C[0, 1] partial orderings:

(i) $h \mathcal{R} f$ means that $h(t) \le f(t)$ for every $t \in [0, 1]$

(ii) $h \mathcal{R}_1 f$ means that $\sup_t h(t) < \sup_t f(t)$.

Hint:

(i) Since $f(t) \le f(t)$ for all $t \in [0, 1]$, $\mathcal{R}$ is reflexive. If $h(t) \le f(t)$ and $f(t) \le h(t)$, $h(t) = f(t)$ for all $t \in [0, 1]$, so that $\mathcal{R}$ is antisymmetric. Finally, $\mathcal{R}$ is obviously transitive, implying that it is a partial ordering.

(ii) It is easy to verify that $\mathcal{R}_1$ is not antisymmetric. It is, however, reflexive and transitive, which makes $\mathcal{R}_1$ a **"preorder."**

2. If $\mathcal{R}_1$ and $\mathcal{R}_2$ are partial ordering and a preordering in a set X, respectively, show that

(i) $\mathcal{R}_1 \cup \mathcal{R}_2 = \mathcal{R}_2$; (ii) $\mathcal{R}_1 \cap \mathcal{R}_2 = \mathcal{R}_1$.

Hint:

(i) Every partially ordered set is a preorder.

(ii) Clearly, since $\mathcal{R}_1 \subset \mathcal{R}_2$.

3. Show that a partially ordered set $(X, <)$ can have at most one smallest and one largest element.

Hint:

Recall that $x_0 \in X$ is smallest if and only if $x_0 < x$ for every $x \in X$. If $x_0' \in X$ is another smallest element then $x_0' < x_0$. But since we have that $x_0 < x_0'$, $=> x_0 = x_0'$. In the same way we prove the uniqueness of the largest element of $(X, <)$.

4. Let $(X, <)$ be partially ordered set and let $Y \sim X$. Show that the set Y can be partially ordered.

Hint:

Since $X \sim Y$, there exists a bijection $h:X \to Y$. For comparable $x_1, x_2 \in X$ such that $x_1 < x_2$ we "stipulate" that $h(x_1) < h(x_2)$. This defines a partial ordering $\lessgtr$ in Y as follows: given any $y_1, y_2 \in Y$, then there exist x' and $x''\in X$ such that $y_1 = h(x')$ and $h_2 = h(x'')$. If x' and x'' are not related, neither are y_1 and y_2. If, on the other hand, x' and x'' are related and $x' < x''$ then $y_1 \lessgtr y_2$ and conversely, if $y_1 \lessgtr y_2$ then necessarily $x' < x''$.

5. Let $\mathscr{C}$ be the collection of all chains in a partially ordered set $(X, <)$. Show that the partially ordered set $(\mathscr{C}, \subset)$ is inductive.

Hint:

Let $\mathbb{C} \subset \mathscr{C}$ be a chain in $(\mathscr{C}, \subset)$, then the union

$$\cup \mathbb{C} = \hat{C}$$

is a chain in $(X, <)$. Indeed, if x, $y \in \hat{C}$ there exist C_1, $C_2 \in \mathscr{C}$ such that $x_1 \subset C_1$ and $x_2 \subset C_2$. But since $\mathscr{C}$ is a chain, either $C_1 \subset C_2$ or $C_2 \subset C_1$. If $C_1 \subset C_2$, then x, $y \in C_2$ and since C_2 is a chain in $(X, <)$ either $y < x$ or $x < y$, proving that $\hat{C}$ is a chain in $(X, <)$.

It follows from the definition of $\hat{C}$ that $\hat{C} = \sup \mathscr{C}$ such that $\hat{C} \in \mathscr{C}$. Consequently, every chain in $(\mathscr{C}, \subset)$ has a least upper bound which makes the partially ordered set $(\mathscr{C}, \subset)$ strictly inductive.

6. (Continuation) Show that the partially ordered set $(\mathscr{C}, \subset)$ has a maximal element, if it has a maximal chain.

Hint:

Since $(\mathscr{C}, \subset)$ is strictly inductive every chain of this set has a least upper bound. Now, let $\mathscr{C}^*$ be a maximal chain of $(\mathscr{C}, \subset)$ and denote by $\hat{C}_* = \cup \mathscr{C}^*$. We claim that $\hat{C}_*$ is a maximal element of $(\mathscr{C}, \subset)$. If it is not, there exists a member $C_0 \in \mathscr{C}$ such that $\hat{C}_* \subset C_0$. But then $\mathscr{C}^* \cup \{C_0\}$ is clearly a chain which properly contains $\mathscr{C}^*$. This, however, contradicts the maximality of $\mathscr{C}^*$. Thus, $\hat{C}_*$ is maximal element of $(\mathscr{C}, \subset)$ and consequently, a maximal chain in $(X, <)$.

7. Let $(X, <)$ be a partially ordered set. Prove that every chain $C \subset X$ is contained in a maximal chain.

Hint:

Let $\mathscr{H}_C$ be the collection of all chains in X which contain C. Clearly $(\mathscr{H}_C, \subset)$ is a partially ordered set. If $\mathscr{C} \subset \mathscr{H}_C$ is a chain then it is clear that

$$\cup \mathscr{C} = H \in \mathscr{H}_C$$

is the least upper bound of $\mathscr{C}$. Thus, every chain in $(\mathscr{H}_C, \subset)$ has a least upper bound in $(\mathscr{H}_C, \subset)$ implying that $(\mathscr{H}_C, \subset)$ has a maximal element, which is a chain and which contains C.

8. Given $(\mathscr{A}, \subset)$ where $\mathscr{A}$ is a class of nonempty sets. A subclass $\mathscr{F} \subset \mathscr{A}$ is said to have the finite intersection property (f.i.p) if every finite subclass $\mathscr{F}_0 \subset \mathscr{F}$ satisfies $\cap \{A; A \in \mathscr{F}_0\} \neq \theta$. Show that $(\mathscr{A}, \subset)$ contains a maximal class having the (f.i.p).

Hint:

Define

$$\overset{0}{\mathscr{H}} = \{\mathscr{C} \subset \mathscr{A}; \mathscr{F} \subset \mathscr{C} \text{ and } \mathscr{C} \text{ has the (f.i.p.)}\}.$$

It is not difficult to verify that $(\overset{0}{\mathscr{H}}, \subset)$ is inductive (see problem 5). Consequently, there is a maximal element, say $\mathscr{M} \in \overset{0}{\mathscr{H}}$ which contains $\mathscr{F}$ and which obviously has the (f.i.p).

9. Let $X = \mathbb{R}^{\mathbb{N}_+}$; order X "lexicographically", i.e., $(x_1, x_2, \ldots) <' (y_1, y_2, \ldots)$ if the first index n where they differ, $x_n \leq y_n$. Show that $(X, <)$ is totally ordered.

Hint:

Clearly, any two elements from X are $<'$-related.

10. Ordering by extension. Let $X \neq \theta$ and denote by $\mathcal{F}$ the collection of all maps

$$f_A : A \to Y, \quad A \in \mathcal{P}(X).$$

Order the collection $\mathcal{F}$ by "extension", i.e., two members f_{A_1} and f_{A_2} of $\mathcal{F}$ are related and $f_{A_1} \langle f_{A_2}$ (f_{A_1} precedes f_{A_2}) if and only if

$$A_1 \subset A_2 \text{ and } f_{A_2} = f_{A_1} \text{ on } A_1$$

(i.e., $f_{A_2} | A_1 = f_{A_1}$ so that f_{A_2} is an extension of f_{A_1}). Show that the relation $\langle$ is a partial ordering.

Hint:

Clearly $f_A \langle f_A$ for all $f_A \in \mathcal{F}$; it is also clear that $f_{A_1} \langle f_{A_2}$ and $f_{A_2} \langle f_{A_1} \Rightarrow A_1 = A_2$. Finally, if

$$f_{A_1} \langle f_{A_2} \text{ and } f_{A_2} \langle f_{A_3} \Rightarrow f_{A_1} \langle f_{A_3}.$$

11. Show that the partially ordered set $(\mathcal{F}, \langle)$ is inductive.

Hint:

Let $C = \{f_{A_t} ; t \in T\} \subset \mathcal{F}$ be a chain. Clearly,

$$f_{\underset{t \in T}{\cup A_t}} \text{ is an extension of each } f_{A_s} \in C,$$

and consequently an upper bound of C.

12. A **"lattice"** is a partially ordered set $(X, <)$ in which any two elements have a least upper bound and greatest lower bound. If every nonempty subset of X has a sup. and an inf., it is called a **"complete lattice."** Show that if $(X, <)$ has a greatest element and every nonempty subset has an inf., it is a complete lattice.

Hint:

We must show that every nonempty set $A \subset X$ has a sup. Let B be the set of upper bounds of A; clearly $B \neq \theta$. Let $b = \inf B$. Then it is clear that $b = \sup A$.

11. THREE MAXIMAL PRINCIPLES OF SET THEORY

Among the milestones of set theory are the

(i) **Hausdorff maximal principle:** Every partially ordered set has a maximal chain.

(ii) **Zorn lemma:** Every inductive partially ordered set has a maximal element.

(iii) **Kuratowski lemma:** Every chain in a partially ordered set is contained in a maximal chain.

In this section we prove the equivalence of these three principles and show that they imply the Axiom of Choice. We begin with some preliminaries which will simplify our presentation. Let $\mathscr{A}$ be a class of sets; for the reason of economy we shall continue to use the notation

$$\cup \mathscr{A} = \cup \{A; \ A \in \mathscr{A} \}.$$

We shall need the next lemma (see also problem 5 of section 10)

Lemma 1.

Let $\mathscr{C}$ be the collection of all chains in a partially ordered set $(X, <)$, then the partially ordered set $(\mathscr{C}, \subset)$ is strictly inductive.

Proof:

We must show that every chain $\mathbb{C}$ in $(\mathscr{C}, \subset)$ has a least upper bound in $(\mathscr{C}, \subset)$. To this end it suffices to prove that

$$\cup \mathbb{C} = \hat{C}$$

is a chain in $(X, <)$. Indeed, if $x, \ y \in \hat{C}$ then $x \in C_1$ and $y_2 \in C_2$, where C_1 and C_2 are elements of $\mathbb{C}$. But $\mathbb{C}$ is a chain so that either $C_1 \subset C_2$ or $C_2 \subset C_1$. Assume $C_1 \subset C_2$, then $x, \ y \in C_2$ and since C_2 is a chain in $(X, <)$, either $x<y$, or $y<x$, proving that the set $\hat{C}$ is a chain in $(X, <)$. Consequently, $\hat{C}$ is an element of $(\mathscr{C}, \subset)$ and obviously $\hat{C} = \sup \mathbb{C}$, which proves that every chain in $(\mathscr{C}, \subset)$ has a least upper bound in $(\mathscr{C}, \subset)$.

Thus, $(\mathscr{C}, \subset)$ is strictly inductive. ♥

Corollary 1.

If it happens that $\mathbb{C}_* \subset \mathscr{C}$ is a maximal chain, then since

$$\cup \mathbb{C}_* = \hat{C}_* \ \text{ is a chain in } (X, <),$$

$\hat{C}_*$ is a maximal element in $(\mathscr{C}, \subset)$ and hence, a maximal chain in $(X, <)$. For, if it is not, there is a chain $C_0 \supset \hat{C}_*$. But then $\mathbb{C}_* \cup \{C_0\}$ is a chain which contains properly $\mathbb{C}_*$. This contradicts the maximality property of $\mathbb{C}_*$.

Proposition 1.

$$(i) <\!=\!> (ii)$$

Proof:

Let $(X, <)$ be inductive. If $C^* \subset X$ is a maximal chain, and $b^* \in X$ is an upper bound of C^*, then b^* is a maximal member in $(X, <)$. For, if it is not, there exists a $b_0 \in X$ such that $b^* < b_0$. But then $C^* \cup \{b_0\}$ is a chain in $(X, <)$ which contains properly C^*. However, this contradicts the maximality of C^*. Thus, b^* is a maximal element. Notice that $b^* \in C^*$, for if it is not, $C^* \subset C^* \cup \{b^*\}$, which is contradiction.

The easiest way to prove the converse is to start with the partially ordered set $(\mathscr{C}, \subset)$, where $\mathscr{C}$ is the class of all chains in $(X, <)$. According to lemma 1, $(\mathscr{C}, \subset)$ is inductive. Then it has a maximal element, which is a maximal chain in $(X, <)$. This completes our proof. ♥

Proposition 2.

$$(ii) <=> (iii)$$

Proof:

Given a partially ordered set $(X, <)$ and a chain $C \subset X$, denote by

$$\mathscr{C}_C = \{C' \in \mathscr{C}, C' \supset C\}.$$

Let $\mathbb{C}$ be a chain in $(\mathscr{C}_C, \subset)$, then the set $H = \cup \mathbb{C}$ is a chain in $(X, <)$; clearly, $H \in \mathscr{C}_C$ (because $C \subset H$).

Thus, every chain in $(\mathscr{C}_C, \subset)$ has an upper bound in $(\mathscr{C}_C, \subset)$. This proves that the set $(\mathscr{C}_C, \subset)$ is inductive and according to proposition 1, it has a maximal element, which is a maximal chain in $(X, <)$ and which contains C.

To prove the converse, let $(X, <)$ be an inductive partially ordered set, whose each chain is contained in a maximal chain. Since $(X, <)$ is inductive every maximal chain has an upper bound which is a maximal element of $(X, <)$. ♥

Remark 1.

The Zorn lemma and the Hausdorff maximal principle are useful devices to establish if a partially ordered set has a maximal element. According to the Hausdorff maximal principle, every partially ordered set has a maximal chain. Now if every chain has an upper bound, the existence of a maximal element is guaranteed by the Zorn lemma. As an application we prove the following:

Proposition 3.

The Zorn lemma implies the Axiom of Choice.

Proof:

Let $\mathscr{M} \neq \theta$ be a class of nonempty sets. There are subclasses (for instance, every finite one) on which we know how to define a choice function. Let $\mathscr{F}$ be the class of all these (choice) functions. Given $h \in \mathscr{F}$, let $\mathscr{D} \subset \mathscr{M}$ be its domain of definition, i.e.,

$$h : \mathscr{D} \to Y = \cup \mathscr{M} \text{ such that } h(A) = x \in A.$$

In terms of ordered pairs (see section 8)

$$h = \{(A, x); A \in \mathscr{D}\} \text{ and } x = h(A).$$

In other words, $h \subset \mathscr{D} \times Y$; thus, $(\mathscr{F}, \subset)$ is a partially ordered set. Notice that $h_1, h_2 \in \mathscr{F}$ are related if either $h_1 \subset h_2$ or $h_2 \subset h_1$. If, for instance,

$$h_1 = \{(A', x'); A' \in \mathscr{D}'\} \quad \text{and} \quad h_2 = \{(A'', x''); A'' \in \mathscr{D}''\},$$

then $h_1 \subset h_2$ if $\mathscr{D}' \subset \mathscr{D}''$ and $h_1 = h_2$ on $\mathscr{D}'$.

Let $\mathbb{C} \subset \mathscr{F}$ be a chain, and set

$$h^* = \cup\, \mathbb{C} = \cup\{h : h \in \mathbb{C}\}.$$

If $\mathscr{D}^*$ is the domain of definition of h^*, then given any $B \in \mathscr{D}^*$ we have that $B \in \mathscr{D}'$ where $\mathscr{D}'$ is the domain of some $h' \in \mathbb{C}$. Thus,

$$h^*(B) = h'(B) \in B,$$

which clearly indicates that h^* is a choice function defined on $\mathscr{D}^*$ and also the least upper bound of $\mathbb{C}$ in $(\mathscr{F}, \subset)$. Thus, $(\mathscr{F}, \subset)$ is inductive and according to Zorn's lemma $(\mathscr{F}, \subset)$ has a maximal element, say f.

The function f is defined everywhere on $\mathscr{M}$ and is a choice function on $\mathscr{M}$. To see this, suppose that $D \in \mathscr{M}$ is a member on which f is not defined. Then $f \cup \{(D, x)\}$, where $x \in D$, is clearly an element of $\mathscr{F}$, containing f, which contradicts the maximality property of f. This proves the proposition.   ♥

We are now in position to prove proposition 3 of section 6, i.e., that $X \sim X \times X$ for any infinite set X.
Proof:

Denote by $\mathscr{F}$ the collection of all bijections

$$f_A : A \to (A \times A), \quad A \subset X.$$

Clearly $\mathscr{F} \neq \theta$ ($A \sim A \times A$ if A is countable). Partially order the collection $\mathscr{F}$ by extension (problems 10 and 11 of section 10). Since $(\mathscr{F}, \langle\,\rangle)$ is inductive it has a maximal element (Zorn's lemma) say f_M,

$$f_M : M \to M \times M, \quad \text{where } M \sim M \times M.$$

To prove the proposition it suffices to show that $M \sim X$. Assume, by the way of contradiction, that $M \nsim X$. Since $M \subset X$, Card $M <$ Card X. Therefore, since $X = M \cup (X-M)$ we have that

$$\text{Card } M < \text{Card } (X-M) \tag{1}$$

(for if Card $M \geq$ Card $(X-M)$ we would have that

$$\text{Card } X \leq \text{Card } M + \text{Card } M = 2\, \text{Card } M = \text{Card } M$$

which is contradiction).

Because of (1) there exists $Y \subset X-M$ such that $Y \sim M$. Then the map

$$f_{M \cup Y} : (M \cup Y) \to (M \cup Y) \times (M \cup Y) = (M \times M) \cup (M \times Y) \cup (Y \times M) \cup (Y \times Y)$$

is a bijection because

$$M \sim M \times M \quad \text{and} \quad Y \sim (M \times Y) \cup (Y \times M) \cup (Y \times Y).$$

Since

$$f_{M \cup Y} \,|\, M = f_M$$

$f_{M \cup Y}$ is an extension of f_M so that $f_M \langle f_{M \cup Y}$, which contradicts the maximality of f_M. Thus,

Card M < Card X is a false assumption. This proves the proposition. ♥

Problems and complements

1. Let X be a partially ordered set. If $C \subset X$ is a maximal chain, show that by adding an element $z \in C^c$ to C we do not obtain a new chain.

    Hint:

    If $C \cup \{z\}$ is chain then it contains C as a proper subset. This contradicts the maximality of C so we must have that $C = C \cup \{z\}$. But then $z \in C$, contradicting the assumption that $z \in C^c$.

2. Let $\mathcal{A} = \{A_1, A_2, \ldots, A_n\}$ be a class of pairwise disjoint nonempty sets. Find a choice function defined on $\mathcal{A}$.

    Hint:

    Since $\mathcal{A}$ is finite we can select one element from each A_i to obtain a set C which consists of n elements. The map

$$h : \mathcal{A} \to \bigcup_1^n A_i \quad \text{defined by } h(A_i) \in C \cap A_i$$

is obviously a choice function on $\mathcal{A}$.

3. Do the same thing for $\mathcal{A}$ if $\{A_i\}_1^n$ are not disjoint.

    Hint:

$$\text{Write } \{A\}_1^n \text{ as } A_1, A_2 - A_1, \ldots, A_n - \bigcap_{i=1}^{n-1} A_i .$$

4. Prove the proposition 3 using the ordering by extension.

    Hint:

    Let $\mathcal{A} \neq \theta$ be a class of nonempty sets. There is a subclass $S \subset \mathcal{A}$ on which is possible to define a choice function, say

$$f_S : S \to \cup\, S.$$

Denote by $\overset{o}{\mathcal{H}}$ the collection of all such classes and by

$$F = \{f_S ; S \in \overset{o}{\mathcal{H}}\}$$

(the class $\overset{o}{\mathcal{H}} \neq \theta$ since it contains finite subclasses of $\mathcal{A}$). The nonempty collection F can be partially ordered by extension. In other words, two $f_{S_1}, f_{S_2} \in F$ are related, and $f_{S_1} < f_{S_2}$ if and only if

$$S_1 \subset S_2 \text{ and } f_{S_1} = f_{S_2} \mid S_1 \qquad (\text{i.e., } f_{S_1} = f_{S_2} \text{ on } S_1).$$

So f_{S_2} is an extension to S_2 of f_{S_1}. Let $\mathcal{C} \subset F$ be a chain and denote by $S^* = \sup \mathbf{N}$, where $\mathbf{N} = \{S, f_S \in \mathcal{C}\}$. Since $f_S \langle f_{S^*}$ for any $f_S \in \mathcal{C}$, it is clear that the map f_{S^*} is an upper bound of $\mathcal{C}$. It is also clear that

$$f_S = f_{S^*} \mid S \text{ for any } f_S \in \mathcal{C},$$

proving that $f_{S^*} \in F$. Thus, the partially ordered set $(F, \langle\,)$ is inductive. Now, according to Zorn's lemma F has a maximal element, say f^*. This function must be defined everywhere on $\mathcal{K}$, for if there is a set $K \in \mathcal{K}$ on which it is not defined, we could choose an element $x \in K$ and put $f^*(K) = x$. But this would be a proper extension of f^*, contradicting its maximality.

12. WELL ORDERING THEOREM OF ZERMELO

One of the crowning achievements of the theory of ordered sets is the theorem due to Zermelo which states that every set can be well ordered, provided that the Axiom of Choice holds. In other words, for each set X there is a well ordering relation in X. **Recall (see section 10), an ordering relation $<$ in a set X is said to be "well ordering" if every nonempty subset of $(X, <)$ has a "smallest" or "first" element.**

There are sets which can be easily well ordered, for instance all finite sets. We can effectively well order all countable sets. Unfortunately, however, no method for well ordering of any uncountable set has been found so far.

An essential feature of every well ordered set is that every one of its elements (except the last one, if it exists) has its "immediate successor." **Thus, every well ordering relation has the proper characteristic of counting.** For instance, if x_1 is the first member of a well ordered set $(X, <)$ then the first members of the well ordered set $(X-\{x_1\}, <)$ is the immediate successor of x_1.

Remark 1

Let $(X, <)$ be a well ordered set and let $y \notin X$. In the set $X^*=X\cup\{y\}$ we can define a well ordering relation $<'$ as follows: $<'=<$ in X and $x<'y$ for every $x\in X$. To show $(X^*, <')$ is well ordered, let $A\subset X^*$. Then either $A=\{y\}$ or $A\cap X\neq\theta$. Since $A\cap X$ has a first element (as a subset of X), that element is also the first element of A in the set $(X^*, <)$. Here, of course, y is the last element of $(X^*, <)$. Note that the element y does not have its immediate predecessor.

We are now ready to prove the Zermelo well ordering (WO) theorem. Using the abbreviation (ZL) for the Zorn lemma we have

Proposition 1

$$(ZL) => (WO).$$

Proof:

Let X be a set to be well ordered. Clearly, there are subsets of X which can be well ordered (for instance, all finite once). Let $\boldsymbol{W}$ be the collection of all such subsets of X, then $(\boldsymbol{W}, <)$ is a partially ordered set where $A_1<A_2$ means: (i) $A_1\subset A_2$; (ii) the well ordering of A_1 is induced by the well ordering of A_2.

Let $\mathcal{C}$ be a chain in $(\boldsymbol{W}, <)$, then clearly

$$\cup\mathcal{C} = \hat{C} \in \boldsymbol{W}.$$

Consequently, $(\boldsymbol{W}, <)$ is inductive and according to Zorn's lemma, it has a maximal element, say M. If M=X the proposition is proved; otherwise, $X-M\neq\theta$ and if $x_0 \in X-M$ then obviously $x_0 \notin M$. Now, according to the remark 1, $M^*=M\cup\{x_0\}$ is also well ordered which contains M properly. This, however, contradicts the maximality of M, so that $M=M^*$. But this implies that $x_0 \in M$. Thus, the assumption that $X-M\neq\theta$ is false, implying that M=X. This completes the proof of the (WO) theorem. ♥

Corollary 1

Every well ordered set X has a choice function. We shall prove this claim by exhibiting such a function. Define a map $h: \mathscr{P}_X \to X$ as follows: for every $A \in \mathscr{P}_X = \mathscr{P}(X) - \{\theta\}$ define:

$$h(A) = \text{the first member of } A.$$

Clearly, $h(A) \in A$.

Remark 2

We now know, if the Zorn lemma is assumed, that every set can be well ordered. Unfortunately, the well ordering theorem is of the existential type. No specific procedure for well ordering of any uncountable set exists at preset time. No well ordering relation of the set of real numbers has been found to this day.

Example 1

Let us show that the lexicographic ordering in $N_+ \times N_+$ is a well ordering. Recall (see problem 9 of section 10) that $(i, j) < '(i_1, j_1)$ if either $i < i_1$ (here $<$ is the usual ordering in R) or else, $i = i_1$ and $j < j_1$. Let $A \subset N_+ \times N_+$ and denote by

$$k_0 = \inf\{k; (k, \ell) \in A\}, \quad k_1 = \inf\{\ell; (k_0, \ell) \in A\}.$$

Clearly, $(k_0, k_1) \in A$ is the $<'$-smallest member of A.

The significance of the "well ordering theorem" lies in the possibility that we can apply the idea and principle of "mathematical induction" to well ordered sets.

Mathematical Induction.

Let $\varphi(n)$ be a statement which one wants to prove to hold for every $n \in N_+$. If

(i) $\varphi(1)$ is proved to be true;

(ii) for each $n \in N_+$ the hypothesis that $\varphi(n)$ is true implies that $\varphi(n+1)$ is true.

Then $\varphi(n)$ is true for every $n \in N_+$.

Proof:

(By contradiction). Suppose that under the conditions (i) and (ii) $\varphi(n)$ does not hold for every $n \in N_+$. The subset of N_+, on which $\varphi(n)$ is false, has a first element (as a subset of well ordered set) say $n_1 > 1$. Thus, $\varphi(n_1)$ is false while $\varphi(n_1 - 1)$ is true, which contradicts (ii). ♥

Let $(X, <)$ be a well ordered set. Given any $x \in X$, the set

$$X_x = \{u \in X; u < x\} \tag{1}$$

is called an "**initial segment**" of $(X, <)$ determined by x. Clearly, $x \notin X_x$.

Transfinite Induction.

Let $(X, <)$ be a well ordered set, and let $K \subset X$ be a subset such that if for an element $x \in X$, $X_x \subset K$ implies that $x \in K$, then $K = X$.

Proof:

If x_0 is the first element of X, then since

$$X_{x_0} = \theta \subset K \qquad \text{it follows that } x_0 \in K.$$

The set K^c does not have the first element (for if it does, and $y \in K^c$ is the first element, then $X_y \subset K \Rightarrow$ that $y \in K$, contradiction) which is in contradiction with the definition of well ordering. Consequently, $K^c = \theta$, proving our contention. ♥

Remark 3.

On $(N_+, <)$ the transfinite induction principle is equivalent to the following: Given a proposition $\varphi(n)$ on N_+ such that $\varphi(1)$ is true. If $\varphi(k)$ is true for every $k \leq n$, then $\varphi(n+1)$ is also true.

Thus, the ordinary mathematical induction and the transfinite induction differ in their formulations. The transfinite induction, instead of passing to each element from its immediate predecessor, passes to each element from the set of all its predecessors. It is obvious that on N_+ these two induction principles are equivalent.

Problems and complements

1. If $(X, <)$ is an ordered set and $Y \sim X$; show that Y can be also ordered.

    Hint:

    Since $X \sim Y$ there is a bijection $f{:}X \to Y$. If $x_1 < x_2$ in X we stipulate that $y_1 = f(x_1)$ is a predecessor of $y_2 = f(x_2)$; this defines an ordering relation, say $<'$ in Y. Actually, we have that

$$x_1 < x_2 \iff f(x_1) <' f(x_2)$$

so that f is an **"isomorphism."**

2. Let $(X, <)$ and $(Y, <')$ be two well ordered sets. Find a well ordering relation in $X \times Y$.

    Hint:

    Lexicographic ordering (see example 1).

3. Let $(X, <)$ be a well ordered and uncountable set. Show that there exists $x_0 \in X$ such that the initial segment X_{x_0} is countable.

    Hint:

    Denote by $B \subset X$ the set of all those elements y for which X_y is uncountable. Let x_0 be the "smallest" element of B. Clearly then X_{x_0} is countable.

4. Let $(X, <)$ and $(Y, <')$ be two ordered sets, such that $X \sim Y$. An "order preserving" bijection $h{:}X \to Y$ (i.e., if $x_1 < x_2$ then necessarily $h(x_1) <' h(x_2)$) is called an **"isomorphism."** If it exists X and Y are said to be **"isomorphic"** and we write $X \cong Y$. If X and Y are well ordered and $h{:}X \to Y$ is an isomorphism, show that h is unique.

    Hint:

    We have for any $x \in X$ that

$$h(X_x) = \{h(u); u < x\} = \{y \in Y; y <' h(x)\} = Y_{h(x)} \, .$$

This obviously implies that $X_x \cong Y_{h(x)}$. If, on the other hand, $h_1{:}X \to Y$ is also an isomorphism we have that for all $x \in X$, $X_x \cong Y_{h_1(x)}$, which is impossible unless $h(x) = h_1(x)$ for every $x \in X$.

5. Let $(X, <)$ be a well ordered set; show that every element of X which is not the last one, has an immediate successor.

    Hint:

    Let $x_0 \in X$ be arbitrary, then the nonempty subset $K_{x_0} = \{x \in X; x_0 < x\}$ has a first element, say x'. Clearly, $x_0 < x'$ and there is no element between x_0 and x'.

6. Let $(X, <)$ be well ordered and $h{:}X \to X$ an order preserving map. Show that $(h(X), <)$ is isomorphic to $(X, <)$.

    Hint:

    For any $x_1, x_2 \in X$ such that $x_1 < x_2$, $h(x_1) < h(x_2)$. Consequently, $h{:}X \to h(X)$ is an order preserving bijection.

7. Let $(Y, <')$ be (totally) ordered set, isomorphic to a well ordered set $(X, <)$. Show that $(X, <')$ is also well ordered.

Hint:

Let $h:X \to Y$ be an isomorphism and $G \subset Y$. Denote by $G' = h^{-1}(G)$, then if $x_0 \in G'$ is a first element of G', $h(x_0) = y_0 \in G$ is a first element of G, because, if there exists $y'_0 \in G$ and $y'_0 < y_0$, then $y'_0 = h(x')$, which then implies that $x' < x_0$, contradiction.

8. Show that every countable set can be well ordered.

Hint:

Use the previous exercise and $(N_+, <)$.

13. (AC) <=> (ZL)

For a partially ordered set $(\mathcal{U}, \subset)$ it was proved (see proposition 1 of section 10), that there is a map $f: \mathcal{U} \to \mathcal{U}$ such that $A \subset f(A)$ for every $A \in \mathcal{U}$ if the Axiom of Choice holds. Let us associate with every $H \in \mathcal{U}$ a subclass $\mathcal{A}_H \subset \mathcal{U}$ defined by

$$\mathcal{A}_H = \{B \in \mathcal{U}; B \supset H\},$$

then from corollary 1 of the section 10 we deduce that

(i) $H \in \mathcal{A}_H$; (ii) $f(\mathcal{A}_H) \subset \mathcal{A}_H$; (iii) $(\mathcal{A}_H, \subset)$ is inductive if $(\mathcal{U}, \subset)$ is inductive.

Let $H \in \mathcal{U}$ be a member which will remain fixed throughout this section. Any subclass $\mathcal{F} \subset \mathcal{U}$ which contains H as an element and satisfies the conditions (ii) and (iii) will be called **"admissible."** The collection $\mathcal{A}$ of all admissible subclasses of $\mathcal{U}$ is obviously nonempty $(\mathcal{A}_H \in \mathcal{A})$. We now prove the existence of a minimal admissible subclass $\mathcal{E}$ containing H.

Proposition 1

Let $(\mathcal{U}, \subset)$ be inductive, then $\cap\{\mathcal{F}; \mathcal{F} \in \mathcal{A}\} = \mathcal{E}$ is admissible.

Proof:

(i) $H \in \mathcal{E}$ because each $\mathcal{F}$ contains H.

(ii) $f(\mathcal{E}) \subset \cap\{f(\mathcal{F}); \mathcal{F} \in \mathcal{A}\} \subset \cap\{\mathcal{F}; \mathcal{F} \in \mathcal{A}\} = \mathcal{E}$

(iii) If $\mathbf{C} \subset \mathcal{E}$ is a chain, then since $\mathbf{C} \subset \mathcal{F} \in \mathcal{A}$, $\cup \mathbf{C} \in \mathcal{F}$ for every $\mathcal{F} \in \mathcal{A}$. Consequently, $\cup \mathbf{C} \in \mathcal{E}$, which proves the proposition. ♥

Next we investigate the structure of the class $\mathcal{E}$. Our goal is to establish that $\mathcal{E}$ is a chain in $(\mathcal{U}, \subset)$. Notice that $\mathcal{E} \subset \mathcal{A}_H$. First, let us examine some basic properties of the elements of $\mathcal{E}$. To this end let us call an element $R \in \mathcal{E}$ "regular" if for any $A \in \mathcal{E}$ for which

$$A \subset R \text{ properly,} \Rightarrow f(A) \subset R. \tag{1}$$

The set H is regular vacuously. An essential feature of any regular set $R \in \mathcal{E}$, to be proved next is: for any $A \in \mathcal{E}$

$$\text{either } A \subset R \text{ properly or } A \supset f(R). \tag{2}$$

In other words, any member A of $\mathcal{E}$ is either a proper subset of a regular set R or it contains R as a subset. To prove this contention consider a regular set $R \in \mathcal{E}$, and denote by $\mathcal{F}_R \subset \mathcal{E}$ the collection of all those sets satisfying (2), that is:

$$\mathcal{F}_R = \{B \in \mathcal{E}; B \subset R \text{ properly or } f(R) \subset B\}.$$

Proposition 2

$$\mathcal{F}_R = \mathcal{E}. \tag{3}$$

Proof:

The idea is to show that $\mathcal{F}_R$ is admissible.

(i) $H \in \mathcal{F}_R$ since $H \subset R$.

(ii) $f(\mathcal{F}_R) = \{f(B); B \subset R \text{ properly or } f(R) \subset B\} \subset \{f(B); f(B) \subset R\} \cup \{f(B); f(R) \subset f(B)\} \subset \{A; A \subset R\} \cup \{A; f(R) \subset A\} = \mathcal{F}_R$.

(iii) If $C \subset \mathcal{F}_R$ is a chain and $C \subset R$ properly for each $C \in C$, then clearly $\cup C \subset R$ properly. If on the other hand, for some $C' \in C$, $C' \not\subset R$ then since $C' \in \mathcal{F}_R$, $C' \supset f(R)$ so that $\cup C \supset f(R)$. This proves that $\mathcal{F}_R \subset \mathcal{E}$ is admissible and since $\mathcal{E}$ is the minimal admissible class, (3) follows. ♥

Corollary 1

From (3) we deduce at once that that for no $E \in \mathcal{E}$ we have that

$$R \subset E \subset f(R) \tag{4}$$

Proposition 3

Every member of $\mathcal{E}$ is regular.

Proof:

Denote by $\mathcal{F}_*$ the class of all regular members of $\mathcal{E}$. To prove that

$$\mathcal{F}_* = \mathcal{E} \tag{5}$$

we must show that $\mathcal{F}_*$ is admissible. Clearly $H \in \mathcal{F}_*$. If $D \in \mathcal{F}_*$ we must show that $f(D) \in \mathcal{F}_*$. Indeed, if $B \in \mathcal{E}$ is such that $B \in f(D)$ properly, it implies that $f(B) \subset f(D)$ or that $B \supset f(D)$. If $B \subset D$ properly, then $f(B) \subset D \subset f(D)$. In the case $B \supset f(D) \supset D$ we would have $D \subset B \subset f(D)$ which due to (4) is not possible. Thus, $D \in \mathcal{F}_*$ implies $f(D) \in \mathcal{F}_*$ and consequently we have that $f(\mathcal{F}_*) \subset \mathcal{F}_*$. Finally, if $C \subset \mathcal{F}_*$ is a chain let $B \subset \cup C$ properly, then there exists a $C \in C$ such that $B \subset C$ properly so that $f(B) \subset C$ and consequently, $f(B) \subset \cup C$. Thus $\cup C \in \mathcal{F}_*$, which proves that $\mathcal{F}_*$ is inductive. Thus $\mathcal{F}_*$ is admissible. ♥

Proposition 4

$(AC) \Rightarrow (ZL)$

Proof:

Any two $R_1, R_2 \in \mathcal{E}$ are regular and according to proposition 2, either $R_1 \subset R_2$ or $R_2 \subset R_1$. Thus $\mathcal{E}$ is a chain such that

$$W = \cup\, \mathscr{C} \in \mathscr{C}.$$

Since $f(\mathscr{C}) \subset \mathscr{C}$ it follows that W is an upper bound of $f(\mathscr{C})$, implying that $f(W) \subset W$, which contradicts the fact that $W \subset f(W)$. Thus

$$f(W) = W,$$

which proves that W is a maximal element of $(\mathscr{U}, \subset)$. This proves the proposition. ♥

Remark 1

The last proposition and proposition 3 of section 11 prove the basic claim of this section.

To summarize, we have established that the minimal admissible class $\mathscr{C}$ is a chain in $(\mathscr{U}, \subset)$. Since $\mathscr{C} \subset \mathscr{C}$ and $\mathscr{C}$ is inductive, $W = \sup \mathscr{C} \in \mathscr{C}$. On the other hand, $f(\mathscr{C}) \subset \mathscr{C}$, implying that the set W is an upper bound of $f(\mathscr{C})$, so that $f(W) \subset W$. From this and proposition1 of section 10, it follows that $f(W) = W$.

With this we have demonstrated that the (AC), (ZL), (HMP), (KL) and (WO) ordering, are equivalent statements. Schematically, we have:

$$\begin{array}{ccc}
\text{(AC)} & \Rightarrow & \text{(ZL)} \\
\Uparrow & & \Downarrow \\
\text{(WO)} & & \text{(KL)} \\
\Uparrow & & \Downarrow \\
\text{(ZL)} & \Leftarrow & \text{(HMP}
\end{array}$$

14. PRINCIPLE OF RECURSIVE DEFINITION

Let $X\neq\theta$ be a set. Recall that by a sequence in X we mean a map $f:N_+\to X$, which is usually given in the form

$$(x_1, x_2, \dots) \quad \text{where} \quad x_n = f(n) \quad \text{for all } n\in N_+.$$

It is often required to construct a sequence in X by specifying $f(1)$, then defining $f(n)$ in terms of $f(n-1)$ and possibly of other values of $f(k)$ for $k<(n-1)$. This is trivial if the sequence $f(\cdot)$ is subject to a "recursive" relation. As an illustration consider the following:

Example 1

In calculating the circumference of the unit circle from inscribed regular polygons of 2^n sides, $n=2, 3, \dots$ one finds that the length $S(2^{n+1})$ of the side of an inscribed 2^{n+1}-polygon is

$$S(2^{n+1}) = \sqrt{2-\sqrt{4-S^2(2^n)}}\,.$$

If we now set

$$x_{n+1}^2 = 4 - S^2(2^n)$$

it follows that

$$S(2^{n+1}) = \sqrt{2-x_{n+1}}\,;$$

then after some simple calculations we obtain the recursion

$$x_{n+1}^2 - x_n = 2\,. \tag{1}$$

Since $S(2^2)=\sqrt{2}$ we have that the initial condition $x_2=0$. With this it is easy to obtain further terms of the sequence $(x_2, x_3, x_4, \dots)$.

In its classical form the principle of recursive definition is formulated as follows: Let $f:N_+\to X$ be a sequence defined recursively by specifying the "initial value" $f(1)$, then defining $f(n+1)$ in terms of n, $f(n)$ and possibly of other values $f(k)$ for $k<n$, by the method of mathematical induction. The method of mathematical induction easily proves the uniqueness of the map $f(\cdot)$ if it exists, but **not its existence.** The next proposition, termed the "recursion theorem" establishes the existence of $f(\cdot)$.

Proposition 1

Let $F:N_+\times X\to X$ be a given map. There exists a uniquely defined sequence $f:N_+\to X$ such that for some $\gamma\in X$

$$f(1) = \gamma \quad \text{and} \quad f(n+1) = F(n, f(n)) \quad \text{for } n\in N_+ \tag{2}$$

or equivalently $x_1=\gamma$ and $x_{n+1}=F(n, x_n)$.

Proof:

Let us first establish the uniqueness of the map $f(\cdot)$ if it exists. If $h: N_+ \to X$ is another sequence satisfying (2), then

$$h(1) = \gamma = f(1), \quad h(2) = F(1, h(1)) = F(1, \gamma) = f(2)$$

and so on. So, if $h(n)=f(n)$ we have that

$$h(n+1) = F(n, h(n)) = F(n, f(n)) = f(n+1).$$

By mathematical induction we conclude that $h(k)=f(k)$ for all $k \in N_+$.

To prove the existence of the map $f(\cdot)$ recall that $f \subset N_+ \times X$ (section 8). Thus, we have to construct a set P of ordered pairs $(k, x_k) \in N_+ \times X$ satisfying:

$$(1, \gamma) \in P; \text{ from } (k, x_k) \in P \Rightarrow (k+1, F(k, x_k)) \in P. \tag{3}$$

In other words, the set P is formed starting from $(1, \gamma)$ by repeated application from (k, x_k) to $(k+1, F(k, x_k))$. Then the set P is the desired function $f(\cdot)$, if we can show that for every $j \in N_+$ there exists exactly one $y_j \in X$, such that $(j, y_j) \in P$.

We shall prove this by contradiction. Assume that for some $i \in N_+$ there exists only one $u \in X$ such that $(i, u) \in P$, but that for $i+1$ we have

$$(i+1, u_1) \in P \quad \text{and} \quad (i+1, u_2) \in P, \ u_1 \neq u_2.$$

Then according to (3) it follows that

$$u_1 = F(i, y_1) \qquad u_2 = F(i, y_2) \in P,$$

which contradicts the assumed uniqueness of (i, u). Thus, we have that $y_1 = y_2$, proving that the set of ordered pairs is a function which satisfies (3). This proves the proposition.   ♥

Remark 1

With the initial condition $f(1)= \gamma$ the other values of the sequence $f(\cdot)$ follow recursively from (2);

$$f(2) = F(1, f(1)) = F(1, \gamma)$$
$$f(3) = F(2, f(2)) = F(2, F(1, \gamma))$$

and so on.

Remark 2

Notice that the previous proposition cannot be proved by simply defining $f: N_+ \to X$ by formula $f(k+1)=F(k, f(k))$; such a definition would be circular in the sense that it defines $f(\cdot)$ in terms of itself. The whole purpose of the proposition is to show that there exists a uniquely defined map $f: N_+ \to X$ having this property.

Denote by $(N_+)_n = \{1, 2, \ldots, n\}$; A more general form of recursive definition is obtained under the assumption that the value $f(n+1)$ depends on $f(1), \ldots, f(n)$, i.e., on

$f|(N_+)_n$. Thus, let $\{F_n\}_1^\infty$ be a given sequence of maps

$$F_n: X^n \to X, \quad n = 1, 2, \ldots$$

where $X^n = X \times \cdots \times X_n$ (the Cartesian product of n copies of X). Replace the condition (2) by

$$f(1) = \gamma, \quad f(n+1) = F_n(f|(N_n)), \quad \text{for all } n \in N_+ \tag{4}$$

and denote by

$$f^*(n) = f|(N_+)_n = (f(1), \ldots, f(n)).$$

It is clear that the map $f(\cdot)$ is uniquely determined by $f^*(\cdot)$. In order to establish the existence and uniqueness of $f(\cdot)$ in (4) we define a map F by

$$F(n, f^*(\cdot)) = (f^*(\cdot), F_n(f(n))).$$

Consequently, since

$$f^*(n+1) = (f^*(n), F_n(f^*(n)))$$

it follows that

$$f^*(n+1) = F_n(f^*(n))) \qquad n = 1, 2, \ldots,$$

which is the required generalization.

The transfinite analogy constructs a function $f(\cdot)$ on a well ordered set $(W, \langle\ )$ whose value at any point $a \in W$ depends on $f|W_a$, where W_a is the initial segment of W determined by a (see (1) of section 12). Given a set $X \neq \theta$, let $\mathcal{F}$ be the collection of all maps φ such that for some $y \in W$, $\varphi: W_y \to X$ (into). Assume also that a map $F: \mathcal{F} \to X$ is given, then we have:

Proposition 2

There exists a unique $f: W \to X$ such that for every $y \in W$

$$f(y) = F(f|W_y) \tag{5}$$

and $f(y_0) = F(\theta)$, where $y_0 \in W$ is the first element.

Proof

We shall prove the uniqueness of $f(\cdot)$ if it exists. If $h: W \to X$ is another such map satisfying (4), we have that $h(y_0) = F(\theta) = f(y_0)$. Next, assume that

$$K = \{u \in W; h(u) \neq f(u)\} \neq \theta$$

and $s \in K$ be its first element. Then since $h|W_s = f|W_s$, it follows from (4) that

$$h(s) = F(h|W_s) = F(f|W_s) = f(s),$$

which is contradiction. Thus, $K = \theta$.

Next, we shall show that a map $f(\cdot)$ satisfying (5) exists. This can be accomplished in a similar way as in proposition 1. Here, we shall use a slightly different approach. Denote by

$\overline{W}_y = W_y \cup \{y\}$, and let $\Gamma \subset W$ be the set of all $u \in W$ such that

$$\varphi(u) = F(\varphi \,|\, W_u) \qquad \text{for all } u \in \overline{W}_y.$$

The map φ is unique and consequently $\varphi = f$ on $\overline{W}_y$. In addition, Γ is obviously a set $\overline{W}_z$ for some $z \in W$. The union of all such φ on W_y, for all $y \in \Gamma$, is well defined function which will also be called $f(\cdot)$. If $X - \Gamma \neq \theta$ and $u \in X - \Gamma$ is the first member, then

$$f(u) \cup \{(u, F(f))\} \text{ is a function on } \overline{W}_y$$

with the desired properties, so that $u \in \Gamma$. This contradiction proves that $f(\cdot)$ exists and since it is unique on every $\overline{W}_y$, it is unique. This proves the proposition. ♥

Problems and complements.

1. Show that the $\lim x_n = 2$ and that

$$\lim_{n \to \infty} 2^n \sqrt{2 - x_n} = \pi$$

where x_n is given by (1).

Hint:

Trivial.

2. The recursions (1) and (2) are clearly the difference equations of order 1 and the initial condition γ. Their solution would give us $f(\cdot)$ as an explicit function of n. This problem, however, is not trivial at all. As a simple exercise, determine $f(\cdot)$ explicitly in terms of n if in equation (2), $X=R$ and $F(k, x)=k \cdot x$.

Hint:

Under the given conditions (2) becomes

$$f(n+1) = n \cdot f(n).$$

By dividing both sides with n! we obtain

$$\frac{f(n+1)}{n!} = \frac{f(n)}{(n-1)!} = \lambda => f(n) = \lambda \cdot (n-1)!$$

3. The book "Liber Abacci" published circa 1202, by the Italian mathematician Fibonacci Leonardo (of Pisa), introduced Hindu-Arabic number system to Europe. The book also contains the famous guinea pig problem which we shall formulate as follows: A pair of guinea pigs born at time "0" creates a further pair in every month of its existence, beginning in the second month, and its descendant behave in the same way. How many pairs will be after n months?

Hint:

If we denote by f(n) this number, it seems clear that (except for the first two terms f(0) and f(1), each term is the sum of the preceding two. The recursive scheme is obviously ($f(0)=0$)

$$f(1) = 1, \quad f(n+1) = f(n) + f(n-1) \tag{*}$$

yielding the "Fibonacci numbers"

$$0, 1, 1, 2, 3, 5, 8, 13, \ldots$$

To determine $f(\cdot)$ as an explicit function of n, we have to solve the difference equation (*) given the initial conditions ($f(0)=0$ and $f(1)=1$. Clearly, (*) is a homogeneous linear difference equation with constant coefficients. The corresponding "secular" equation

$$\lambda^2 - \lambda - 1 = 0$$

has two distinct roots $\lambda_1 = (1 + \sqrt{5}) / 2$ and $\lambda_2 = (1 - \sqrt{5}) / 2$, so that the general solution of (*) is

$$f(n) = C_1 \left(\frac{1+\sqrt{5}}{2}\right)^n + C_2 \left(\frac{1-\sqrt{5}}{2}\right)^n.$$

By virtue of the initial conditions we obtain that $C_1 + C_2 = 0$ and $(C_1 + C_2) + 5(C_1 - C_2) = 2$ yielding $C_1 = 1/\sqrt{5}$ and $C_2 = -1/\sqrt{5}$. Thus,

$$f(n) = \frac{1}{\sqrt{5}}\left[\left(\frac{1+\sqrt{5}}{2}\right)^n - \left(\frac{1-\sqrt{5}}{2}\right)^n\right].$$

15. ISOMORPHISM AND ORDER TYPE

Let $(X, <)$ and $(Y, <')$ be ordered sets. If $X \sim Y$, there exists a bijection $h{:}X \to Y$. The map h is called an **"isomorphism"** if it is "order preserving", i.e., if for any $x_1, x_2, \in X$ such that $x_1 < x_2$, necessarily $h(x_1) <' h(x_2)$ (clearly $h(x_1) <' h(x_2)$ implies that $x_1 < x_2$ for if $x_2 < x_1$ we would have that $h(x_2) <' h(x_1)$). We then say that the ordered sets $(X, <)$ and $(Y, <')$ are **"isomorphic" or "similar"** (in symbol, $X \cong Y$).

Remark 1

If $(X, <)$ and $(Y, <')$ are isomorphic and one of them has a first (last) element, so does the other. For, if $x_0 \in X$ is a first element then $h(x_0)$ is a first element of $(Y, <')$, because if it is not, there exists $y_0 \in Y$ such that $y_0 <' h(x_0)$. But then, since $y_0 = h(x_0')$, we would have that $x_0' < x_0$, which is contradiction.

Example 1

The ordered sets $(N_+, <)$ and $(M, <)$, where $<$ is the "usual" ordering relation on the real line and $M = \{n / (n+1)\}_1^{\infty}$, are similar since $N_+ \sim M$ and the map $h{:}N_+ \to M$ defined by

$$h(k) = \frac{k}{k+1}$$

is an isomorphism. In other words, $N_+ \cong M$. However, the sets $(N_+, <)$ and $(\overline{M}, <)$, where

$$(\overline{M}, <) = \{\frac{1}{2}, \frac{2}{3}, \ldots, \frac{n}{n+1}, \ldots, 1\}$$

are not similar, although $N_+ \sim \overline{M}$. The reason for this is that 1 is the last element of $(\overline{M}, <)$ while $(N_+, <)$ does not have the last number (see remark 1).

Remark 2

Similarity $\cong$ of ordered sets is an equivalence relation, since

(i) $X \cong X$ (reflexivity)

(ii) If $X \cong Y$ then $Y \cong X$ (symmetry) for if $h{:}X \to Y$ is an isomorphism and $y_1, y_2 \in Y$, then there exists (a unique pair) $x_1, x_2 \in X$ such that $y_1 = h(x_1)$ and $y_2 = h(x_2)$. If $y_1 <' y_2$ then necessarily $x_1 < x_2$.

(iii) If $X \cong Y$ and $Y \cong Z$ then $X \cong Z$ (transitivity) because the composition of two isomorphisms is an isomorphism.

It is often convenient, when it is possible, to write an ordered set $(A, <)$ in the form

$$(A, <) = (a_1, a_2, \cdots),$$

where the order in which the elements are written corresponds to the ordering relation $<$

(i.e., $a_1 < a_2 < \cdots$).

Since the concept of "similarity" is an equivalence relation in a collection of ordered sets, such a collection can be partitioned into "equivalence" classes of isomorphic ordered sets. All members of an equivalence class are said to have the same "order type." Thus, just as the concept of equivalence of sets led to that of cardinal, the notion of similarity leads to that of **"order type."**

The order type of the empty set θ is 0. Every singleton has order type 1 and the order type of every ordered finite set consisting of n elements is n. Thus, the order type of any finite set is its cardinal number. The order type of the ordered set $(N_+, <)$, and of any other ordered set similar to it is commonly denoted by the symbol ω. The order type of $(M, <)$ is also ω but not of $(\overline{M}, <)$. As we shall see later its order type is $\omega+1$. The ordered set

$$(\cdots, 4, 3, 2, 1) \qquad (\cdots < 4 < 3 < 2 < 1)$$

is not similar to $(N_+, <)$ (it has the last element but not the first one). We use the symbol $*\omega$ for its order type. In general, if the order type of a set $(X, <)$ is μ, then $*\mu$ denotes the order type of the ordered set obtained from $(X, <)$ when the order of succession of the elements is reversed. Clearly, $*n=n$. It is also clear that $*(*\omega)=\omega$ and in general that $*(*\mu)=\mu$ for any order type of μ. The symbol μ is commonly used for the order type of $(Q, <)$, where Q is the set of rational numbers and $<$ is the usual ordering relation on the real line R. The symbol λ is used for the order type of $(R, <)$ and π for

$$(Z, <) = (\cdots, -3, -2, -1, 0, 1, 2, 3, \cdots)$$

Clearly, $*\pi=\pi$, $*\mu=\mu$ and $*\lambda=\lambda$.

Let $(X, <)$ and $(Y, <')$ be ordered sets such that $X \cap Y = \theta$. One way to order the union $X \cup Y$ is as follows:

(i) For any $x \in X$ and $y \in Y$, we set $x \langle y$.

(ii) On X, $\langle = <$ and on Y, $\langle = <'$. With this ordering relation we write $(X \cup Y, \langle) = X+Y$. Note that the order of X and Y matters here (we can express this by writing $X \langle Y$). In a similar way we define the sum $Y+X$. In general, $X+Y$ and $Y+X$ are not similar, i.e., $X+Y \not\cong Y+X$. However,

$$N_+ + M = (1, 2, \cdots; \frac{1}{2}, \frac{2}{3}, \cdots)$$

$$M + N_+ = (\frac{1}{2}, \frac{2}{3}, \cdots; 1, 2, \cdots),$$

and it is not difficult to show (see problem 11) that $N_+ + M \cong M+N_+$.

If $(X, <)$, $(Y, <')$ and $(Z, <'')$ are disjoint ordered sets we define

$$X + Y + Z = (X + Y) + Z.$$

If the order type of $(X, <)$ is v_1 and v_2 is the order type of $(Y, <)$, then the order type $X+Y$ is v_1+v_2, and the order type of $Y+X$ is v_2+v_1. Clearly, $v_1+v_2 \neq v_2+v_1$. If the order type of $(Z, <")$ is v_3, the order type of $Y+X+Z=(X+Y)+Z$ is $(v_1+v_2)+v_3$.

For example, the order type of $(\overline{M}, <)$ is $\omega+1$ because $(\overline{M}, <)=(\frac{1}{2}, \frac{2}{3}, \cdots, 1)=M+\{1\}$. But the order type of $(0, 1, 2, \cdots)$ is clearly ω and since $(0, 1, 2, \cdots) = \{0\}+N_+$, we have that $1+\omega=\omega$. It follows from the definition of the order types ω and $*\omega$, that $*\omega+\omega=\pi$. Clearly $\omega+*\omega \neq *\omega+\omega$, because the first one is the order type of the ordered set which contains both the first and the last element, while the second one is the order type of an ordered set which contains neither. Finally, it is not difficult to show that for any order type v_1 and v_2

$$*(v_1+v_2) = *v_2 + *v_1.$$

Notice that $n+\omega=\omega$; since $*(n+\omega)=*\omega$ we have that $*\omega+n=*\omega$ $((\cdots 3, 2, 1) \cong (\cdots, n+2, n+1) + (n, n-1, \cdots, 1)$, so that $*\omega=*\omega+n)$. However, $*\omega$, $1+*\omega$, $\cdots$, $n+*\omega$ are all distinct order types. Thus, the operation "addition" of order types is not commutative. However it is associative, i.e., if v_1, v_2, v_3 are three order type, their sum is defined by

$$v_1+v_2+v_3 = (v_1+v_2) + v_3 = v_1+ (v_2+v_3).$$

From this it follows that

$$*\omega + n + \omega = *\omega + (n + \omega) = *\omega + \omega$$
$$\omega + n + \omega = \omega + (n + \omega) = \omega + \omega$$
$$\omega + *\omega + n = \omega + (*\omega + n) = \omega + *\omega.$$

To summarize, after $0, 1, 2, \cdots$ comes ω; after ω comes $\omega+1$, $\omega+2$, $\cdots$, $\omega+\omega$ and then comes $\omega+\omega+\omega$, and so on.

The operation "addition" can be extended to arbitrary many order types as follows: Let $(T, <)$ be an ordered set and let $(A_t, <_t)$, $t \in T$ be a family of disjoint ordered sets, each $(A_t, <_t)$ having the order type μ_t. Set $S=\cup\{A_t ; t \in T\}$; we order the union S to obtain $(S, \langle)$ where the ordering relation $\langle$ is specified as follows: For any $t_1, t_2 \in T$ such that $t_1<t_2$ we have that $A_{t_1} \langle A_{t_2}$, meaning that every $x \in A_{t_1}$ precedes every $y \in A_{t_2}$, i.e., $x \langle y$, and for any $x_1, x_2 \in S$ which belong to the same $(A_t, <_t)$, $x_1 \langle x_2$ if $x_1 <_t x_2$. The ordered union is usually denoted by

$$\sum_{t \in T} A_t$$

and the order type μ is then $\mu = \sum_{t \in T} \mu_t$.

Example 2

If $(T, <)=(N_+, <)$ and Card $A_t = t$ for each $t \in N_+$ we have that

$$\mu = \sum_{t=1}^{\infty} t = \omega,$$

because the cardinal and the order type of every finite ordered set consisting of n elements is n. Notice also, since Card = ordinal type, the finite ordinal types are comparable while the transfinite are not.

To define the product of two order types, consider the ordered sets $(X, <)$ and $(Y, <')$ (not necessarily disjoint) with the order types μ_1 and μ_2, respectively. The "ordered product" of X and Y, denoted by $X \cdot Y$, is ordered set

$$(X \times Y, \langle)$$

where $\langle$ is the "lexicographic" ordering relation $X \times Y$ (problem 9 of section 10). The order type of $X \cdot Y$ is then $\mu_2 \cdot \mu_1$ (not $\mu_1 \cdot \mu_2$).

As an example consider the ordered sets $(A, <)=(a_1, a_2)$ and $(B, <')=(b_1, b_2, b_3)$, then we have:

$$A \cdot B = ((a_1, b_1), (a_1, b_2), (a_1, b_3), (a_2, b_1), (a_2, b_2), (a_2, b_3))$$

$$B \cdot A = ((b_1, a_1), (b_1, a_2), (b_2, a_1), (b_2, a_2), (b_3, a_1), (b_3, a_2))$$

Thus, the order type of $A \cdot B$ and $B \cdot A$ is 6. On the other hand,

$$A \cdot N_+ = ((a_1, 1), (a_1, 2), \cdots; (a_2, 1), (a_2, 2), \cdots)$$

is $\omega + \omega = \omega \cdot 2$, while the order type of

$$N_+ \cdot A = <(1, a_1), (1, a_2), (2, a_1), (2, a_2), \cdots >$$

is clearly $2\omega = \omega$. In general we have that

$$\omega \cdot n = \omega + \omega + \cdots \omega \qquad \text{(n-times)}$$

$$n \cdot \omega = n + n + \cdots = \omega$$

and if μ_1 and μ_2 are two order type we have (somewhat loosely) that

$$\mu_1 \cdot \mu_2 = \mu_1 + \mu_1 + \cdots \qquad (\mu_2\text{-times})$$

$$\mu_2 \cdot \mu_1 = \mu_2 + \mu_2 + \cdots \qquad (\mu_1\text{-times})$$

It seems clear from these examples that the product of order types is not "commutative", however, it is associative, i.e.,

$$\mu_1 \cdot \mu_2 \cdot \mu_3 = (\mu_1 \cdot \mu_2) \cdot \mu_3 = \mu_1 \cdot (\mu_2 \cdot \mu_3).$$

The distributive law does not hold in general either. It is only valid when the second factor in the product is as a sum. i.e.,

$$\mu_1 \cdot (\mu_2 + \mu_3) = \mu_1 \cdot \mu_2 + \mu_1 \cdot \mu_3 ;$$

while

$$(\mu_2 + \mu_3) \cdot \mu_1 = (\mu_2 + \mu_3) + (\mu_2 + \mu_3) + \cdots \quad (\mu_2\text{-times}).$$

For instance,

$$2 \cdot (\omega + n) = 2\omega + 2n = \omega + 2n; \text{ but}$$

$$(\omega + n) \cdot 2 = (\omega + n) + (\omega + n) = \omega \cdot 2 + 2n.$$

We also have that

$$\omega^2 = \omega \cdot \omega = \omega + \omega + \cdots$$

$$\omega \cdot (1+\omega) = \omega + \omega^2 = \omega + \omega + \cdots = \omega^2$$

$$\omega \cdot (\omega+1) = \omega^2 + \omega$$

$$(1+\omega)\,\omega = (1+\omega) + (1+\omega) + \cdots = \omega + \omega^2 = \omega^2 .$$

As we have seen, the symbol λ is used to denote the order type of $(R, <)$, where $<$ is the usual ordering relation in the real line R. Since any open interval $(a, b) \sim R$, the order type of $((a, b), <)$ is also λ. On the other hand, since $[a, b) = \{a\} \cup (a, b)$, the order type of $\{a\}+(a, b)$ is $1+\lambda$ and of $(a, b)+\{b\}$ is $\lambda+1$. The order type of $([a, b], <)$ is $1+\lambda+1$ since $[a, b] = \{a\} \cup (a, b) \cup \{b\}$. However, $(a, b] \cup (b, c) = (a, c)$, so that the order type of $((a, c), <) = ((a, b]+(b, c), <)$ is

$$\lambda = \lambda + 1 + \lambda$$

and so on.

As we have seen, the symbol λ is used to denote the order type of $(R, <)$, where $<$ is the usual ordering relation in the line R. Since any open interval $(a, b) \sim R$, the order type of $((a, b), <)$ is also λ. On the other hand, since $[a, b) = \{a\} \cup (a, b)$, the order type of $\{a\}+(a, b)$ is $1+\lambda$. The order type of $([a, b], <)$ is $1+\lambda+1$ since $[a, b] = \{a\} \cup (a, b) \cup \{b\}$. However, $(a, b] \cup (b, c) = (a,c)$, so that order type of $((a, c), <) = ((a, b]+(b, c), <)$ is

$$\lambda = \lambda + 1 + \lambda,$$

and so on.

Problems and complements

1. Show that $(N_+, <)$ and $(n, n+1, n+2, \cdots)$ are isomorphic.

   Hint:

   The map $h:N_+ \to \{k\}_n^\infty$ defined by $h(k)=k+(n-1)$ is an isomorphism.

2. Let $(X, <)$ be an ordered set such that $X \sim Y$. Find an ordering relation in Y so that $X \cong Y$.

   Hint:

   Since $X \sim Y$, there exists a bijection $h:X \to Y$ (which may not be easy to construct). If $x_1, x_2 \in X$ such that $x_1 < x_2$, and $y_1 = h(x_1)$, $y_2 = h(x_2)$ then an ordering relation $<'$ in Y can be induced by stipulating that $y_1 <' y_2$. Then the map h is an isomorphism since for any y', $y'' \in Y$ with $y' <' y''$ and $y' = h(x')$, $y'' = h(x'')$, necessarily $x' < x''$ (for if $x'' < x'$ we would have that $y'' < y'$).

3. Show that the composition of two isomorphisms is an isomorphism.

   Hint:

   Let $(X, <)$, $(Y, <')$ and $(Z, <'')$ be ordered sets such that $h_1:X \to Y$ and $h_2:Y \to Z$ are isomorphisms. Denote by $h = h_2 \circ h_1$, then if $x_1 < x_2$

   $$h(x_1) = h_2(h_1(x_1)) <'' h_2(h(x_2)) = h(x_2).$$

This proves that $h:X \to Z$ is order preserving.

4. If $(X, <)$ and $(Y, <')$ are ordered sets and the map $h:X \to Y$ an order preserving surjection, show that the map $h(\cdot)$ is an isomorphism.

   Hint:

   If $x_1 < x_2$ in X, then $h(x_1) <' h(x_2)$. Thus for no $x_1 \neq x_2$, $h(x_1) \neq h(x_2)$, proving that h is an isomorphism.

5. Construct a bijective map $h:(0, 1) \to (0, 1]$ (which exists since $(0, 1) \sim (0, 1]$).

   Hint:

   Define

   $$h(x) = \sum_{i=1}^{\infty} (3 \cdot 2^{-i} - x) I_{[2^{-i}, 2^{-i+1})}(x);$$

The map is obviously a bijection ($h(1/2)=1$).

6. Show that $\pi = *\pi$.

   Hint:

   Recall that the "inverse" of an ordered set, say $(X, <) = (\cdots, x, y, z, \cdots)$, is obtained by replacing the relation $<$ with $>$ throughout the set so that we obtain

   $$*(\cdots x, y, z, \cdots) = (\cdots z, y, x, \cdots).$$

Thus, if $(\cdots, -3, -2, -1, 0, 1, 2, 3, \cdots)$, its inverse is $(\cdots, 3, 2, 1, 0, -1, -2, -3, \cdots)$. These two sets are clearly isomorphic.

7. Let $(X, <)$ and $(Y, <')$ be two disjoint ordered sets. If $X'\cong X$, $Y'\cong Y$ and $X'\cap Y'=\theta$, show that $X'+Y'\cong X+Y$.

Hint:

Let $\varphi_1:X'\to X$ and $\varphi_2:Y'\to Y$ be isomorphisms, then the map $\varphi:X'+Y'\to X+Y$, defined by

$$\varphi = \begin{cases} \varphi_1 & \text{on} \quad X' \\ \varphi_2 & \text{on} \quad Y' \end{cases}$$

and $\varphi(x)<\varphi(y)$ if $x\in X'$ and $y\in Y'$, is clearly an isomorphism.

8. Show that the order type of

    (i) $N_+ +\{\alpha\}$ is $\omega+1$

    (ii) $\{\alpha\}+N_+$ is ω

    (iii) $N_+ + (\alpha_1, \alpha_2, \cdots, \alpha_n)$ is $\omega+n$

    (iv) $(\alpha_1, \alpha_2, \cdots, \alpha_n)+N_+$ is $n+\omega = \omega$

    (v) $(1, 3, 5, \cdots)+(\cdots, 6, 4, 3, 2)$ is $\omega+{}^*\omega$

    (vi) $(1, 3, 5, \cdots)+(2, 4, 6, \cdots)$ is $\omega 2$

9. Verify if the following holds:

    (i) $\omega + \omega + \omega + \cdots = \omega^2$ where $\omega\cdot\omega = \omega^2$

    (ii) $\omega + \omega^2 = \omega (1+\omega) = \omega\cdot\omega = \omega^2$

    (iii) $\omega^2+\omega = \omega (\omega+1)$

    (iv) $\omega\cdot(\omega+\omega) = \omega\cdot(\omega\cdot2) = (\omega\cdot\omega)\cdot2 = \omega^2 \cdot2$

    (v) $(\omega+\omega)\cdot\omega = (\omega\cdot2) \omega = \omega\cdot(2\omega) = \omega^2$

10. Prove the following:

    (i) $1+\lambda = (1+\lambda) \cdot n = (1+\lambda) \cdot \omega$

    (ii) $\lambda+1 = (\lambda+1)n = (\lambda+1)\cdot {}^*\omega$

Hint:

    $[0, 1) \cong [0, n) = [0, 1) \cup [1, 2) \cup \cdots \cup [n-1, n)$.

We also have that $[0, 1) \cong [0, \infty)$.

11. Prove that $N_+ +M \cong M+N_+$.

Hint:

Both sets are countable so that $N_+ +M \sim M+N_+$. Next, the map $\varphi:N_+ \to M+N_+$ defined by

$$\varphi = \begin{cases} \dfrac{k}{k+1} & \text{on} \quad N_+ \\[2mm] \dfrac{x}{1-x} & \text{on} \quad M \end{cases}$$

is clearly an isomorphism. For instance, if $n \in N_+$ and $\dfrac{k}{k+1} \in M$ we have that in $N_+ + M$

$$n < \frac{k}{k+1} \; ; \text{ since } \varphi(n) = \frac{n}{n+1} \text{ and } \varphi(\frac{k}{k+1}) = k, \; \frac{n}{n+1} < k.$$

16. ORDINALS AND ORDINAL NUMBERS

Recall that **an ordinal is well ordered set**. The Zermelo proof that every set can be well ordered (proposition 1 of section 12), assuming of course the axiom of choice, is a compelling reason to investigate properties of this particular class of ordered sets in some detail.

If an ordinal is isomorphic to a proper part of itself, the following fundamental proposition characterizes the corresponding isomorphic map.

Proposition 1

Let $(W, <)$ be an ordinal and consider a subset $B \subset W$ such that $B \sim W$. Then, for any isomorphic map $h:W \to B$, we have that

$$x < h(x) \quad \text{for all} \quad x \in W. \tag{1}$$

Proof:

If (1) is false then the (well ordered) set K defined below is nonempty, i.e.,

$$K = \{x \in W;\ h(x) < x\} \neq \theta.$$

If $y_0 \in K$ is the smallest member of K, we have that

$$z_0 = h(y_0) < y_0 \quad (\text{note that } z_0 \notin K).$$

Since $h(\cdot)$ is order preserving, it is clear that

$$h(z_0) < h(y_0) = z_0,$$

which is contradiction. Thus, $K = \theta$. ♥

Corollary 1

Recall that given $y \in W$ the set $W_y = \{x \in W;\ x < y\}$ is called an "initial segment" of $(W, <)$, determined by y. An ordinal can never be isomorphic to any of its initial segment. For if $W \cong W_y$ for some $y \in W$ and $h:W \to W_y$ is and isomorphism, we would have that

$$h(x) < y \text{ for all } x \in W,$$

which contradicts (1).

Corollary 2

From the definition of "initial segment," it seems clear that for any $y,\ z \in W$ we have that, $(W_z)_y = W_y$ if $y < z$. In other words, W_y is an initial segment of W_z for every $y < z$.

This and corollary 1 imply that no two initial segments of an ordinal are similar.

Proposition 2

Let $(X, <)$ and $(Y, <')$ be isomorphic ordinals, then the isomorphic map $h:X \to Y$ is unique.

Proof:

We shall first show that for any $x \in X$

$$h(X_x) = Y_{h(x)}. \tag{2}$$

If $y_0 \in h(X_x)$ there exists $x_0 \in X_x$ such that $y_0 = h(x_0)$. But $h(x_0) <' h(x)$, implying that $y_0 \in Y_{h(x)}$. Thus, $h(X_x) \subset Y_{h(x)}$. On the other hand, if $y' \in Y_{h(x)}$, $y' < h(x)$ and if $x' \in X$ is such that $y' = h(x')$, we have that $x' < x$. Hence, $y' \in h(X_x)$ and consequently, $Y_{h(x)} \subset h(X_x)$ (see problem 4 of section 12) proving (2). Now, if there is another isomorphic map $h_1 : X \to Y$, then for every $x \in X$ we have

$$Y_{h(x)} \cong X_x \text{ and } X_x \cong Y_{h_1(x)}$$

implying that $Y_{h(x)} \cong Y_{h_1(x)}$ which, according to corollary 2 is not possible if $h \neq h_1$. This completes the proof. $\qquad\qquad\heartsuit$

The next proposition, known as the "**comparability theorem**" for ordinals, is fundamental.

Proposition 3

Let $(X, <)$ and $(Y, <')$ be ordinals, then either $X \cong Y$ or one of the ordinals is similar to an initial segment of the other ordinal.

Proof:

Let $\overset{o}{X} \subset X$ consist of those $x \in X$ for which there exists $y \in Y$ such that $X_x \cong Y_y$.

Denote by φ this correspondence and by $\varphi(\overset{o}{X}) = \overset{o}{Y}$. Next, assuming that φ is an isomorphism, $\overset{o}{X} \cong \overset{o}{Y}$. Thus, if $\overset{o}{X} = X$, and if $\overset{o}{Y} = Y$ we would have that $X \cong Y$. If, on the other hand, $\overset{o}{X} = X$ but $Y - \overset{o}{Y} \neq \theta$ and $y_0 \in Y - \overset{o}{Y}$ is the smallest member of this set, we would have that $\overset{o}{Y} = Y_{y_0}$. Consequently, it would follow that $X \cong Y_y$ as we have contended.

To complete the proof it remains to show first, that φ is an isomorphism and second, that $\overset{o}{Y} = Y_{y_0}$ for some $y_0 \in Y$. Let $x_1, x_2 \in \overset{o}{X}$ such that $x_1 < x_2$, and set $y_1 = \varphi(x_1)$, $y_2 = \varphi(x_2)$. Then since $X_{x_1} \cong Y_{y_1}$ and $X_{x_2} \cong Y_{y_2}$ and

$$X_{x_1} = (X_{x_2})_{x_1} \cong (Y_{y_2})_{y'} \text{ where } y' <' y_2,$$

it follows that $X_{x_1} \cong Y_{y'}$. Hence, $y' = \varphi(x_1) = y_1 <' y_2$, which proves that φ is an isomorphism. In other words, $\overset{o}{X} \cong \overset{o}{Y}$.

To show that $\overset{o}{Y} = Y_{y_0}$ notice that for any $y \in \overset{o}{Y}$, $y <' y_0$, implying that $y \in Y_{y_0}$;

consequently, $\overset{0}{Y} \subset Y_{y_0}$. If $y \in Y_{y_0}$, then $y<'y_0$, due to the fact that y_0 is the smallest

member of Y that does not belong to $\overset{0}{Y}$. Consequently $y \in \overset{0}{Y}$ and hence we have that

$Y_{y_0} \subset \overset{0}{Y}$, proving that $\overset{0}{Y} = Y_{y_0}$. ♥

Corollary 3

If $A \subset W$, where (W, <) is an ordinal, then either $A \cong W$ or $A \cong W_x$ for some $x \in W$. This is a simple consequence of the last proposition since A is also well ordered set.

Corollary 4

Let X and Y be sets and assume that there exists an injection $f_1 : X \rightarrow Y$ and also an injection $f_2 : Y \rightarrow X$. By well ordering of X and Y we would have, according to corollary 3, that either $f_1(X) \cong Y$ or that $f_1(X) \cong Y_y$, for some $y \in Y$. Thus, Card $X \le$ Card Y. In the same fashion we obtain by means of the injective map f_2 that Card $Y \le$ Card X, yielding that CardX = Card Y (Bernstein's equivalence theorem).

Given two ordinals (X, <) and (Y, <'), then if $X \cong Y_y$ for some $y \in Y$, we shall say (temporarily) that (X, <) is "shorter" then (Y, <').

Proposition 4

Every family O of non- isomorphic ordinals contains a shortest member.

Proof:

Choose $(X, <) \in O$. If the (X, <) is the shortest ordinal the proposition is proved. If it is not, there are members of O shorter than (X, <). Each of these is isomorphic to an initial segment of (X, <). Denote by $X_0 \subset X$ the set of their determine points. Let $x_0 \in X_0$ be the least element of X_0, and let $(Z, <') \in O$ satisfy $Z \cong X_{x_0}$, then (Z, <') is the shortest element of

O. Indeed, for any ordinal (Y, <") shorter than (X, <') we have that $Y \cong X_x$ for some $x \in X_0$. Clearly, $x_0 < x$ and X_{x_0} is an initial segment of X_x. This proves that (Z, <') is shorter than (Y, <"). ♥

Proposition 5

Let (T, <) be an ordinal and let

$$\{(X_t, <_t); t \in T\}$$

be a family of disjoint ordinals, then the ordered sum

$$S = \sum_{t \in T} X_t$$

is an ordinal.

Proof:

Let $S_0 \subset S$ and denote by $T_0 \subset T$ the set of those $t \in T$ for which

$$X_t \cap S_0 \neq \theta.$$

If t_* is the first member of T_0, it is also the first member of S_0. This proves that S is an ordinal. ♥

As an example consider the well ordered sets $<1, 3, 5, \cdots >$ and $<2, 4, 6, \cdots >$. The ordered sum

$$(N_+, \langle o) = <1, 3, 5, \cdots > + <2, 4, 6, \cdots >$$

where $\langle o$ is defined on N_+ as follows: $2n-1 < 2$ for all $n=1, 2, \cdots$ and $\langle o = < $ on each of the two disjoint ordinals, is an ordinal, since any $K \subset N_+$ has a first element. The order type of $(N_+, \langle o)$ is obviously $\omega + \omega = \omega \cdot 2$.

We now introduce the concept of an **"ordinal number."** In the classical Cantor theory, the order type of an ordinal is called an "ordinal number." In the case of finite sets, cardinals, the order types and the ordinal numbers coincide. For instance, if a finite set F consists of n elements its cardinal n is at the same time an order type and an ordinal number. This, however, is not the case with the transfinite ordinals.

The set $(N_+, <)$, (where $<$ is the usual ordering relation on the real line) is well ordered and since its order type is ω, this is its ordinal number. On the other hand, $*\omega$ is not an ordinal number (it is not the order type of a well ordered set). For the same reason, the order types η, λ and π are not ordinal numbers. The set $(\overline{M}, <)$ in example 1 of section 15, is an ordinal, whose ordinal number is $\omega + 1$.

Consider the ordered set $(Q_+, <)$ where Q_+ is the set of positive rationales and $<$ is the usual ordering relation on the real line. The well ordering version of this set, denote by the symbol $(Q_+, \lessgtr)$, is given by (example 3 of section 10)

$$<1, 2, 3, \cdots; \frac{1}{2}, \frac{3}{2}, \frac{5}{2}, \cdots; \frac{1}{3}, \frac{2}{3}, \frac{4}{3}, \cdots >$$

$$= <1, 2, 3, \cdots > + < \frac{1}{2}, \frac{3}{2}, \frac{5}{2}, \cdots > + < \frac{1}{3}, \frac{2}{3}, \frac{5}{3} \cdots > + \cdots .$$

The ordinal number of $(Q_+, \lessgtr)$ is then

$$\omega + \omega + \cdots = \omega \cdot \omega = \omega^2.$$

The transfinite order types are not comparable in general. It is only clear when two order type are equal. The relation "less than" does not exist. However, it is possible to define an ordering relation in a set of ordinal numbers. This is what we will do next.

Definition 1

Let $(X, <')$ and $(Y, <'')$ be ordinals with ordinal numbers α_1 and α_2, respectively. We say that α_1 precedes α_2 (or that α_1 is "smaller" than α_2) and we write $\alpha_1 < \alpha_2$ if $X \cong Y_y$ for some $y \in Y$.

It is clear from this definition that for any couple of ordinal numbers, say γ_1 and γ_2, only one of the three relations

$$\gamma_1 < \gamma_2, \gamma_1 = \gamma_2; \text{ and } \gamma_2 < \gamma_1 \tag{3}$$

is possible. If, for instance, we have that simultaneously $\gamma_1 < \gamma_2$ and $\gamma_1 = \gamma_2$, it would follow that $\gamma_1 < \gamma_1$. In other words, we would have that for some $x \in X$, $X_x \cong X$, which is clearly impossible (corollary 1). From this we deduce that any two ordinal numbers are comparable.

Proposition 6

Every set of ordinal numbers, if ordered by the "magnitude" of its elements, is an ordinal.

Proof:

Let $(O, <)$ be an ordered set of ordinal numbers. With each $\alpha \in O$ we can associate an ordinal whose ordinal number is α. According to proposition 4 there exists the shortest among these ordinals. Its ordinal number precedes every other ordinal number in O. In other words, the set O has a least element. ♥

Let $\beta > 0$ be ordinal number; denote by W_β the set of all ordinal numbers which are "less" than β (which precede β). According to proposition 6, W_β is an ordinal. For instance, $W_n = <0, 1, 2, \cdots, n-1>$ (we take that $W_0 = \theta$) and $W\omega = <0, 1, 2, \cdots>$, $W\omega+1 = <0, 1, 2, \cdots, \omega>$. Let us prove the following:

Proposition 7

The ordinal number of W_β is β.

Proof:

Let M be an ordinal with ordinal number β, and denote by $\mathcal{H}$ the class of all initial segments of M. Any $\alpha \in W_\beta$ is obviously ordinal number of an element of $\mathcal{H}$. Conversely, the ordinal numbers of members of $\mathcal{H}$ are elements of W_β. This correspondence defines a map

$$h : W_\beta \to \mathcal{H}$$

which is obviously a bijection (no two distinct initial segments of M are similar). Thus, to every $\alpha \in W_\beta$ there corresponds a unique member of $\mathcal{H}$. The set $(\mathcal{H}, \langle)$, with ordering relation $\langle$ defined as follows: any two M_{m_1}, $M_{m_2} \in \mathcal{H}$ satisfy $M_{m_1} \langle M_{m_2}$ if the corresponding $\alpha_1, \alpha_2 \in W_\beta$ satisfy $\alpha_1 < \alpha_2$, is obviously an ordinal, and the map $h(\cdot)$ is an isomorphism. Consequently, $W_\beta \cong \mathcal{H}$ and since $\mathcal{H} \cong M$, we have that

$$W_\beta \cong M,$$

which proves our contention. ♥

Corollary 5

Let $(X, <)$ be an ordinal with ordinal number γ, then we have:

$$(X, <) = (x_0, x_1, \cdots, x_\alpha, \cdots) \qquad (\alpha < \gamma).$$

Here, the index of every element in $(X, <)$ is the order type of the set of elements preceding it.

Remark 1

As an illustration we will list several ordinal numbers. First come $0, 1, 2, \cdots$; then comes ω and after ω comes $\omega+1$, $\omega+2$, $\cdots$; then comes $\omega \cdot 2$. After comes $\omega \cdot 2 + 1$, and then $\omega \cdot 2 + 2$, $\cdots$. Next comes $\omega \cdot 3$, $\omega \cdot 3 + 1$, $\cdots$ and after that comes $\omega \cdot 4$, and so on. Then comes ω^2, $\omega^2 + 1$, $\omega^2 + 2$, $\cdots$; $\omega^2 + \omega$, $\omega^2 + \omega + 1$, $\cdots$; $\omega^2 + \omega \cdot 2$, $\omega^2 + \omega \cdot 2 + 1$, $\cdots$; $\omega^2 + \omega \cdot 3$, $\omega^2 + \omega \cdot 3 + 1$, $\cdots$; $\omega^2 \cdot 2$, $\cdots \omega^2 \cdot 3$, $\cdots$, ω^3, $\cdots$, ω^4, $\cdots$, ω^ω, $\cdots$.

Is there a set W of all ordinal numbers? If the answer is yes, W is well ordered, with an ordinal number γ. Now, according to proposition 7, W_γ has the ordinal number γ. Consequently, W and its initial segment W_γ are isomorphic, which contradicts corollary 1. This paradox is known as the **Burali-Forti antinomy**.

Example 1

The ordinal number of the set $(N_+, <)$ is ω. Since

$$(N_+, <) = (1, 2, \cdots, n) + (n+1, n+2, \cdots)$$

it follows that $\omega = n + \omega$, for every $n = 0, 1, \cdots$. Note that various reordering of the set N_+ yields various order types. For instance, the order type of $(2, 4, 6, \cdots) + (1, 3, 5, \cdots)$ is $\omega + \omega = \omega \cdot 2$, $(n, n+1, n+2, \cdots; 1, 2, \cdots, n-1) = (n, n+1, n+2, \cdots) + (1, 2, \cdots, n-1)$ has the order type $\omega + (n-1)$. Clearly,

$$\omega < \omega + 1 < \omega + 2 < \cdots.$$

The order type of

$$(\cdots, n+1, n, \cdots, 6; 1, 2, 3, 4, 5) \text{ is } {}^*\omega + 5.$$

The set N_+ can be partitioned in countably many countable subsets. For instance, consider the following array (the numbers are arranged diagonally from upper write to the lower left)

$$
\begin{array}{llllll}
1 & 2 & 4 & 7 & 11 & 16\ \ldots\ldots \\
3 & 5 & 8 & 12 & 17\ldots\ldots\ldots \\
6 & 9 & 13 & 18\ldots\ldots\ldots \\
10 & 14 & 19\ldots\ldots\ldots \\
15 & 20\ldots\ldots\ldots \\
21\ldots\ldots\ldots \\
\ \ \ \ \ldots\ldots\ldots
\end{array}
$$

yielding

$$(N_+, <') = (1, 2, 4, \cdots) + (3, 5, 8, \cdots) + (6, 9, 13, \cdots) + \cdots .$$

Consequently, the ordinal number of $(N_+, <')$ is

$$\omega + \omega + \cdots = \omega \cdot \omega = \omega^2 .$$

Note that all ordered sets listed here were countable.

It is possible to define "ordinal number" within the framework of set theory. Here we outline the basic ideas of this approach. Recall (section 1) that every natural number is a set: for instance,

$$0 = \theta;\ 1 = \{\theta\};\ 2 = \{\theta, \{\theta\}\};\ 3 = \{\theta, \{\theta\}, \{\theta, \{\theta\}\}\}$$
$$4 = \{\theta, \{\theta\}, \{\theta, \{\theta\}\}, \{\theta, \{\theta\}, \{\theta, \{\theta\}\}\}\}$$

and so on. From this we deduce that

$$0 = \theta;\ 1 = \{0\};\ 2 = \{0, 1\};\ 3 = \{0, 1, 2\},\ \cdots$$

and that in general

$$n = \{0, 1, 2, \cdots, n-1\}. \tag{4}$$

Thus, every natural number is a finite set. Consequently, its cardinal and ordinal number coincide, so the ordinal number of set n is the set n.

It follows from (4) that every element of the set n is also its subset (if $k \in n$, then $k \subset n$, since $k = \{0, 1, 2, \cdots, k-1\}$). This gives rise to the following:

Definition 5

A set X is called **"transitive"** if each of its members is also a subset of X, or equivalently, if $x \in y$ and $y \in X$ then $x \in X$.

Definition 6

An ordinal number is a set α which is well ordered by the membership relation $\in$ and which is transitive.

The successor α' of an ordinal number α is defined by

$$\alpha' = \alpha \cup \{\alpha\}.$$

We contend that α' is also an ordinal number. To prove this we have to show that for any $y \in \alpha'$, if $x \in y$ then $x \in \alpha'$. For, since $y \in \alpha'$, $y \in \alpha$ or $y = \alpha$. In the first case $y \subset \alpha$ so that $x \in y \subset \alpha$

and thus $x \in \alpha \cup \{\alpha\} = \alpha'$. In the second case, $x \in y = \alpha$ so that again $x \in \alpha \cup \{\alpha\} = \alpha'$ proving that α' is transitive. If $\{x, y\} \subset \alpha' \Rightarrow \{x, y\} \subset \alpha$ or $x \in \alpha$, $y = \alpha$ or $y \in \alpha$ and $x = \alpha$. In any case we see that either $x \in y$ or that $t \in x$.

As an example let us show that

$$\omega = (\ 0,\ 1,\ 2,\ 3,\ \cdots\ )$$

is an ordinal number. It is clear that ω is well ordered. To show that ω is transitive denote by $S \subset \omega$ the set of all transitive elements of ω, i.e., $n \in S$ if and only if $n \in \omega$, and for every $x \in n$, $x \in \omega$. Clearly, $0 \in S$ ($0 = \theta$). If $n \in S$ then $n' \in S$ so that by the principle of mathematical induction $S = \omega$. This proves that ω is an ordinal number since it is obviously an ordinal.

The set ω is an example of an ordinal number which is not a natural number. If we name $\omega + 1 = \omega'$ and then $\omega + 2 = (\omega')' = \omega' \cup \{\omega'\} = (\omega + 1) \cup \{\omega + 1\}$, and so on, we have:

$$\omega + 1 = (\ 0,\ 1,\ 2,\ \cdots,\ \omega\ )$$
$$\omega + 2 = (\ 0,\ 1,\ 2,\ \cdots,\ \omega,\ \omega + 1\ ).$$

Problems and complements

1. Show that any two finite ordered sets A and B such that A~B, are isomorphic.

Hint:

Set $A=(a_1, a_2, \cdots, a_n)$ and $B=(b_1, b_2, \cdots, b_n)$, then the map $h:A\to B$ defined by $h(a_i)=b_i$, $i=1, 2, \cdots, n$ is an isomorphism.

2. Let $(X, <)$ be an ordinal; show that each $z\in X$, if it is not the last one, has an immediate successor.

Hint:

Let y be the first element of the nonempty $\{x\in X; z<x\}$. Clearly, $z<y$ and there are no elements of X between z and y.

3. Let $(X, <)$ be an ordinal and $h:X\to X$ order preserving. Show that $(h(X), <)$ is isomorphic to $(X, <)$.

Hint:

Since for any $x_1<x_2$ from X, $h(x_1)<h(x_2)$, the map $h:X\to h(X)$ is a bijection.

4. Show that every transfinite ordinal $(X, <)$ contains a subset whose ordinal number is ω.

Hint:

Let $x_0 \in X$ be the first element. Let x_1 be the immediate successor of x_0; let x_2 be the immediate successor of x_1 and so on. The ordinal

$$(a_0, a_1, a_2, \cdots)$$

has obviously the ordinal number ω.

5. Let $(X, <)$ be an ordinal; show that each proper subset $B\subset X$ such that $y\in B$ and $z<y$ implies that $z\in B$, is an initial segment.

Hint:

Let $x\in X-B$ be its least element, then clearly if $y<x$, $y\in B$. Thus, $y\in X_x$ and $B\subset X_x$. On the other hand, if $y\in X_x => y<x$ so that $y\in B$. Consequently $X_x\subset B$.

6. Let $(X, <)$ be an ordinal. Show that the intersection (union) of any family of initial segments of X is an initial segment.

Hint:

Let $x\in \bigcap_t X_t$ and $y<x$. It follows from this that $x\in X_t$ for all t and consequently $y\in X_t$ for all t, which implies that $y\in \bigcap_t X_t$.

7. Let $(X, <)$ be an ordinal and denote by $\mathcal{H}$ the collection of all initial intervals of X. Show that the map $h:X\to\mathcal{H}$ defined by $h(x)=X_x$ is an isomorphism, if $\mathcal{H}$ is ordered by inclusion.

Hint:

Clearly $x_1 < x_2 \Rightarrow X_{x_1} \subset X_{x_2}$ and $x_1 \neq x_2 \Rightarrow X_{x_1} \neq X_{x_2}$. Thus, h is bijection and obviously order preserving. Thus, $X \cong \mathcal{H}$ and consequently $\mathcal{H}$ is an ordinal.

8. Let $(X, <)$ and $(Y, <')$ be ordinals and f_1 and f_2 the maps defined in corollary 4. Show that

$$f_1 = f_2^{-1} \text{ and } f_2 = f_1^{-1}.$$

Hint:

Since $X \cong Y$, f_1 and f_2 are isomorphism. But $f_1^{-1}: Y \to X$ is also an isomorphism. Thus, according to proposition 2, f_2 is unique, so that

$$f_2 = f^{-1}.$$

9. Let $(X, <)$ and $(Y, <')$ be an ordinal and an ordered set, respectively. If $X \cong Y$, show that $(Y, <')$ is also an ordinal.

Hint:

Let $h: X \to Y$ be an isomorphism and let $Y_0 \subset Y$. Let $X_0 \subset X$ be such that $Y_0 = h(X_0)$. If $x_0 \in X_0$ is the first member of X_0 then $h(x_0) \in Y_0$ is the first element of Y_0, for if not, there exists $x' \in X_0$ such that $y' = f(x')$. Now, since $f(x') <' f(x_0)$ it follows that $x' < x_0$, which is contradiction.

10. If $(X, <)$ and $(Y, <')$ are two disjoint ordinals define $X+Y$ as the ordered set $(X \cup Y, \langle)$ where $x \langle y$ if $x \in X$ and $y \in Y$ or $x, y \in X$ and $x < y$ or $x, y \in Y$ and $x <' y$. Show that $X+Y$ is also an ordinal.

Hint:

(Proposition 5).

11. Given a set of ordinal numbers O show that there exists an ordinal number γ greater than any $\alpha \in O$.

Hint:

Clearly, $\alpha < \alpha+1$. Consequently, if O contains a largest member, say β, then $\alpha < \beta+1$ for every $\alpha \in O$. Assume now that O does not have the greatest member and let $(O, <)$ be well ordered (proposition 6), then the sum

$$s = \sum_{\alpha \in O} \alpha$$

is an ordinal greater than any $\alpha \in O$. Indeed, we associate with each $\alpha \in O$ and ordinal X_α whose ordinal number is α. We may assume that all X_α are disjoint. The ordered sum

$$\overset{o}{X} = \sum_{\alpha \in O} X_\alpha$$

is an ordinal (proposition 5). Clearly, each $X_\alpha \subset \overset{o}{X}_t$, where t is the least member of the set $X_{\alpha *}$ with $\alpha < \alpha^* \in O$. Since no well ordered set is isomorphic to any of its initial segments we clearly have that $\alpha < \mathbf{s}$ for all $\alpha \in O$.

12. (Continuation). Show that an immediate successor of any $\alpha \in O$ is $\alpha + 1$.

    Hint:

    If β is an immediate successor of an $\alpha \in O$ then

$$O_\beta = O_\alpha + \{\alpha\}.$$

By proposition 7 the ordinal number of O_β is so that

$$\beta = \alpha + 1.$$

13. Starting from the fact that $1 + \omega = \omega$ show that

    (i) $1 + \omega + \omega^2 + \cdots + \omega^n = \omega^n \qquad (n = 1, 2, \cdots)$

    (ii) $1 + \omega + \omega^2 + \cdots = \omega^\omega$

    Hint:

$$1 + \omega + \omega^2 = \omega + \omega^2 = \omega(1 + \omega) = \omega \cdot \omega.$$

14. Let $(X, <)$ be an ordinal with Card $X \geq \mathbf{c}$, and the last element $\Omega \in X$. Show that there is an $y_0 \in X$ such that Card $X y_0 \leq \aleph_0$.

    Hint:

    Let $Y \subset X$ be the subset of all $y \in X$ such that X_y is uncountable. $Y \neq \theta$ because $\Omega \in Y$.

Since $Y \subset X$ it is well ordered and consequently contains the smallest element, say $y_0 \in Y$. Then

$$X y_0 = \{x \in X;\ x < y_0\}$$

is countable. If X does not have a last element, take an element $z \notin X$, replace X by $X \cup \{z\}$, and extend the order $<$ by setting $x < z$ for all $x \in X$.

Chapter II

MEASURE THEORY

17. MEASURES

Here as usual the set of real numbers is denoted by R and $\overline{R} = [-\infty, \infty] = R \cup \{-\infty, \infty\}$ is extended real number system. The objects $-\infty$ and $+\infty$ are not real numbers. The usual ordering on R is extended to $\overline{R}$ by setting $-\infty < x < +\infty$.

Let X be a nonempty set and let $\mathcal{K} \subset \mathcal{P}(X)$ be a subclass such that $\theta \in \mathcal{K}$. A map

$$\varphi : \mathcal{K} \to \overline{R}$$

is called a "**set function**." The map φ is said to be "**finitely additive**" (on $\mathcal{K}$) if $\varphi(\theta)=0$ and if for any finite disjoint sequence $\{A_k\}_1^n \subset \mathcal{K}$, such that $\bigcup_1^n A_k \in \mathcal{K}$ we have that

$$\varphi\left(\bigcup_1^n A_k\right) = \sum_1^n \varphi(A_k). \tag{1}$$

Notice that the condition $\varphi(\theta)=0$ is almost redundant since if there exists only one $B \in \mathcal{K}$ such that $|\varphi(B)| < \infty$, from $\varphi(B)=\varphi(B \cup \theta)=\varphi(B)+\varphi(\theta)$ we have that $\varphi(\theta)=0$.

In order that the sum on the right-hand side of (1) is well defined, the sequence $\{A_k\}_1^n$ must not contain members A_i and A_j such that $\varphi(A_i)=-\infty$ and $\varphi(A_j)=+\infty$. For this reason we shall assume from now on that the range of any set function is in $(-\infty, +\infty]$. If for every infinite sequence of disjoint sets $\{M_k\}_1^\infty$ from $\mathcal{K}$ such that $\bigcup_1^\infty M_k \in \mathcal{K}$, we have that

$$\varphi\left(\bigcup_1^\infty M_k\right) = \sum_1^\infty \varphi(M_k), \tag{2}$$

the set function φ is said to be countably additive (or σ-additive) on $\mathcal{K}$.

A non-negative countably additive set function is called a "**measure**." Natural domain of definition of a measure is a σ-algebra. Recall that, **given a set X, an algebra $\mathcal{A}_0 \subset \mathcal{P}(X)$ is a class which contains θ, X and which is closed under the formation of finite union and complement. An algebra closed under the formation of countable union is a σ-algebra $\mathcal{A}$. Note that $\mathcal{A}_0 \neq \theta$.**

Let $\mathcal{A} \subset \mathcal{P}(X)$ be a σ-algebra; the couple $(X, \mathcal{A})$ is termed a "**measurable space**", and every element of $\mathcal{A}$ is called a "**measurable set**." Let μ be a measure on $\mathcal{A}$. **The system $(X, \mathcal{A}, \mu)$ is called a "measure space."** If $\mu(X)<\infty$ we say that μ is **finite**. If $\mu(X)=+\infty$, but there exists a sequence $\{E_n\}_1^\infty$ from $\mathcal{A}$ such that $X=\bigcup_1^\infty E_n$ and $\mu(E_n)<\infty$ for all n=1, 2, $\cdots$, then we say that μ **is "σ-finite."**

Any set $N \in \mathcal{A}$ such that $\mu(N)=0$ is called a **"μ-null"** set (or simply a null set if no other measures are involved). A set $F \in \mathcal{A}$ such that $\mu(F^c)=0$ is often called a set of **"full measure."** One of the key concepts in measure theory is that of **"almost everywhere"**, abbreviated (a.e). Let π be a property of some elements of X. We say that π holds (a.e) on X if it holds at every point of a subset of full measure.

An **"atom"** of a measure space $(X, \mathcal{A}, \mu)$ is a (μ-equivalence class) set $A \in \mathcal{A}$ such that $0<\mu(A)<\infty$, and if $C \in \mathcal{A} \cap A$ then either $\mu(C)=0$ or $\mu(A-C)=0$. Any two atoms A_1 and A_2 of $(X, \mathcal{A}, \mu)$ are called distinct if we have that $\mu(A_1 \cap A_2)=0$. A measure space which does not have atoms is called **"nonatomic."**

As an example, consider a measurable space $(X, \mathcal{A})$, where X is a finite or countably infinite set. In such a case the σ-algebra $\mathcal{A}=\mathcal{P}(X)$. A measure μ on X is then usually given in the form $p(x)=\mu(\{x\})$ and for any nonempty set $A \subset X$

$$\mu(A) = \sum_{x \in A} p(x).$$

Here every singleton $\{x\}$, such that $0< p(x)<\infty$, is an atom. The cardinality of $\mathcal{A}$ is clearly 2^n if Card $X = n$ and, if Card $X=\aleph_0$, Card $\mathcal{A} = c$. Cardinality of a σ-algebra is never $\aleph_0$.

The idea to define a measure as a set function is technically compelling. It is also very natural one from an intuitive geometric point of view. To see how it arises in this way, consider the problem of formulating the concept of measure on a set X as an abstraction of the notion of length, volume, mass, probability, etc. Formally, it is a function that assigns to a "suitable" subset $A \subset X$ a number $\mu(A) \geq 0$, representing its "measure", and having the properties characteristic of a measure. But which subsets are suitable and which properties are characteristics? Now, one property of the familiar examples of measure is that, so to speak, the whole is the sum of its parts. This leads to the additivity of μ. This, on the other hand, requires that the unions of "suitable sets' are suitable sets.

In the framework of the real line R, the most natural suitable sets are intervals. The class $\mathcal{S}$ of all intervals of the form (a, b] is a **"semiring"** (i.e., a class such that $\theta \in \mathcal{S}$ and if A, $B \in \mathcal{S}$ then $A \cap B \in \mathcal{S}$. In addition, if K, $M \in \mathcal{S}$, there exists a finite subclass $\{C_i\}_1^n \subset \mathcal{S}$ of disjoint sets such that $M-K= \bigcup_{i=1}^{n} C_i$). A **"semialgebra"** $\mathcal{S}^*$ is a semiring which contains R as an element. The collection $\mathcal{A}_0$ of all finite unions of disjoint elements from $\mathcal{S}^*$ is an algebra. For example, while the class of all intervals of the form (a, b], where $-\infty<a \leq b<\infty$, is a semiring, the class of all intervals of the form $(-\infty, a], (\alpha, \beta], (b, \infty)$ is a semialgebra.

Remark 1

The concept of a "measure" on R was first defined by Borel. This is essentially simply the "length function" which we will denote by λ. It assigns to an interval (a, b] the "measure" $\lambda((a, b])=b-a$. The length function λ is countably additive on the semiring $\mathcal{S}$. Its completion (see remark 1 of the next section) is called **the "Lebesgue measure" and is denoted by** $\bar{\lambda}$.

Remark 2

In general, a nonempty class $\mathcal{S} \subset \mathcal{P}(X)$ **is a semiring if** $\theta \in \mathcal{S}$**, and if for any A, B$\in \mathcal{S}$, A$\cap$B$\in \mathcal{S}$; while A-B may not belong to** $\mathcal{S}$ **, there exists a finite sequence of disjoint sets** $\{C_i\}_1^n$ **from** $\mathcal{S}$ **such that** $A\text{-}B = \bigcup_{i=1}^{n} C_i$ **. A semialgebra is a semiring which contains X as an element.**

Problems and complements.

1. Let μ be an additive set function defined on an algebra $\mathcal{A}_0$. Assuming that $A, B \in \mathcal{A}_0$, show that if $A \subset B$ and if $\mu(A) < \infty$, then

$$\mu(B-A) = \mu(B) - \mu(A).$$

Hint:

$B-A \in \mathcal{A}_0$ and $B = A \cup (B-A)$; hence

$$\mu(B) = \mu(A) + \mu(B-A).$$

2. If μ is a measure on an algebra $\mathcal{A}_0$ and if $\{A_i\}_1^n$ and $\{B_i\}_1^n$ are sequences of $\mathcal{A}_0$, show that

$$\mu\left(\bigcup_1^n A_i - \bigcup_1^n B_i\right) \le \sum_1^n \mu(A_i - B_i).$$

Hint:

$$\bigcup_1^n A_i - \bigcup_1^n B_i = \bigcup_1^n A_i \cap \left(\bigcup_1^n B_i\right)^c \subset \bigcup_1^n (A_i - B_i).$$

3. Let μ be a measure on an algebra $\mathcal{A}_0$. Show that for any $A_1, A_2 \in \mathcal{A}_0$

$$\left|\mu(A_1) - \mu(A_2)\right| \le \mu(A_1 \triangle A_2).$$

Hint:

$$\mu(A_1) - \mu(A_2) = \mu(A_1) - \mu(A_1 \cap A_2) - \mu(A_2) + \mu(A_1 \cap A_2)$$

$$= \mu(A_1 - A_2) - \mu(A_2 - A_1).$$

Thus,

$$\left|\mu(A_1) - \mu(A_2)\right| \le \mu(A_1 - A_2) + \mu(A_2 - A_1) = \mu(A_1 \triangle A_2).$$

4. Let $\mathcal{A}_0$ be an algebra and $\varphi : \mathcal{A}_0 \to [0, \infty)$ a finitely additive set function. Show that φ is countably additive if and only if for any monotone sequence $\{M_n\}_1^\infty$ from $\mathcal{A}_0$ such that $M_n \downarrow \theta$, then necessarily $\varphi(M_n) \to 0$.

Hint:

If φ is σ-additive then

$$\varphi(M_1) = \sum_{k=1}^\infty \varphi(M_k - M_{k+1}) < \infty$$

Since the series converges and $M_n = \bigcup_{i=n}^\infty (M_i - M_{i+1})$, we have that

$$\sum_{k=n}^\infty \varphi(M_k - M_{k+1}) \to 0 \text{ as } n \to \infty => \varphi(M_n) \to 0.$$

Let $\{G_n\}_1^\infty \subset \mathscr{A}_0$ be a disjoint sequence such that $G = \bigcup\limits_{k=1}^\infty G_k \in \mathscr{A}_0$ and set $B_1 = G$, $B_n = G$

$-\bigcup\limits_1^{n-1} G_k$. Clearly, each $B_n \in \mathscr{A}_0$ and $B_n \downarrow \theta$ as $n \to \infty$. Since

$$\varphi(B_n) = \varphi(G) - \sum\limits_{k=1}^{n-1} \varphi(G_k)$$

and $\varphi(B_n) \to 0$ as $n \to \infty$, the proof is completed.

5. Let $(X, \mathscr{A}, \mu)$ be a measure space such that $\mu(X) < \infty$. Show that the collection of all distinct atoms is at

most countable.

Hint:

Let $\mathscr{C} \subset \mathscr{A}$ be the collection of all distinct atoms of μ. For each $n=1, 2, \cdots$, denote by

$$\mathscr{C}_n = \{A \in \mathscr{C}; \mu(A) \geq \mu(X) / n \}.$$

Since for every $A_1, A_2 \in \mathscr{C}$ we have that $\mu(A_1 \cap A_2) = 0$, the number of atoms in $\mathscr{C}_n$ is at most $n \cdot \mu(X)$.

Now, since $\mathscr{C} = \bigcup\limits_1^\infty \mathscr{C}_n$ the assertion follows.

6. Let $\{A_k\}_1^\infty$ be a sequence from an algebra, $\mathscr{A}_0$. Show that there exists a disjoint sequence

$\{G_k\}_1^\infty \subset \mathscr{A}_0$ such that

$$\bigcup\limits_{k=1}^\infty A_k = \bigcup\limits_{k=1}^\infty G_k .$$

Hint:

Set $G_1 = A_1$ and $G_n = A_n \bigcap\limits_1^{n-1} A_i^c$, $n = 2, 3, \cdots$. Clearly, $G_i \cap G_j = \theta$ for every $i \neq j$. It is also clear that

$\bigcup\limits_1^\infty G_k \subset \bigcup\limits_1^\infty A_k$. Now, if $\omega \in \bigcup\limits_1^\infty A_k$ it must belong to at least one A_k. Let n be the least index such that

$\omega \in A_n$ (i.e. $\omega \notin A_i$, $i=1, 2, \cdots, n-1$ and $\omega \in A_n$). In other words,

$$\omega \in A_n \bigcap\limits_1^{n-1} A_i^c \Rightarrow \omega \in G_n.$$

7. Recall **that a class $\boldsymbol{\mathcal{S}} \subset \boldsymbol{\mathcal{P}}(X)$ closed under finite intersection is a "semiring" if $\theta \in \boldsymbol{\mathcal{S}}$ and if $\{A, B\} \subset \boldsymbol{\mathcal{S}}$,**

there exists a finite sequence of disjoint sets $\{C_k\}_1^n \subset \boldsymbol{\mathcal{S}}$ such that

$$A - B = \bigcup\limits_1^n C_k .$$

If A and $\{A_k\}_1^n$ are from $\boldsymbol{\mathcal{S}}$ show that

$$D = A - \bigcup_{k=1}^{n} A_k = \bigcup_{i=1}^{m} B_i$$

where $\{B_i\}_1^m \subset \mathcal{S}$ are disjoint.

Hint:

By induction; for $n=1$ the claim is obvious by the definition of a semiring. Assume that $A_{n+1} \in \mathcal{S}$, then

$$A - \bigcup_{k=1}^{n+1} A_k = D - A_{n+1} = \bigcup_{i=1}^{m} B_i - A_{n+1} = \bigcup_{i=1}^{m} (B_i - A_{n+1}).$$

But each

$$B_i - A_{n+1} = \bigcup_{i=1}^{m_i} C_j^i \quad \text{where} \quad \{C_j^i\}_1^{m_i} \subset \mathcal{S}$$

are disjoint. Consequently,

$$A - \bigcup_{k=1}^{n+1} A_k = \bigcup_{i=1}^{m} \left(\bigcup_{j=1}^{m_i} C_j^i \right).$$

8. A class $\mathcal{O} \subset \mathcal{P}(X)$ is called a "ring" if it is closed under the formation of set theoretic difference and finite union. If $\mathcal{S} \subset \mathcal{P}(X)$ is a semiring, show that the class of all finite disjoint unions of elements from $\mathcal{S}$ is a ring.

Hint:

Let $\mathcal{O}$ be the class of all finite unions of disjoint elements from $\mathcal{S}$. Clearly $\mathcal{S} \subset \mathcal{O}$. Now consider A, $B \in \mathcal{O}$ and suppose that

$$A = \bigcup_{1}^{n} A_i , \quad B = \bigcup_{1}^{m} B_j .$$

Let us show that $A \cap B \in \mathcal{O}$. Denote by

$$A_i \cap B_j = C_{ij} \in \mathcal{S}.$$

Clearly then

$$A \cap B = \bigcup_{i=1}^{n} \bigcup_{j=1}^{m} C_{ij} \in \mathcal{O},$$

proving that the class $\mathcal{O}$ is closed under finite intersection. Next, consider

$$A - B = \bigcup_{i=1}^{n} \left(A_i - \bigcup_{j=1}^{m} B_j \right).$$

But according to the previous problem

$$A_i - \bigcup_{j=1}^{m} B_j = \bigcup_{\ell=1}^{r_i} E_{i\ell} ,$$

where for every $i=1, 2, \cdots, n$ fixed, $\{E_{i\ell}\}_1^{r_i} \subset \mathscr{S}$ are disjoint. Clearly,

$$(A_v - \bigcup_{j=1}^{m} B_j) \cap (A_k - \bigcup_{j=1}^{m} B_j) = \theta \ \ \text{if} \ v \neq k$$

so that

$$A - B = \bigcup_{i=1}^{n} (\bigcup_{\ell=1}^{r_i} E_{i\ell}) \in \mathscr{C}.$$

Since $A \cup B = B \cup (A\text{-}B) \in \mathscr{C}$, it follows that $A \cup B \in \mathscr{C}$, proving that $\mathscr{C}$ is a ring.

9. Let $\mathscr{C} \subset \mathscr{P}(X)$ be an arbitrary class. Show that the intersection of all rings containing $\mathscr{C}$, denoted by $\mathscr{O}(\mathscr{C})$, is a ring (generated by $\mathscr{C}$). Show also that every element of $\mathscr{O}(\mathscr{C})$ can be covered by a finite union of elements from $\mathscr{C}$.

Hint:

If $A, B \in \mathscr{O}(\mathscr{C})$, then A and B belong to every ring which contains $\mathscr{C}$; clearly then $A \cap B$ also belongs to every ring which contains $\mathscr{C}$, so that $A \cap B \in \mathscr{O}(\mathscr{C})$, etc.

Let $\mathscr{K} \subset \mathscr{O}(\mathscr{C})$ be the class of all those elements which can be covered by a finite union of elements from $\mathscr{C}$. It is not difficult to see that $\mathscr{K}$ is a ring (which contains $\mathscr{C}$). Thus, $\mathscr{O}(\mathscr{C}) \subset \mathscr{K}$.

10. Let $\mathscr{C} \subset \mathscr{P}(X)$ be a subclass such that $X \in \mathscr{C}$. Let $\mathscr{A}_0(\mathscr{C})$ be the least (with respect to inclusion) algebra containing $\mathscr{C}$ (i.e., the intersection of all algebras containing $\mathscr{C}$). Does every member of $\mathscr{A}_0(\mathscr{C})$ can be covered with a finite union of element from $\mathscr{C}$?

Hint:

The answer is yes in the case when $X \in \mathscr{C}$, because $\mathscr{O}(\mathscr{C}) = \mathscr{A}_0(\mathscr{C})$. Otherwise the answer is no. As a counter example let X be an infinite set and let $\mathscr{C}$ be the collection of all its singletons. Then the algebra $\mathscr{A}_0(\mathscr{C})$ consists of all finite unions of elements from $\mathscr{C}$ and their complements. Clearly, no infinite member from $\mathscr{A}_0(\mathscr{C})$ can be covered by a finite union of members from $\mathscr{C}$.

11. Let $\mathscr{A}_0$ be an algebra. The least (with respect to inclusion) σ-algebra containing $\mathscr{A}_0$ is usually denoted by $\sigma\{\mathscr{A}_0\}$ (i.e., the intersection of all σ-algebras containing $\mathscr{A}_0$). Show that every member of $\sigma\{\mathscr{A}_0\}$ can be covered by a finite or countable union of elements from $\mathscr{A}_0$.

Hint:

Denote by $\mathscr{K} \subset \sigma\{\mathscr{A}_0\}$ the collection of sets with this property. Clearly, $\mathscr{A}_0 \subset \mathscr{K}$ and since $\mathscr{K}$ is a σ-algebra we have that $\sigma\{\mathscr{A}_0\} \subset \mathscr{K}$.

12. Let X be an uncountably infinite set and denote by $\hat{S} \subset \mathscr{P}(X)$ the collection of all singletons of X. Show that $\sigma\{\hat{S}\}$ consists of those subsets $F \subset X$ such that F or F^c is countable.

Hint:

It is clear that $\sigma\{\hat{S}\}$ contains such sets. It remains to show that the class of all such sets forms a σ-algebra. If $\{F_n\}_1^\infty$ is a sequence of such sets their union is countable if each F_n is. If the complement of one of them is countable then the complement of their union will be, a fortiori

13. Recall that a class $\mathscr{S}^* \subset \mathscr{P}(X)$ closed under finite intersections is called a semialgebra if $\theta, X \in \mathscr{S}^*$ and if, given any $K \in \mathscr{S}^*$, then K^c is the union of a finite family of disjoint members of $\mathscr{S}^*$. Show that the collection of all finite unions of disjoint members of $\mathscr{S}^*$ is $\mathscr{A}_0(\mathscr{S}^*)$, i.e., the least (with respect to inclusion) algebra which contains $\mathscr{S}^*$.

Hint:

It follows from problem 8 and the fact that $\mathscr{S}^*$ is a semiring. The following is an independent proof.

Denote by $\mathscr{L}$ the class of such unions. Clearly, $\theta, X \in \mathscr{L}$ and if $\bigcup_1^n A_i$, $\bigcup_1^m B_j \in \mathscr{L}$ we have that

$$\left(\bigcup_1^n A_i \cap \bigcup_1^m B_j\right) = \bigcup_{i=1}^n \bigcup_{j=1}^m (A_i \cap B_j) \in \mathscr{L}.$$

The class $\mathscr{L}$ is closed under complementation because

$$\left(\bigcup_{i=1}^n A_i\right)^c = \bigcap_{i=1}^n A_i^c, \text{ and } A_i^c \in \mathscr{L}.$$

It is now clear that $\mathscr{L} = \mathscr{A}_0(\mathscr{S}^*)$.

18. PROPERTIES OF MEASURES

Let $(X, \mathcal{A})$ be a measurable space. Recall that a set function

$$\mu : \mathcal{A} \to [0, \infty] \quad \text{with} \quad \mu(\theta) = 0,$$

is said to be a measure if it is countably additive. We say that a measure μ is "finite" if $\mu(X) < \infty$. If μ is not

finite but there exists a sequence $\{A_k\}_1^\infty$ from $\mathcal{A}$ such that $X = \bigcup_1^\infty A_k$ and $\mu(A_k) < \infty$ for every $k = 1, 2, \cdots$,

the measure μ is said to be σ-finite.

Next we outline some simple properties of μ. For any $\{A, B\} \subset \mathcal{A}$

$$\mu(A\text{-}B) = \mu(A) - \mu(A \cap B);$$

this follows from $A = (A\text{-}B) \cup (A \cap B)$ and additivity of μ. If $B \subset A$ we obtain from this that

$$\mu(A\text{-}B) = \mu(A) - \mu(B),$$

and hence that $\mu(B) \leq \mu(A)$ (monotonicity of μ).

Proposition 1

If $\{M_k\}_1^\infty$ is an increasing sequence from $\mathcal{A}$, then

$$\mu(\lim_{k \to \infty} M_k) = \lim_{k \to \infty} \mu(M_k).$$

Proof:

Set $M_0 = \theta$, then (see section 2)

$$\mu(\lim_{k \to \infty} M_k) = \mu(\bigcup_{k=1}^\infty M_k) = \mu(\bigcup_1^\infty (M_k - M_{k-1}))$$

$$= \sum_{k=1}^\infty \mu(M_k - M_{k-1}) = \lim_{n \to \infty} \sum_{k=1}^n \mu(M_k - M_{k-1}) = \lim_{n \to \infty} \mu(M_n). \qquad \heartsuit$$

Corollary 1

If $\{D_k\}_1^\infty$ is a decreasing sequence from $\mathcal{A}$ such that for some n $\mu(D_n) < \infty$, then $\mu(\lim_{k \to \infty} D_k) =$

$\lim_{k \to \infty} \mu(D_k)$. Indeed, we have that

$$\mu(\lim_{k \to \infty} (D_n - D_k)) = \mu(D_n) - \mu(\lim_{k \to \infty} D_k)$$

and since $\{D_n - D_k\}_{n+1}^\infty$ increases

$$\mu(\lim_{k \to \infty} (D_n - D_k)) = \lim_{k \to \infty} \mu(D_n - D_k) = \mu(D_n) - \lim_{k \to \infty} \mu(D_k),$$

which proves our claim.

Proposition 2

For any $\{B_k\}_1^\infty$ from a σ-algebra $\mathcal{A}$ we have:

$$\mu\left(\bigcup_{k=1}^{\infty} B_k\right) \le \sum_{k=1}^{\infty} \mu(B_k)$$

Proof:

Since (see section 2)

$$\bigcup_{1}^{\infty} B_k = B_1 \cup (B_2 \cap B_1^{c}) \cup (B_3 \cap B_1^{c} \cap B_2^{c}) \cup \cdots,$$

it follows that

$$\mu\left(\bigcup_{1}^{\infty} B_k\right) = \mu(B_1) + \mu(B_2 \cap B_1^{c}) + \mu(B_3 \cap B_1^{c} \cap B_2^{c}) + \cdots$$

$$\le \mu(B_1) + \mu(B_2) + \mu(B_3) + \cdots, \text{ as claimed.} \qquad \heartsuit$$

Recall that an element $N \in \mathscr{A}$ is called **a "null" set** if $\mu(N)=0$. A subset $\Gamma \subset N$ is said **to be "negligible." The measure space $(X, \mathscr{A}, \mu)$ is said to be "complete" if the σ-algebra $\mathscr{A}$ contains every negligible subset of X**. Since the union of every countable family of null sets is a null set, the union of every countable family of negligible sets is a negligible set. Thus, the class $\mathscr{E}$ of all negligible sets is closed under the operation of countable union.

Proposition 3

Let $(X, \mathscr{A}, \mu)$ be a measure space. Denote by $\overline{\mathscr{A}}$ the class of all sets of the form $A \cup L$ where $A \in \mathscr{A}$ and L is negligible, then

$$\overline{\mathscr{A}} = \sigma\{\mathscr{A} \cup \mathscr{E}\}.$$

Proof:

We clearly have that $\{\mathscr{A} \cup \mathscr{E}\} \subset \overline{\mathscr{A}} \subset \sigma\{\mathscr{A} \cup \mathscr{E}\}$. But $\overline{\mathscr{A}}$ is a σ-algebra because if $\{(A_i \cup L_i)\}_{1}^{\infty}$ is a sequence from $\overline{\mathscr{A}}$

$$\bigcup_{1}^{\infty}(A_i \cup L_i) = A \cup L \in \overline{\mathscr{A}}$$

where $A = \bigcup_{1}^{\infty} A_i \in \mathscr{A}$ and $L = \bigcup_{1}^{\infty} L_i \in \mathscr{E}$. Also, if $(A \cup L) \in \overline{\mathscr{A}}$

$$(A \cup L)^{c} = (A^{c} \cap L^{c}) \cap (N^{c} \cup N) = (A \cup N)^{c} \cup N \cap (A \cup L)^{c} \in \overline{\mathscr{A}}$$

where $L \subset N$ and $\mu(N)=0$. Thus, $\overline{\mathscr{A}}$ is closed under the countable union and complementation. This proves our contention. $\qquad \heartsuit$

Remark 1

The set function $\overline{\mu}$ defined on $\overline{\mathscr{A}}$ by

$$\overline{\mu}(A \cup L) = \mu(A)$$

is complete measure on $\overline{\mathscr{A}}$. For, if L is a negligible set

$$\bar{\mu}(L) = \bar{\mu}(\theta \cup L) = \mu(\theta).$$

The set function $\bar{\mu}$ is "well defined", for if

$$A_1 \cup L_1 = A_2 \cup L_2 \text{ where } \{A_1, A_2\} \subset \mathcal{A} \text{ and } \{L_1, L_2\} \subset \mathcal{C},$$

we obtain that

$$(A_1 \triangle A_2) \subset L_1 \cup L_2.$$

Thus, $\mu(A_1 \triangle A_2) = 0$ and consequently $\mu(A_1) = \mu(A_2)$ so that

$$\bar{\mu}(A_1 \cup L_1) = \mu(A_1) = \mu(A_2) = \bar{\mu}(A_2 \cup L_2).$$

Finally, if $\{(A_k \cup L_k)\}_1^{\infty} \subset \overline{\mathcal{A}}$ are disjoint we have that

$$\bar{\mu}(\bigcup_1^{\infty}(A_k \cup L_k)) = \bar{\mu}((\bigcup_1^{\infty} A_k) \cup L) = \mu(\bigcup_1^{\infty} A_k) = \sum_1^{\infty} \mu(A_k) = \sum_1^{\infty} \bar{\mu}(A_k \cup L_k).$$

Thus $\bar{\mu}$ is a measure on $\overline{\mathcal{A}}$; this proves that the measure space $(X, \overline{\mathcal{A}}, \bar{\mu})$ is complete.

The genesis of modern measure theory can be found in the works of Borel and Lebesgue. The first measure was **the Lebesgue measure** on the real line which is essentially an extension of the "length" function. This function assigns to every interval (a, b) the "measure" (b-a). A useful generalization is the concept of **the "Lebesgue-Stieltjes" measure** where (b-a) is replaced by the increment $F(b)-F(a)$ of a nondecreasing function $F:R \rightarrow [0,\infty)$, say left continuous, i.e., such that $F(x-0)=F(x)$ at every $x \in R$.

The collection $\mathcal{S}$ of intervals

$$\mathcal{S} = \{(a, b]; -\infty < a \leq b < \infty\}$$

is a semiring. The set function μ_F defined on $\mathcal{S}$ by

$$\mu_F (a, b] = F(b) - F(a)$$

is called the Lebesgue-Stieltjes measure induced by F. To deserve this name we must show that μ_F is countably additive on $\mathcal{S}$. First of all $\mu_F(\theta)=0$. Let $\{(a_i, a_{i+1}]\}_1^{\infty} \subset \mathcal{S}$ be a collection of disjoint intervals such that

$$\bigcup_1^{\infty}(a_i, a_{i+1}] = (\alpha, \beta] \in \mathcal{S}.$$

We have to show that

$$\mu_F(\bigcup_1^{\infty}(a_i, a_{i+1}]) = \sum_1^{\infty} \mu_F(a_i, a_{i+1}].$$

Indeed,

$$\sum_{i=1}^{\infty} \mu_F(a_i, a_{i+1}] = \lim_{n \to \infty} \sum_{i=1}^{n} \mu_F(a_i, a_{i+1}]$$

$$= \lim_{n \to \infty} [F(a_{n+1}) - F(a_1)] = F(\beta) - F(\alpha) = \mu_F(\alpha, \beta]$$

where $\lim_{n\to\infty} F(a_n)=F(\beta-0)=F(\beta)$.

Next we take up and solve the problem of extending a measure μ_0 on a semialgebra $\mathcal{S}^*$ of subsets of a set X to a measure on the algebra $\mathcal{A}_0(\mathcal{S}^*)$ (generated by $\mathcal{S}^*$). Recall that θ, $X\in\mathcal{S}^*$ and that $\mathcal{S}^*$ is closed under finite intersection. In addition, if $C\in\mathcal{S}^*$, then C^c is the union of a finite family of disjoint elements of $\mathcal{S}^*$. The algebra $\mathcal{A}_0(\mathcal{S}^*)$ is the collection of all finite unions of disjoint elements of $\mathcal{S}^*$ (see problem 13 of section 17).

Proposition 5

Let μ_0 be a measure on a semialgebra $\mathcal{S}^*\subset\mathcal{P}(X)$. There exists a unique extension μ of μ_0 to the algebra $\mathcal{A}_0(\mathcal{S}^*)$.

Proof:

Any $A\in\mathcal{A}_0(\mathcal{S}^*)$ is of the form

$$A = \bigcup_1^n E_i, \text{ where } \{E_i\}_1^n \text{ are from } \mathcal{S}^* \text{ and } E_j\cap E_k \text{ if } j\neq k.$$

Let μ be a set function on $\mathcal{A}_0(\mathcal{S}^*)$ defined by

$$\mu(A) = \sum_1^n \mu_0(E_i). \tag{1}$$

Clearly, $\mu=\mu_0$ on $\mathcal{S}^*$. To show that μ is well defined on $\mathcal{A}_0(\mathcal{S}^*)$, suppose that there exists another finite sequence of disjoint sets $\{F_j\}_1^m\subset\mathcal{S}^*$ such that $A=\bigcup_1^m F_j$. We must show that

$$\sum_{i=1}^n \mu_0(E_i) = \sum_{j=1}^m \mu_0(F_j).$$

To this end denote by $H_{jk} = F_j\cap E_k$; Clearly, $\{H_{jk}\}\subset\mathcal{S}^*$ is a finite disjoint sequence such that

$$E_k = E_k\cap \bigcup_{j=1}^m F_j = \bigcup_{j=1}^m H_{jk}$$

$$F_j = F_j\cap \bigcup_{k=1}^n E_k = \bigcup_{k=1}^n H_{jk}.$$

Thus,

$$\sum_{k=1}^n \mu_0(E_k) = \sum_{k=1}^n\sum_{j=1}^m \mu_0(H_{jk}) = \sum_{k=1}^m \mu_0(F_j)$$

as claimed. The finite additivity of μ is obvious. To show that μ is countably additive on $\mathcal{A}_0(\mathcal{S}^*)$, let $\{G_\ell\}_1^\infty\subset\mathcal{A}_0(\mathcal{S}^*)$ be a disjoint sequence such that

$$G = \bigcup_1^{\infty} G_i \in \mathcal{A}_0(\boldsymbol{S}^*).$$

Then, since every member of $\mathcal{A}_0(\boldsymbol{S}^*)$ is a finite union of disjoint elements from $\boldsymbol{S}^*$ we have that

$$G = \bigcup_{i=1}^{n} C_i \quad \text{where } \{C_i\}_1^n \text{ are from } \boldsymbol{S}^* \text{ and } C_i \cap C_j = \theta \ \text{ if } i \neq j.$$

Now for every $k = 1, 2, \cdots$

$$G_k = \bigcup_{i=1}^{r_k} G_{ki} \quad \text{where } \{G_{ki}\}_1^{r_k} \subset \boldsymbol{S}^* \text{ and } G_{ki} \cap G_{kj} = \theta \text{ if } i \neq j$$

and for every $i = 1, 2, \cdots, n$

$$C_i = C_i \cap G = \bigcup_{\ell=1}^{\infty}(C_i \cap C_\ell) = \bigcup_{\ell=1}^{\infty} \bigcup_{j=1}^{r_\ell}(C_i \cap G_{\ell j}).$$

Taking into account that $C_i \cap G_{\ell j} \subset \mathcal{S}^*$ and that μ_0 is a measure on $\boldsymbol{S}^*$ we have that

$$\mu_0(C_i) = \sum_{\ell=1}^{\infty} \sum_{j=1}^{r_\ell} \mu_0(C_i \cap G_{\ell j}).$$

Thus,

$$\mu(G) = \sum_{i=1}^{n} \mu_0(C_i) = \sum_{i=1}^{n} \sum_{\ell=1}^{\infty} \sum_{j=1}^{r_\ell} \mu_0(C_i \cap G_{\ell j})$$

$$= \sum_{\ell=1}^{\infty} \sum_{j=1}^{r_\ell} \mu_0(G_{\ell j}) = \sum_{\ell=1}^{\infty} \mu(G_\ell),$$

which proves that μ is a measure on $\mathcal{A}_0(\boldsymbol{S}^*)$.

To show that the measure μ given by (1) is unique, assume that μ_1 is another extension of μ_0 to $\mathcal{A}_0(\boldsymbol{S}^*)$ which is a measure. Since for any $A \in \mathcal{A}_0(\boldsymbol{S}^*)$ we have that

$$A = \bigcup_1^{n} D_k \quad \text{where } \{D_k\}_1^n \subset \boldsymbol{S}^* \text{ and } D_i \cap D_j = \theta \text{ if } i \neq j,$$

it follows that

$$\mu_1(A) = \sum_1^{n} \mu_1(D_k) = \sum_1^{n} \mu_0(D_k) = \mu(A).$$

This completes the proof of our contention. ♥

Remark 2

Let $(X, \mathcal{A}, \mu)$ be a measure space and $A, B \in \mathcal{A}$ such that $A \not\subset B$. **If $\mu(A-B)=0$ we say that $A \subset B$** **"almost everywhere" (a.e.) and we write $A \overset{a.e.}{\subset} B$.** The (a.e) inclusion $\overset{a.e.}{\subset}$ is a "preorder" in $\mathcal{A}$ (see problem 1 of section 10) because it is reflexive, $\mu(A\text{-}A)=0$, and transitive (if A, B, C$\in\mathcal{A}$ and if A $\overset{\text{a.e.}}{\subset}$ B,

B $\overset{\text{a.e.}}{\subset}$ C then

$$\mu(A\text{-}C) = \mu(A\cap B\cap C^{c}) + \mu(A\cap B^{c}\cap C^{c}) \leq \mu(A\text{-}B) + \mu(B\text{-}C) = 0 \ ).$$

Thus, $(\mathcal{A}, \overset{\text{a.e.}}{\subset})$ is a preordered set. Recall that a preordered set is said to be **"inductive"** if every chain $\mathcal{C}\subset$

$\mathcal{A}$ has an upper bound in $(\mathcal{A}, \overset{\text{a.e.}}{\subset})$.

Remark 3

Recall that by a sequence in a class $\mathcal{A}$ of sets we mean a map $f{:}N_{+}\to\mathcal{A}$, usually given in the form

$(A_{1}, A_{2}, \cdots)$ where $f(n)=A_{n}$ for each $n\in N_{+}$. We sometimes write $\{B_{k}\}_{1}^{\infty}\subset\mathcal{A}$ if each $B_{k}\in\mathcal{A}$ and

$B_{i}\cap B_{j}=\emptyset$ if $i\neq j$.

Problems and complements

1. Construct an example of a decreasing sequence of measurable sets $E_1 \supset E_2 \supset \cdots$ such that

$$\mu\left(\bigcap_1^\infty E_k\right) \neq \lim_{k\to\infty} \mu(E_k).$$

Hint:

Let μ be the Lebesgue measure and define $E_n = \{x \in R; n < x < \infty\}$. Clearly, then

$$\bigcap_{n=1}^\infty E_n = \theta \quad \text{while} \quad \mu(E_n) = +\infty \text{ for all } n = 1, 2, \cdots.$$

2. Let μ be a σ-finite measure on an algebra $\mathcal{A}_0 \subset \mathcal{P}(X)$. Given any $x > 0$, show that there exists a set

$B \in \mathcal{A}_0$ such that $\mu(B) \geq x$.

Hint:

Since μ is σ-finite there exists a sequence of disjoint sets $\{A_k\}_1^\infty \subset \mathcal{A}_0$ such that $\mu(A_k) < \infty$ for

each $k = 1, 2, \cdots$ and such that $X = \bigcup_1^\infty A_k$. Set

$$n = \inf_k \left\{ \sum_{i=1}^k \mu(A_k) \geq x \right\},$$

then clearly $B = \bigcup_1^n A_k$.

3. Let X be uncountable and let $\mathcal{R} \subset \mathcal{P}(X)$ be a class defined by

$$\mathcal{R} = \{G \subset X; G \text{ or } G^c \text{ is at most countable}\}.$$

Show that the $\mathcal{R}$ is a σ-algebra and that the map $\mu: \mathcal{R} \to \{0, 1\}$ specified by

$$\mu(E) = \begin{cases} 0 & \text{if} \quad E \text{ is at most countable} \\ 1 & \text{if} \quad E^c \text{ is at most countable} \end{cases}$$

is a measure on $\mathcal{R}$ (see problem 12 of section 17)

Hint:

The class $\mathcal{R}$ is the σ-algebra generated by the collection of singletons $\{x\}$ of X. Indeed, θ, $X \in \mathcal{R}$. If

$\{B_n\}_1^\infty$ is a sequence from $\mathcal{R}$ and if each B_k is at most countable so is $\bigcup_1^\infty B_k \in \mathcal{R}$. On the other hand, if

for some k, B_k is not countable, then B_k^c is countable and the inclusion

$$\left(\bigcup_1^\infty B_k\right)^c \subset B_k^c \implies \bigcup_1^\infty B_k \in \mathcal{R}.$$

If elements of a sequence $\{E_k\}_1^\infty \subset \mathscr{A}$ are disjoint and each E_k is at most countable so is $E = \bigcup_1^\infty E_k$ and

$$\mu(E) = 0, \qquad \mu(\bigcup_1^\infty E_k) = \sum_1^\infty \mu(E_k) = 0.$$

On the other hand, if E_n^c is at most countable for some n, then since $E_i \subset E_n^c$ for each $i \neq n$, each E_i is at most countable. Consequently,

$$1 = \mu(E_n) = \mu(\bigcup_1^\infty E_k) = \sum_1^\infty \mu(E_k) = \mu(E).$$

This proves that μ is countably additive.

4. Let $(X, \mathscr{A}, \mu)$ be a measure space. Show that for any $\{A_n\}_1^\infty$ from $\mathscr{A}$

$$\mu(A_*) \leq \liminf \mu(A_n) \leq \limsup \mu(A_n) \leq \mu(A^*)$$

where $A_* = \liminf A_n$, $A^* = \limsup A_n$.

Hint:

Since $H_n = \bigcap_{k=n}^\infty A_k \uparrow A_*$ we have that

$$\lim_{n \to \infty} \mu(H_n) = \mu(\lim H_n) = \mu(A_*).$$

On the other hand, $H_n \subset A_n$ for each n=1, 2, $\cdots$, so that

$$\liminf \mu(H_n) \leq \liminf \mu(A_n).$$

Similarly, if we set $C_n = \bigcup_{k=n}^\infty A_k$, we clearly have that $C_n \downarrow A^*$. Hence,

$$\lim_{n \to \infty} \mu(C_n) = \mu(\lim C_n) = \mu(A^*).$$

Since $A_n \subset C_n$ for each n=1, 2, $\cdots$ we have;

$$\limsup \mu(A_n) \leq \limsup \mu(C_n) = \mu(A^*).$$

5. Let $(X, \mathscr{A}, \mu)$ be a nonatomic measure space. If $A \in \mathscr{A}$ satisfies $0 < \mu(A) < \infty$, then given any $\varepsilon > 0$ there

exists a measurable subset $G \subset A$ such that $0 < \mu(G) < \varepsilon$.

Hint:

Let $B \subset A$ be measurable and such that $0 < \mu(B) < \mu(A)$, then either $\mu(A-B) \geq \mu(B)$ or $\mu(A-B) \leq$

$\mu(B)$. Thus,

$$\mu(B) \leq \mu(A)/2 \quad \text{or} \quad \mu(B) > \mu(A)/2.$$

In the later case we have that

$$\mu(A-B) < \mu(A)/2.$$

In either case, for any measurable subset of $B_1 \subset B$, either

$$\mu(B_1) \leq \mu(B)/2 \leq \mu(A)/2^2 \quad \text{or} \quad \mu(B\text{-}B_1) \leq \mu(B)/2 \leq \mu(A)/2^2.$$

Continuing in this fashion, we obtain B_n such that

$$0 < \mu(B_n) \leq \mu(A)/2^{n+1},$$

where B_n is a measurable subset either of B_{n-1} or of B_{n-2} -B_{n-1}. For n so large that $\mu(A)/2^n < \varepsilon$, the measurable subset G is B_n.

6. Let μ_0 be a non-negative finitely additive set function on a semiring $\mathscr{S}$. Show that there exists a unique finitely additive extension μ of μ_0 to the ring $\mathscr{O}(\mathscr{S})$ (the least ring containing $\mathscr{S}$).

    Hint:

    See problem 8 of section 17.

7. Let $\mathscr{S}$ be a semiring and $\mu:\mathscr{S} \to [0, \infty]$ such that $\mu(\theta)=0$. Show that the set function μ is measure if and only if: (i) given any disjoint sequence $\{A_i\}_1^\infty$ from $\mathscr{S}$ such that $\bigcup_1^\infty A_i \subset A \in \mathscr{S}$, then $\sum_1^n \mu(A_i) \leq \mu(A)$ for every n=1,2, . . . ;

(ii) Given $B \in \mathscr{S}$ and a disjoint sequence $\{B_k\}_1^\infty$ from $\mathscr{S}$ such that $\bigcup_1^\infty B_k \supset B$ then

$$\sum_1^\infty \mu(B_k) \geq \mu(B).$$

    Hint:

    Let $\{M_k\}_1^\infty \subset \mathscr{S}$ be a disjoint sequence such that $M = \bigcup_1^\infty M_k \in \mathscr{S}$. If the conditions (i) and (ii) hold we have that

$$\sum_1^n \mu(M_k) \leq \mu(M) \leq \sum_1^\infty \mu(M_k) \quad \text{for every n=1, 2, } \cdots,$$

which proves sufficiency of the conditions. To prove necessity, assume that μ is a measure on $\mathscr{S}$. Since

$$M - \bigcup_1^n M_k = \bigcup_1^m B_j, \ \{B_j\}_1^m \subset \mathscr{S} \ \text{and} \ B_i \cap B_j = \theta \ \text{if} \ i \neq j$$

$$M = (\bigcup_1^n M_k) \cup (\bigcup_1^m B_j) \in \mathscr{S},$$

we have that

$$\mu(M) \leq \sum_1^n \mu(M_k) + \sum_1^m \mu(B_j)$$

proving (i). Now, from $M = \bigcup_1^\infty M_k = \bigcup_1^\infty (M_k \cap M)$ we have that

$$\mu(M) = \sum_1^\infty \mu(M_k \cap M) \le \sum_1^\infty \mu(M_k)$$

proving (ii) and the assertion when $\{M_j\}_1^\infty \subset \mathcal{S}$ are disjoint. In the general case when $\{M_k\}_1^\infty$ are not

disjoint, we can write

$$\bigcup_1^\infty M_k = M_1 \cup (M_2 \cap M_1^c) \cup (M_3 \cap M_1^c \cap M_2^c) \cup \cdots$$

But $M_n \cap M_1^c \cap \cdots \cap M_{n-1}^c = M_n - \bigcup_1^{n-1} M_i = \bigcup_1^\infty C_i^n$ where $C_i^n \cap C_j^n = \emptyset$ if $i \ne j$ and all but finitely many $C_i^n = \emptyset$.

Since

$$\bigcup_1^m C_i^n \subset M_n \implies \sum_1^m \mu(C_i^n) \le \mu(M_n)$$

by (i). Thus,

$$\mu(M) = \sum_1^\infty \mu(M \cap M_n \cap M_1^c \cap \cdots \cap M_{n-1}^c)$$

$$= \sum_1^\infty \sum_1^\infty \mu(C_i^n \cap M) \le \sum_1^\infty \mu(M_k),$$

which proves (ii).

8. Denote by $\mathcal{A}_t = \{A \in \mathcal{A}; \mu(A) \le t\}$ where $0 \le t < \infty$. Show that the preordered set $(\mathcal{A}_t, \overset{\text{a.e.}}{\subset})$ is inductive (see

remark 2) .

    Hint:

    We must show that given any chain $\mathcal{C} \subset \mathcal{A}_t$ there exists a set $B \in \mathcal{A}_t$ such that if $C \in \mathcal{C}$ we have

that $C \overset{\text{a.e.}}{\subset} B$. Denote by $\overset{0}{\mathcal{M}}$ the collection of all at most countable subclasses of the chain $\mathcal{C}$. We shall show

that the supremum

$$m = \sup \{\mu(\cup \mathcal{C}'); \ \mathcal{C}' \in \overset{0}{\mathcal{M}}\}$$

is attained for some $\mathcal{C}_0 \in \overset{0}{\mathcal{M}}$. To this end choose a sequence $\{\mathcal{C}_n\}_1^\infty$ from $\overset{0}{\mathcal{M}}$ such that for every $k = 1, 2,$

$$0 \le m - \mu(\cup \mathcal{C}_k) \le 1/k$$

and set $\mathcal{C}_0 = \mathcal{C}_1 \cup \mathcal{C}_2 \cup \cdots$. Since for each $k = 1, 2, \ldots, \mathcal{C}_k$ is at most countable, $\mathcal{C}_0 \in \overset{0}{\mathcal{M}}$, so that

$$0 \le m - \mu(\cup \mathcal{C}_0) \le 1/n, \qquad n = 1, 2, \cdots,$$

which obviously implies that

$$m = \mu(\cup \mathcal{C}_0).$$

We claim that $B = \cup \mathcal{E}_0$ is an upper bound for the chain $\mathcal{C} \subset \mathcal{K}_t$ (B is measurable since $\mathcal{E}_0$ is countable).

If $K \in \mathcal{C} - \mathcal{E}_0$ we have, due to the "maximality" of $\mathcal{E}_0$, that

$$\mu(B \cup K) = \mu\left(\bigcup_{H \in \mathcal{E}_0 \cup \{K\}} H \right) \leq m \qquad (\mathcal{E}_0 \cup \{K\} \in \overset{0}{\mathcal{M}}).$$

Thus, for any $\Gamma \in \mathcal{C}$

$$m \geq \mu(B \cup \Gamma) = \mu(B) + \mu(\Gamma\text{-}B) = m + \mu(\Gamma\text{-}B).$$

Hence, $\mu(\Gamma\text{-}B) = 0 \Rightarrow \Gamma \overset{a.e.}{\subset} B$ and consequently, B is an upper bound of the chain $\mathcal{C}$ in $(\mathcal{K}_t, \overset{a.e.}{\subset})$.

9. Show that a measure space $(X, \mathcal{A}, \mu)$ is nonatomic if and only if the mapping

$$\mu : \mathcal{A} \to [0, \infty)$$

is surjective.

Hint:

Assume that $(X, \mathcal{A}, \mu)$ is nonatomic. To prove that μ is surjective if suffices to show that for every $t \in [0, \infty)$ there exists at least one $\Gamma \in \mathcal{A}$ such that $\mu(\Gamma)=t$. Consider $\mathcal{K}_t$ from the previous problem. Since $(\mathcal{K}_t, \overset{a.e.}{\subset})$ is inductive, according to Zorn's lemma it has a maximal element, say $M \in \mathcal{K}_t$. If $\mu(M)=t$ the assertion is true. If $\mu(M) < t$, set

$$\eta = t - \mu(M) \leq \mu(M^c).$$

Since by assumption μ is nonatomic, there exists $D \in \mathcal{A} \cap M^c$ such that $0 < \mu(D) < \eta$. Hence,

$$\mu(M \cup D) = \mu(M) + \mu(D) \leq \mu(M) + \eta = t,$$

which implies that $M \cup D \in \mathcal{K}_t$. This, however, contradicts the maximality of M, proving that the assumption $\mu(M) < t$ is fals.

19. OUTER MEASURES AND MEASURABLE SETS

Let $\mathscr{S}^*$ be a semialgebra of subsets of a set X. The problem of extension of a measure μ defined on $\mathscr{S}^*$ to the algebra $\mathscr{A}_0(\mathscr{S}^*)$ (generated by $\mathscr{S}^*$), was discussed in some detail in proposition 5 of section 18. In this section we shall extend a measure μ defined on a semiring $\mathscr{S} \subset \mathscr{P}(X)$ to a much larger class of sets; at the very least to the σ-algebra $\sigma\{\mathscr{S}\}$. To this end we need the concept of "**outer measure**" introduced by C. Carathéodory.

Let μ be a measure on a semiring $\mathscr{S} \subset \mathscr{P}(X)$. An **outer measure** μ^*, induced by μ, is the set function $\mu^*:\mathscr{P}(X) \to [0, \infty]$ defined as follows: for any $C \in \mathscr{P}(X)$,

$$\mu^*(C) = \inf\left\{ \sum_1^\infty \mu(A_i) ;\; \{A_i\}_1^\infty \text{ from } \mathscr{S},\; C \subset \bigcup_1^\infty A_i \right\}. \tag{1}$$

Remark 1

The union of any sequence $\{E_k\}_1^\infty$ from $\mathscr{S}$ can be written as the union of disjoint elements from $\mathscr{S}$ since

$$\bigcup_1^\infty E_k = E_1 \cup (E_2 - E_1) \cup \cdots \cup \left(E_n - \left(\bigcup_1^{n-1} E_i\right)\right) \cup \cdots ,$$

and according to problem 7 of section 17, for every $n=2, 3, \cdots$

$$E_n - \bigcup_1^{n-1} E_i = \bigcup_{j=1}^{i_n} F_{nj} ,\quad \text{where } \{F_{nj}\} \subset \mathscr{S} \text{ and } F_{ni} \cap F_{nj} = \theta \text{ if } i \neq j.$$

Consequently, in (1) we can assume, without any loss of generality, that $\{A_k\}_1^\infty$ from $\mathscr{S}$ is a disjoint sequence.

Remark 2

By the definition of μ^*, given any $C \in \mathscr{P}(X)$ and any $\varepsilon > 0$, there exists a sequence $\{A_i\}_1^\infty$ from $\mathscr{S}$ such that $C \subset \bigcup_1^\infty A_i$ and $\sum_1^\infty \mu(A_i) \leq \mu^*(C) + \varepsilon$.

Next we establish some properties of the set function μ^*. First we prove:

Proposition 1

The restriction of μ^* to $\mathscr{S}$ is equivalent to μ, i.e.

$$\mu^* \mid \mathscr{S} = \mu \tag{2}$$

Proof:

For any $A \in \mathscr{S}$, $A \subset A \cup \theta \cup \theta \cup \cdots$ which yields $\mu^*(A) \leq \mu(A)$; if $A = \theta$ we deduce from this that

$$\mu^*(\theta) = 0.$$

For any disjoint sequence $\{A_k\}_1^\infty$ and A from $\mathcal{S}$ such that $A \subset \bigcup_1^\infty A_k$, we have that

$$\mu(A) = \mu(A \cap \bigcup_1^\infty A_k) = \sum_1^\infty \mu(A \cap A_k) \leq \sum_1^\infty \mu(A_k),$$

which implies that $\mu(A) \leq \mu^*(A)$. This together with the previous inequality gives

$$\mu^*(A) = \mu(A) \quad \text{for every } A \in \mathcal{S}. \qquad \blacktriangledown$$

Proposition 2

The outer measure μ^* is monotone on $\mathcal{P}(X)$.

Proof:

Consider $\{A, B\} \subset \mathcal{P}(X)$ such that $A \subset B$. Let $\{B_k\}_1^\infty$ be a sequence from $\mathcal{S}$ such that $B \subset \bigcup_1^\infty B_i$,

then obviously

$$\mu^*(A) \leq \sum_1^\infty \mu(B_k).$$

From this, after a simple argument, we deduce that

$$\mu^*(A) \leq \mu^*(B) \qquad \blacktriangledown$$

as claimed. Generally speaking, an outer measure is not a measure on $\mathcal{P}(X)$, because it may fail to be additive. However, we have :

Proposition 3

An outer measure μ^* is "**subadditive.**" In other words, given any $\{C_k\}_1^\infty$ from $\mathcal{P}(X)$ we have that

$$\mu^*(\bigcup_1^\infty C_k) \leq \sum_1^\infty \mu^*(C_k). \qquad (3)$$

Proof:

Given any $\varepsilon > 0$, there is, according to remark 2, for each C_n a sequence $\{E_{ni}\}_1^\infty$ from $\mathcal{S}$ such that

$C_n \subset \bigcup_1^\infty E_{ni}$ and

$$\sum_{i=1}^\infty \mu(E_{ni}) \leq \mu^*(C_n) + \varepsilon/2^n.$$

Since

$$\bigcup_{n=1}^\infty C_n \subset \bigcup_{n=1}^\infty \bigcup_{i=1}^\infty E_{ni},$$

we have that

$$\mu^*(\bigcup_{1}^{\infty} C_n) \le \sum_{n=1}^{\infty} \sum_{i=1}^{\infty} \mu(E_{ni}) \le \sum_{n=1}^{\infty} \mu^*(C_n) + \varepsilon,$$

which proves (3) since $\varepsilon > 0$ is arbitrary. ♥

The following definition is due to C. Carathéodory:

Definition 1

A set $M \subset X$ is said to be Carathéodory measurable (or **μ^*-measurable**) if it splits every other subset $C \subset X$ "μ^*- additively." In other words, if

$$\mu^*(C) = \mu^*(M \cap C) + \mu^*(M^c \cap C). \tag{4}$$

Denote by $\mathcal{M} \subset \mathcal{P}(X)$ the collection of all μ^*-measurable sets. The fundamental result of this section is that the class $\mathcal{M}$ is a complete σ-algebra such that $\mathcal{S} \subset \mathcal{M}$ and such that the restriction of μ^* to $\mathcal{M}$ is a measure, which we will denote by $\bar{\mu}$, i.e.

$$\bar{\mu} = \mu^* \mid \mathcal{M}. \tag{5}$$

The miracle is that the class $\mathcal{M}$ exists and that $\sigma\{\mathcal{S}\} \subset \mathcal{M}$, which we will prove next. To this end we need several auxiliary results which are given below. It follows from (4) that $\theta, X \in \mathcal{M}$, since when $M = \theta$ in (4) we have that

$$\mu^*(C) = \mu^*(C \cap \theta) + \mu^*(C \cap X)$$

so that $\theta \in \mathcal{M}$. From the symmetry in (4) we conclude that $X \in \mathcal{M}$. In addition, any μ^*-null set $N \in \mathcal{P}(X)$ is measurable since for any $C \in \mathcal{P}(X)$

$$\mu^*(C) = \mu^*((C \cap N) \cup (C \cap N^c)) \le \mu^*(C \cap N) + \mu^*(C \cap N^c) \le \mu^*(C),$$

so that $N \in \mathcal{M}$. From this we conclude that the class $\mathcal{M}$ is complete.

Proposition 4

$$\mathcal{S} \subset \mathcal{M} \tag{6}$$

Proof:

We have to show that every $B \in \mathcal{S}$ is measurable, i.e., that $\mu^*(C \cap B) + \mu^*(C \cap B^c) = \mu^*(C)$ for every $C \in \mathcal{P}(X)$. Given $C \subset X$ and $\varepsilon > 0$ there exists a sequence $\{E_k\}_1^{\infty}$ from $\mathcal{S}$ such that $C \subset \bigcup_1^{\infty} E_k$ and

$$\sum_1^{\infty} \mu(E_k) \le \mu^*(C) + \varepsilon$$

(see remark 2). Since E_n and B are from $\mathcal{S}$, $(E_n - B) = \bigcup_1^{m} B_j$, where $\{B_j\}_1^{m}$ is a disjoint sequence from $\mathcal{S}$. Since

$$(E_n \cap B) \cup B_1 \cup \cdots \cup B_m = E_n,$$

it follows that

$$\mu^*(C) \leq \mu^*(C \cap B) + \mu^*(C \cap B^C) \leq \sum_1^\infty [\mu^*(E_k \cap B) + \mu^*(E_k - B)]$$

$$\leq \sum_1^\infty [\mu(E_k \cap B) + \sum_1^m \mu(B_j)] = \sum_1^\infty \mu(E_k) \leq \mu^*(C) + \varepsilon.$$

Since $\varepsilon > 0$ is arbitrary the assertion follows. ♥

Proposition 5

The class $\mathcal{M}$ is an algebra.

Proof:

From the symmetry in (4) we see that $A \in \mathcal{M}$ implies that $A^C \in \mathcal{M}$. It rests to show that $\mathcal{M}$ is closed under finite union. Let $\{A, B\} \subset \mathcal{M}$; since μ^* is subadditive we have for any $C \in \mathcal{P}(X)$ that

$$\mu^*(C) \leq \mu^*(C \cap (A \cup B)) + \mu^*(C \cap (A \cup B)^C)$$

$$= \mu^*(C \cap (A \cup B) \cap A) + \mu^*(C \cap (A \cup B) \cap A^C) + \mu^*(C \cap A^C \cap B^C)$$

$$= \mu^*(C \cap A) + \mu^*(C \cap A^C \cap B) + \mu^*(C \cap A^C \cap B^C)$$

$$= \mu^*(C \cap A) + \mu^*(C \cap A^C) = \mu^*(C) .$$

In other words, $\mu^*(C) = \mu^*(C \cap (A \cup B)) + \mu^*(C \cap (A \cup B)^C)$ which proves that $A \cup B \in \mathcal{M}$. Thus, the union of two measurable sets is measurable. By way of induction it follows that the union of any finite collection of measurable sets is measurable, which proves our contention. ♥

Corollary 1

If $\{A, B\} \subset \mathcal{M}$ are disjoint, it follows from (4), after replacing M with A and C with $(A \cup B) \cap C$, that

$$\mu^*(C \cap (A \cup B)) = \mu^*(C \cap (A \cup B) \cap A) + \mu^*(C \cap (A \cup B) \cap A^C)$$

$$= \mu^*(C \cap A) + \mu^*(C \cap B).$$

By induction (see problem 7) we have for any finite disjoint sequence $\{A_k\}_1^n \subset \mathcal{M}$, that

$$\mu^*(C \cap \bigcup_1^n A_i) = \sum_1^n \mu^*(C \cap A_i). \tag{7}$$

The next result is fundamental.

Proposition 6.

The class $\mathcal{M}$ of μ^*-measurable subsets of X is a σ-algebra.

Proof:

Let $\{M_k\}_1^\infty$ be a disjoint sequence from $\mathcal{M}$ and denote by $M_* = \bigcup_1^\infty M_k$, then due to the subadditivity property of μ^* we have:

$$\mu^*(C\cap M_*) \le \sum_1^\infty \mu^*(C\cap M_k),$$

and since μ^* is monotone it follows that

$$\mu^*(C\cap M_*) \ge \mu^*(C\cap \bigcup_1^n M_k) = \sum_1^n \mu^*(C\cap M_k)$$

for every $n=1, 2, \cdots$. By letting $n\to\infty$ this yields

$$\mu^*(C\cap M_*) \ge \sum_1^\infty \mu^*(C\cap M_k),$$

proving that for any $C\subset X$,

$$\mu^*(C\cap \bigcup_1^\infty M_k) = \sum_1^\infty \mu^*(C\cap M_k). \tag{8}$$

Consequently, on $\mathcal{M}$ the outer measure μ^* is countably additive and since $\mathcal{M}$ is an algebra, $\bigcup_1^n M_k \in \mathcal{M}$.

Next, from (4) we have for every $n=1, 2. \cdots$, that:

$$\mu^*(C) = \mu^*(C\cap \bigcup_1^n M_k) + \mu^*(C\cap(\bigcup_1^n M_k)^c)$$

$$\ge \sum_1^n \mu^*(C\cap M_k) + \mu^*(C\cap(\bigcup_1^\infty M_k)^c).$$

By letting $n\to\infty$ it follows that

$$\mu^*(C) \ge \sum_1^\infty \mu^*(C\cap M_k) + \mu^*(C\cap(\bigcup_1^\infty M_k)^c)$$

$$\ge \mu^*(C\cap(\bigcup_1^\infty M_k)) + \mu^*(C\cap(\bigcup_1^\infty M_k)^c).$$

Since the inequality in opposite direction always holds, we have just proved that $\bigcup_1^\infty M_k \in \mathcal{M}$. For any sequence $\{M_k\}_1^\infty$ from $\mathcal{M}$ which is not disjoint see remark 1. ♥

Corollary 2

Since $\mathcal{M}$ is a σ-algebra which contains $\mathcal{S}$, it follows from (6) that $\sigma\{\mathcal{S}\}\subset\mathcal{M}$. If we put $C=X$ in (8) we obtain that

$$\mu^*(\bigcup_1^\infty M_k) = \sum_1^\infty \mu^*(M_k).$$

In other words, on the class $\mathcal{M}$ the outer measure μ^* is a measure. Denote this measure by $\bar\mu$, i.e.,

$$\bar\mu = \mu^* \mid \mathcal{M}, \tag{9}$$

Then, according to proposition 1, $\mu=\mu^*\,|\,\mathcal{S}=(\mu^*\,|\,\mathcal{M})\,|\,\mathcal{S}=\bar\mu\,|\,\mathcal{S}$; thus, the measure $\bar\mu$ is an extension of μ on $\mathcal{S}$ to the σ-algebra $\mathcal{M}$, which is clearly complete. For the most purposes, however, an extension of μ to a measure on $\sigma\{\mathcal{S}\}\subset\mathcal{M}$ will suffice.

Is the extension $\bar\mu$ unique? The answer is no, in general. However, it is yes if the measure μ is σ-finite as the following proposition shows:

Proposition 7

Let μ_1 be a measure on $\sigma\{\mathcal{S}\}$ such that $\mu_1=\mu$ on $\mathcal{S}$. If μ is σ-finite then

$$\mu_1 = \mu^* \mid \sigma\{\mathcal{S}\}$$

Proof:

Let $C\in\mathcal{P}(X)$ and $\varepsilon>0$ be arbitrary, then (see remark 2) there exists $\{A_k\}_1^\infty$ from $\mathcal{S}$ such that

$$C\subset A = \bigcup_1^\infty A_k \ \text{ and } \ \sum_1^\infty \mu(A_k) \le \mu^*(C)+\varepsilon.$$

Let μ_1^* be the outer measure induced by μ_1 and assume that μ is finite. It is clear that

$$\mu_1^*(C) \le \sum_1^\infty \mu_1(A_k) = \sum_1^\infty \mu(A_k) \le \mu^*(C)+\varepsilon, \text{ from which we deduce that}$$

$$\mu_1^*(C) \le \mu^*(C) \text{ for all } C\in\mathcal{P}(X). \tag{10}$$

Next, assume that $C\in\sigma\{\mathcal{S}\}$ and that $\mu^*(C)<\infty$. Assume also that $\{A_k\}_1^\infty$ is a disjoint sequence, then since

$$C\subset A= \bigcup_1^\infty A_k \ \text{ and } \ \sum_1^\infty \mu(A_k)\le \mu^*(C)+\varepsilon, \text{ we have from (10) that}$$

$$\mu_1^*(A\text{-}C) \le \mu^*(A\text{-}C) = \mu^*(A)-\mu^*(C) < \varepsilon,$$

since μ^* is a measure on $\sigma\{\mathcal{S}\}$. Thus,

$$\mu^*(C) \le \mu^*(A) = \sum_1^\infty \mu(A_k) = \sum_1^\infty \mu_1(A_k) = \mu_1^*(A)$$

$$\le \mu_1^*(C) + \mu_1^*(A\text{-}C) \le \mu_1^*(C)+\varepsilon$$

which implies that $\mu^*(C) \le \mu_1^*(C)$. This and (10) then yield that

$$\mu^*(C) = \mu_1^* (C) \quad \text{for all } C\in\sigma\{\mathcal{S}\},$$

provided that μ^* is finite.

If μ^* is not finite but it is σ-finite, then there exists a disjoint sequence $\{Z_k\}_1^\infty$ from $\sigma\{\mathcal{S}\}$ such that

$X=\bigcup_1^\infty Z_k$ and such that $\mu^*(Z_n)<\infty$ for every $n=1, 2, \cdots$. Now, for any $C\in\sigma\{\mathcal{S}\}$ we have that

$$\mu^*(Z_n \cap C) = \mu_1(Z_n \cap C) < \infty \qquad n = 1, 2, \cdots .$$

Consequently,

$$\mu^*(C) = \mu^*\left(C \cap \bigcup_1^\infty Z_n\right) = \sum_1^\infty \mu^*(C \cap Z_n)$$

$$= \sum_1^\infty \mu_1(C \cap Z_n) = \mu_1\left(C \cap \bigcup_1^\infty Z_n\right) = \mu_1(C),$$

which proves the proposition. ♥

Proposition 8

Let $\mathcal{S}_0$ and $\mathcal{S}_1$ be semirings such that $\mathcal{S}_0 \subset \mathcal{S}_1 \subset \mathcal{P}(X)$. If μ_0 and μ_1 are measures on $\mathcal{S}_0$ and $\mathcal{S}_1$, respectively, such that $\mu_1 = \mu_0$ on $\mathcal{S}_0$ then

$$\mu_1 = \mu_0^* \, | \, \mathcal{S}_1$$

if μ_0 is finite.

Proof:

For any $C \subset X$

$$\mu_0^*(C) = \inf\left\{ \sum_1^\infty \mu_0(A_i); \ C \subset \bigcup_1^\infty A_i , \ \{A_i\}_1^\infty \text{ from } \mathcal{S}_0 \right\}$$

$$= \inf\left\{ \sum_1^\infty \mu_1(A_i); \ C \subset \bigcup_1^\infty A_i , \ \{A_i\}_1^\infty \text{ from } \mathcal{S}_0 \right\}$$

$$\geq \inf\left\{ \sum_1^\infty \mu_1(A_i); \ C \subset \bigcup_1^\infty A_i , \ \{A_i\}_1^\infty \text{ from } \mathcal{S}_1 \right\} = \mu_1^*(C).$$

Thus, for every $C \subset X$, $\mu_1^*(C) \leq \mu_0^*(C)$. If $C \in \mathcal{S}_1$ we have that $\mu_1(C) \leq \mu_0^*(C)$. Let $C \in \mathcal{S}_1$, then for any $\varepsilon > 0$ there is a sequence $\{A_i\}_1^\infty$ from $\mathcal{S}_0$ such that

$$C \subset \bigcup_1^\infty A_i \ \text{ and } \mu_1(C) = \mu_1^*(C) \geq \sum_1^\infty \mu_1(A_i) - \varepsilon = \sum_1^\infty \mu_0(A_i) - \varepsilon$$

From this we deduce that $\mu_1(C) \geq \mu_0^*(C)$ for all $C \in \mathcal{S}_1$. This and the previous inequality prove the assertion. ♥

Remark 3

The proposition holds if μ_0 is σ-finite which can be proved in the same way as in proposition 7. It also clear that the μ_0^* is the only extension of μ_0 to $\mathcal{M}$.

Proposition 9

The infimum in (1) is attained.

Proof

We have to prove that there is $B \in \mathcal{M}$ such that $\bar{\mu}(B) = \mu^*(C)$. To this end notice that if $\{B_n\}_1^\infty$ is from $\mathcal{M}$ such that $C \subset B_n$ and $\bar{\mu}(B_n) \to \mu^*(C)$, then $B = \bigcap_1^\infty B_n \in \mathcal{M}$, contains C and from $C \subset \bigcap_1^\infty B_i \subset B_n$ it follows that $\mu^*(C) \leq \bar{\mu}(\bigcap_1^\infty B_i) \leq \bar{\mu}(B_n)$. By letting $n \to \infty$ we have that

$$\mu^*(C) = \mu(\bigcap_1^\infty B_n).$$

♥

Problems and complements

1. Show that a set $A \subset X$ is measurable if and only if for each $C \subset X$

$$\mu^*(C) \geq \mu^*(C \cap A) + \mu^*(C \cap A^c).$$

   Hint:

$$\mu^*(C) = \mu^*((C \cap A) \cup (C \cap A^c)).$$

2. Given $\{A, B\} \subset \mathscr{P}(X)$, show that $\mu^*(A) - \mu^*(A \cap B) \leq \mu^*(A-B)$.

   Hint:

$$A = (A \cap B) \cup (A \cap B^c)$$

3. If $\{N, A\} \subset \mathscr{P}(X)$ and N is μ^*-negligible, show that

$$\mu^*(A) = \mu^*(A \cup N) = \mu^*(A-N).$$

   Hint:

$$\mu^*(A) \leq N^*(A \cup N) \leq \mu^*(A) + \mu^*(N) = \mu^*(A).$$

$$A = N \cup (A-N)$$

4. Show that for any $M \in \mathscr{M}$ and $C \in \mathscr{P}(X)$

$$\mu^*(M \cup C) = \mu^*(M) + \mu^*(C) - \mu^*(M \cap C)$$

   Hint:

$$\mu^*(M \cup C) = \mu^*((M \cup C) \cap M) + \mu^*((M \cup C) \cap M^c)$$

$$= \mu^*(M) + \mu^*(C \cap M^c) + \mu^*(C \cap M) - \mu^*(C \cap M)$$

5. Show that $|\mu^*(A) - \mu^*(B)| \leq \mu^*(A \,\Delta\, B)$

   Hint:

$$A \subset B \cup (A \,\Delta\, B), \qquad B \subset A \cup (A \,\Delta\, B).$$

$$\mu^*(A) \leq \mu^*(B) + \mu(A \,\Delta\, B), \qquad \mu^*(B) \leq \mu^*(A) + \mu^*(A \,\Delta\, B)$$

6. Show that an outer measure μ^* on $\mathscr{P}(X)$ is countably additive if and only if for any finite disjoint

sequence $\{A_k\}_1^n \subset \mathscr{P}(X)$

$$\mu^*\left(\bigcup_1^n A_k\right) = \sum_1^n \mu^*(A_k) \qquad n = 1, 2, \cdots.$$

   Hint:

Let $\{A_k\}_1^\infty \subset \mathscr{P}(X)$ be disjoint, then

$$\mu^*\left(\bigcup_1^\infty A_k\right) = \mu^*\left(\bigcup_1^n A_k \cup \bigcup_{n+1}^\infty A_k\right) = \sum_1^n \mu^*(A_k)$$

$$+ \mu^*\left(\bigcup_{n+1}^\infty A_k\right) \to \sum_1^\infty \mu^*(A_k) + \alpha \text{ as } n \to \infty$$

where $\alpha \geq 0$. If $\alpha = 0$, μ^* is countably additive. If $\alpha > 0$ we have that

$$\mu^*\left(\bigcup_1^\infty A_k\right) \ge \sum_1^\infty \mu^*(A_k).$$

7. Let $\{M_k\}_1^n \subset \mathcal{M}$ be a disjoint sequence. Show that for any $C \subset X$

$$\mu^*\left(C \cap \bigcup_1^n M_k\right) = \sum_1^n \mu^*(C \cap M_k).$$

Hint:

By induction; for $n=1$ the equality holds. Assume that it holds for $n=m$, then for $n=m+1$ we have

$$\mu^*\left(C \cap \bigcup_1^{m+1} M_k\right) = \mu^*\left((C \cap \bigcup_1^{m+1} M_k) \cap M_{m+1}\right) + \mu^*\left((C \cap \bigcup_1^{m+1} M_k) \cap M_{m+1}^c\right)$$

$$= \mu^*(C \cap M_{m+1}) + \mu^*\left(C \cap \bigcup_1^m M_k\right) = \sum_1^{m+1} \mu^*(C \cap M_k).$$

8. Show that a set $E \subset X$ is measurable if and only if $M \cap E \in \mathcal{M}$ for every $M \in \mathcal{M}$.

Hint:

If $E \in \mathcal{M}$ then necessarily $M \cap E \in \mathcal{M}$. Suppose now that $M \cap E \in \mathcal{M}$, then for any $C \subset X$

$$\mu^*(C) = \mu^*(C \cap (M \cap E)) + \mu^*(C \cap (M \cap E)^c).$$

Since this equation holds for every $M \in \mathcal{M}$, choose $M \in \mathcal{M}$ such that $C \subset M$, then we have:

$$\mu^*(C) = \mu^*(C \cap E) + \mu^*(C \cap E^c).$$

9. Show that a set $E \subset X$ is measurable if and only if for every $A \in \mathcal{M}$

$$\mu^*(A) = \mu^*(A \cap E) + \mu^*(A \cap E^c).$$

Hint:

If E is measurable the condition necessarily holds for every $A \in \mathcal{P}(X)$. Assume now that the condition holds, and let $C \in \mathcal{P}(X)$. Then

$$\mu^*(C) \le \mu^*(C \cap E) + \mu^*(C \cap E^c).$$

Given $\varepsilon > 0$ there exists $M \in \mathcal{M}$ such that $C \subset M$ and $\mu^*(M) - \mu^*(C) < \varepsilon$. Then, since

$$\mu^*(C) \le \mu^*(C \cap E) + \mu^*(C \cap E^c)$$

$$\le \mu^*(M \cap E) + \mu^*(M \cap E^c) = \mu^*(M)$$

$$\le \mu^*(C) + \varepsilon.$$

10. Show that a set $A \subset X$ is measurable if and only if for every $\varepsilon > 0$ there exists a set $M \in \mathcal{M}$ such that $M \subset A$ and $\mu^*(A - M) < \varepsilon$.

Hint:

Assume that the condition holds, then for every $n = 1, 2, \cdots$, there exists an $M_n \in \mathcal{M}$ such that $M_n \subset A$ and $\mu^*(A - M_n) \le 1/n$. Clearly,

$$M = \bigcup_1^\infty M_n \subset A \quad \text{and} \quad M \in \mathcal{M},$$

we also have that

$$\mu^*(A\text{-}M) \le \mu^*(A\text{-}M_n) \le 1/n. \qquad n=1, 2, \cdots ,$$

From this it follows that $\mu^*(A\text{-}M)=0$. Thus, A-M is measurable as a null set and from

$$A = M \cup (A\text{-}M)$$

the measurability of A follows, which proves sufficiency of the condition. If A is measurable then A=M satisfies the condition for each $\varepsilon > 0$.

11. If $M \in \mathcal{M}$ is such that $\mu^*(M) > 0$, show that for any $A \subset M$ and $A \notin \mathcal{M}$, $\mu^*(M\text{-}A) > 0$.

    Hint:

    If $\mu^*(M\text{-}A)=0$, then from A=M-(M-A) we would have that $A \in \mathcal{M}$, contradiction.

12. Let $\mu^*(X) < \infty$; show that under this condition a set $E \subset X$ is measurable if and only if

$$\mu^*(E) + \mu^*(E^c) = \mu^*(X).$$

    Hint:

    If $E \in \mathcal{M}$ the condition holds trivially. Now assume that the condition holds, than for any $M \in \mathcal{M}$

$$\mu^*(M) + \mu^*(M^c) = \mu^*(X) = \mu^*(E) + \mu^* (E^c)$$

$$= \mu^*(E \cap M) + \mu^*(E \cap M^c) + \mu^*(E^c \cap M) + \mu^*(E^c \cap M^c)$$

$$\ge \mu^*(M^c) + \mu^*(M \cap E) + \mu^*(M \cap E^c).$$

Thus,

$$\mu^*(M) \ge \mu^*(M \cap E) + \mu^*(M \cap E^c) \ge \mu^*(M).$$

This and problem 9 prove the claim.

13. Let μ be a σ-finite measure on an algebra $\mathcal{A}_0 \subset \mathcal{P}(X)$. Show that for any $B \in \sigma\{\mathcal{A}_0\}$ and $\varepsilon > 0$ there exists $B_0 \in \mathcal{A}_0$ such that

$$\mu^*(B \,\Delta\, B_0) < \varepsilon.$$

    Hint:

    Bearing in mind that an algebra is semiring there exists a sequence $\{A_k\}_1^\infty$ from $\mathcal{A}_0$ such that

$$B \subset \bigcup_1^\infty A_k \quad \text{and}$$

$$\mu^*\left(\bigcup_1^\infty A_k\right) \le \mu^*(B) + \varepsilon/2$$

(see remark 2). Since μ^* is a measure on $\sigma\{\mathcal{A}_0\}$

$$\lim_{n\to\infty} \mu^*\left(\bigcup_1^n A_k\right) = \mu^*\left(\bigcup_1^\infty A_k\right).$$

Then there exists an index $n_0 = n_0(\varepsilon)$ such that

$$\mu^*\left(\bigcup_1^\infty A_k\right) \le \mu^*\left(\bigcup_1^{n_0} A_k\right) + \varepsilon/2.$$

Set $B_0 = \bigcup_1^{n_0} A_k$ then

$$\mu^*(B\text{-}B_0) \le \mu^*\left(\bigcup_1^\infty A_k - B_0\right) \le \varepsilon/2$$

$$\mu^*(B_0\text{-}B) \le \mu^*\left(\bigcup_1^\infty A_k - B\right) \le \varepsilon/2$$

which is the required proof.

14. Show that for any $C \subset \mathscr{P}(X)$

$$\mu^*(C) = \inf\{\,\bar{\mu}(M);\ C \subset M \in \mathscr{M}\,\}$$

$$= \inf\{\,\bar{\mu}(G);\ C \subset G \in \sigma\{\mathscr{S}\}\,\}.$$

Hint:

Since $\mu = \bar{\mu}$ on $\mathscr{S}$, we have (we can assume that $\{A_i\}_1^\infty$ are disjoint).

$$\mu^*(C) = \inf\left\{\sum_1^\infty \bar{\mu}(A_i);\ C \subset \bigcup_1^\infty A_i,\ \{A_i\}_1^\infty \subset \mathscr{S}\right\}$$

$$\ge \inf\left\{\sum_1^\infty \bar{\mu}(A_i);\ C \subset \bigcup_1^\infty A_i,\ \{A_i\}_1^\infty \subset \sigma\{\mathscr{S}\}\right\}.$$

Set $A = \bigcup_1^\infty A_i$; t hen from the last inequality we have:

$$\mu^*(C) \ge \inf\{\,\bar{\mu}(A);\ C \subset A,\ A \in \sigma\{\mathscr{S}\}\,\}$$

$$\ge \inf\{\,\bar{\mu}(A);\ C \subset A,\ A \in \mathscr{M}\,\}$$

$$= \inf\{\mu^*(A);\ C \subset A,\ A \in \mathscr{M}\} \ge \mu^*(C).$$

In other words, the outer measures defined by starting from a semiring $\mathscr{S} \subset \mathscr{P}(X)$ and a measure μ, or the class of measurable sets $\mathscr{M}$ and $\bar{\mu}$, respectively, are identical.

15. Let in proposition 8, $\mu_1 = \mu_0^* \mid \mathscr{S}$; show that $\mu_0^* = \mu_1^*$.

Hint:

Given $C \subset X$, let $\{A_k\}_1^\infty$ from $\mathscr{S}_1$ satisfy $C \subset \bigcup_1^\infty A_k$, then

$$\mu_0^*(C) \le \sum_1^\infty \mu_0^*(A_k) = \sum_1^\infty \mu_1(A_k).$$

Consequently,

$$\mu_1^*(C) = \inf\{\sum_1^\infty \mu_1(A_k); \ C\subset\bigcup_1^\infty A_k \ ; \ \{A_k\}_1^\infty \text{ from } \mathcal{S}_1\} \geq \mu_0^*(C).$$

Next, assume that $\mu_0^*(C) < \infty$, then for any $\varepsilon > 0$ there exists $\{A_n\}_1^\infty$ from $\mathcal{S}_0$ such that $C\subset\bigcup_1^\infty A_n$ and

$\sum_1^\infty \mu_0(A_n) \leq \mu_0^*(C) + \varepsilon$. But since $\mu_1^*(C)\leq\sum_1^\infty \mu_1(A_n)=\sum_1^\infty \mu_0(A_n)$ we get that $\mu_1^*(C)\leq \mu_0^*(C)$; this proves

our claim.

16. Let μ_i be a measure on a semiring $\mathcal{S}_i\subset\mathcal{P}(X)$, $i = 0, 1$. Show that:

$$\mu_0^* = \mu_1^* \ <=> \ \mu_0^* = \mu_1 \text{ on } \mathcal{S}_1 \text{ and } \mu_1^* = \mu_0 \text{ on } \mathcal{S}_0.$$

Hint:

Assume that $\mu_0^* = \mu_1^*$, then since $\mu_0 = \mu_0^*$ on $\mathcal{S}_0$ we have that $\mu_0 = \mu_1^*$ on $\mathcal{S}_0$ and similarly, $\mu_1 = \mu_1^*$

on $\mathcal{S}_1$ implies that $\mu_1 = \mu_0^*$ on $\mathcal{S}_1$. This proves necessity of the condition.

Let $C\subset X$ satisfy $\mu_1^*(C) < \infty$; given $\varepsilon > 0$ there exists (see remark 2) a sequence $\{A_k\}_1^\infty\subset\mathcal{S}_1$ such

that $C\subset\bigcup_1^\infty A_k$ and such that

$$\mu_1^*(C) \geq \sum_1^\infty \mu_1(A_n)-\varepsilon/2 = \sum_1^\infty \mu_0^*(A_n)-\varepsilon/2$$

Next, there exists $\{E_{ni}\}_1^\infty\subset\mathcal{S}_0$ such that $A_n\subset\bigcup_{i=1}^\infty E_{ni}$ and such that $\mu_0^*(A_n)\geq \sum_{i=1}^\infty \mu_0(E_{ni})-\varepsilon/2^n$.

Consequently,

$$\mu_1^*(C) > \sum_{n=1}^\infty \sum_{i=1}^\infty \mu_0(E_{ni})-\varepsilon > \mu_0^*(C)-\varepsilon$$

Since $\varepsilon > 0$ is arbitrary we have that

$$\mu_0^*(C) \leq \mu_1(C).$$

Similarly, assuming that $\mu_0^*(C)<\infty$, given any $\varepsilon>0$ there exists $\{B_n\}_1^\infty$ from $\mathcal{S}_0$ such that

$C\subset\bigcup_1^\infty B_n$ and

$$\mu_0^*(C) \geq \sum_1^\infty \mu_0(B_n)-\varepsilon/2 = \sum_1^\infty \mu_1^*(B_n)-\varepsilon/2$$

Now, for each B_n there exists $\{F_{ni}\}\subset\mathcal{S}_1$ such that $B_n\subset\cup F_{ni}$ and $\sum_1^\infty \mu_1(F_{ni}) < \mu_1(B_n)+\varepsilon/2^n$. Thus,

$$\mu_0^*(C) \geq \sum_{n=1}^{\infty} \sum_{i=1}^{\infty} \mu_1 (F_{ni}) - \varepsilon \geq \mu_1^*(C) - \varepsilon$$

implying that

$$\mu_0^*(C) \geq \mu_1(C).$$

This proves the assertion.

17. Let $C \subset X$ be fixed; show that the map $\varphi_C(\cdot)$ on algebra $\mathcal{A}_0$ defined by

$$\varphi_C(A) = \mu(A) - \mu^*(A-C)$$

is monotone in $A \in \mathcal{A}_0$.

Hint:

Let $A, B \in \mathcal{A}_0$ and assume that $A \subset B$; if $C \subset A$ then

$$B - C = (B \cap C^c) \cap (A \cup A^c) \subset (A-C) \cup (B-A).$$

Hence,

$$\mu^*(B-C) \leq \mu^*(A-C) + \mu(B-A)$$

$$\mu(B) = \mu(A) + \mu(B-A).$$

By combining these two expressions we obtain

$$\mu(B) - \mu^*(B-C) \geq \mu(A) - \mu^*(A-C).$$

In other words, $\varphi_C(A) \leq \varphi_C(B)$ if $A \subset B$.

20. MEASURABLE COVERS

Let $(X, \mathcal{A}, \mu)$ be a measure space. The outer measure induced by μ is, as usual, denoted by μ^*. As we have seen (problem 14 of section 19), for any $C \subset X$

$$\mu^*(C) = \inf\{\mu(A); \ C \subset A \in \mathcal{A}\}. \tag{1}$$

In this section we are concerned with the relationship between μ and μ^*. One of the principal results here is that the completion $\tilde{\mathcal{A}} = \mathcal{M}$ if the measure μ is σ-finite.

Given set $C \subset X$, we shall say that a set $E \in \mathcal{A}$ is a "**measurable cover**" of C in case: (i) $C \subset E$, and (ii) if $H \subset E-C$ and $H \in \mathcal{A}$, then $\mu(H)=0$. **Each $B \in \mathcal{A}$ is a measurable cover of itself.**

Proposition 1

If E_1 and E_2 are measurable covers of a set $C \subset X$ so is $E = E_1 \cap E_2$; in addition, $\mu(E_1 \Delta E_2)=0$.

Proof:

Clearly, $C \subset E$ and $E-C \subset E_i-C$, $i=1, 2$. Thus, if $H \subset E-C$, necessarily $H \subset E_i-C$, $i=1, 2$, so that if $H \in \mathcal{A}$ we have that $\mu(H)=0$. On the other hand, since $E_i-E \subset E_i-C$ it follows that $\mu(E_i-E)=0$. ♥

Proposition 2

For a set $C \subset X$ a set $E \in \mathcal{A}$ is a measurable cover if $C \subset E$ and if $\mu^*(C) = \mu(E) < \infty$.

Proof:

Let $H \subset E-C$ be measurable, then since $C \subset E-H$ and $\mu^*(C)=\mu(E)$ we obtain that

$$\mu(E) = \mu^*(C) \le \mu^*(E-H) = \mu(E)-\mu(H).$$

Thus by the finiteness of $\mu(E)$, $\mu(H)=0$. ♥

The next proposition is of existence type. It also shows that the infimum μ^* is attained.

Proposition 3

If $\mu^*(C) < \infty$, there exists a measurable cover E of C and $\mu^*(C)=\mu(E)$.

Proof:

From (1) it follows that there exists a sequence $\{G_n\}_1^\infty$ from $\mathcal{A}$ such that $C \subset G_n$ for all $n=1, 2, \cdots$ and

$$\mu(G_n) \le \mu^*(C) + 1/n.$$

Set $E = \bigcap_1^\infty G_n$; since $C \subset E \in \mathcal{A}$ we have:

$$\mu^*(C) \le \mu(E) \le \mu(G_n) \le \mu^*(C) + 1/n$$

for all $n=1, 2, \cdots$. By the finiteness assumption we deduce that $\mu^*(C)=\mu(E)$ and that $\mu^*(C) = \lim_{n \to \infty} \mu(G_n)$.

From this and the previous proposition the assertion follows. ♥

Remark 1

As a matter of fact, if $\mu^*(C) < \infty$, and E_0 is any other measurable cover of C, then $\mu^*(C) = \mu(E_0)$, for

$\mu(E)=\mu^*(C) \le \mu^*(E_0)=\mu(E_0)=\mu(E)$ according to proposition 1.

The following is the principal result of this section.

Proposition 4

If the measure μ is σ-finite, then any $C \subset X$ has a measurable cover.

Proof:

Since by assumption μ is σ-finite so is μ^*. Thus, there exists a sequence $\{D_k\}_1^\infty$ from $\mathcal{P}(X)$ such

that $C \subset \bigcup_1^\infty D_k$ and $\mu^*(D_k)<\infty$ for every $k=1, 2, \cdots$. Clearly,

$$C = \bigcup_1^\infty (C \cap D_k)$$

and according to proposition 3, each $C \cap D_k$ has a measurable cover, say E_k. We claim that the set $E=\bigcup_1^\infty E_k$

is a measurable cover of C. Now, obviously,

$$E - C = \bigcup_1^\infty E_k - \bigcup_1^\infty (C \cap D_k) \subset \bigcup_1^\infty (E_k - C \cap D_k).$$

If $H \subset E$-C is measurable then clearly

$$H \cap E_k \subset (E\text{-}C) \cap E_k \subset E_k\text{-}C \subset E_k\text{-}C \cap D_k.$$

Since E_k is a measurable cover of $C \cap D_k$, we have that $\mu(H \cap E_k)=0$ for every $k=1, 2, \cdots$. From this and the

fact that $H \subset E$ we have that

$$\mu(H) \le \sum_1^\infty \mu(H \cap E_k) = 0,$$

which proves that E is a measurable cover of C. ♥

Proposition 5

If the measure μ is σ-finite then for any $C \subset X$ there exists $A, B \in \mathcal{A}$ such that

$$A \subset C \subset B \quad \text{and} \quad \mu(B\text{-}A) = 0.$$

Proof:

Let B be a measurable cover of C, and let G be a measurable cover of B-C. Denote by $A=B$-G;
because $A \subset C$ we have;

$$C\text{-}A = C\text{-}(B\text{-}G) = C \cap G \subset (C \cap G) \cup (G\text{-}B)$$

$$= G \cap (C \cup B^c) = G\text{-}(B\text{-}C).$$

Since μ is σ-finite, $\mu(G) = \mu^*(B\text{-}C)$; consequently, $\mu^*(C\text{-}A)=0$. On the other hand,

$$B\text{-}A = (B\text{-}A) \cap (C \cup C^c) = (B\text{-}C) \cup (C\text{-}A),$$

from which we deduce that $\mu(B\text{-}A)=0$. This proves the proposition. ♥

The completion $(X, \widetilde{\mathcal{A}}, \overline{\mu})$ of a measure space $(X, \mathcal{A}, \mu)$ was discussed in proposition 3 of section

18. Let μ^* be the outer measure induced by μ and denote by $\mathcal{M}$ the class of μ^*-measurable sets. It is clear from proposition 3 of section 18 that $\tilde{\mathcal{A}} \subset \mathcal{M}$. In the next proposition we demonstrate that the converse inclusion also holds if the measure μ is σ-finite.

Proposition 6

Let $(X, \mathcal{A}, \mu)$ be a σ-finite measure space, then (see proposition 3, section 18)

$$\tilde{\mathcal{A}} = \mathcal{M} \quad \text{and} \quad \bar{\mu} = \mu^* \mid \mathcal{M}.$$

Proof:

Clearly, $\tilde{\mathcal{A}} \subset \mathcal{M}$; to prove the converse inclusion consider a set $M \in \mathcal{M}$. Since by hypothesis the measure μ is σ-finite there exists a measurable cover S of M (see proposition 4). Let $H \subset S - M$ be measurable and write $B = S - H$, then

$$Z = M - B \subset S - B = S - (S - H) = H.$$

Thus, $\mu^*(Z) = 0$ and since

$$M = B \cup (M\text{-}B) \in \tilde{\mathcal{A}}$$

(because M-B$\subset$H), we have that $\mathcal{M} \subset \tilde{\mathcal{A}}$, which proves the proposition.   ♥

Problems and complements

1. Let E_k be a measurable cover of C_k, $k=1, 2, \cdots$. Show that $E=\bigcup_1^\infty E_k$ is a measurable cover of $C=\bigcup_1^\infty C_k$.

  Hint:

  We must prove that every measurable $H\subset E\text{-}C$ is a null set. Since

$$H\subset(E\text{-}C) \Rightarrow H\cap E_k \subset (E\text{-}C)\cap E_k = E_k\text{-}C \subset E_k\text{-}C_k,$$

we have that $\mu(H\cap E_k)=0$. Now

$$\mu(H) = \mu(H\cap E) \le \sum_1^\infty \mu\,(H\cap E_k) = 0.$$

2. Let E be a measurable cover of C, where $\mu^*(C)<\infty$. Assuming that μ is σ-finite show that:

(i) If $C\in\mathcal{M}$, then necessarily $\mu^*(E\text{-}C)=0$.

(ii) If $\mu(E)>0$ and $C\notin\mathcal{M}$, then $\mu^*(E\text{-}C)>0$.

  Hint:

  Clearly

$$C = E\text{-}(E\text{-}C);$$

(i) Since μ is σ-finite we have that $\mu^*(E\text{-}C)=0$.

(ii) Since $C\notin\mathcal{M}$, $E\text{-}C\notin\mathcal{M}$ so that necessarily $\mu^*(E\text{-}C)>0$, for if $\mu^*(E\text{-}C)=0$, $E\text{-}C$ is a null set so that $C\in\mathcal{M}$.

3. Let $(X, \mathcal{B}, \mu)$ be a measure space with $\mu(X)<\infty$ and $\mathcal{B}=\{A, A^C, \theta, X\}$. Define μ^* and find a measurable cover of a set $C\subset X$.

  Hint:

  For any $D\subset X$, $\mu^*(D)=\mu(A)$ if $D\subset A$, $\mu^*(D)=\mu(A^C)$ if $D\subset A^C$ and $\mu^*(D)=\mu(X)$ in all other cases. Clearly, A is the measurable cover of any subset $C\subset A$, since the only measurable set $H\subset A\text{-}C$ is θ. If $\mu(A)>0$ then from $D=A\text{-}(A\text{-}D)$ it follows that $\mu^*(A\text{-}D)=0$. Notice that $\mu(\theta)=0$ and that $\mathcal{M}=\mathcal{B}$.

4. Show that a set $C\subset X$ is measurable if and only if there exists $E\in\mathcal{A}$ such that $C\subset E$, $\mu(E)<\infty$ and

$$\mu(E) = \mu^*(C) + \mu^*(E\text{-}C).$$

  Hint:

  If $C\in\mathcal{M}$ then the condition necessarily holds. Assume now that the condition holds. Since $\mu^*(C)<\infty$, according to proposition 4 there exists a measurable cover, say F of C so that $\mu^*(C)=\mu(F)$. Clearly

$$C \subset E\cap F = F_0 \subset E.$$

Let G be a measurable cover of $E\text{-}C$ then we have that $\mu(G)=\mu^*(E\text{-}C)$. From this and the condition we obtain

$$\mu(F) = \mu^*(C) + \mu^*(E\text{-}C) = \mu(F) + \mu(G).$$

Let us show that $\mu(F_0\cap G)=0$. Since $F_0\cup G=E$

$$\mu(F) = \mu(F_0\cup G) = \mu(F_0) + \mu(G)-\mu(F_0\cap G)$$

$$= \mu(E)-\mu(F_0 \cap G) => \mu(F_0 \cap G) = 0.$$

Next, since $C=(E-G)\cup(C\cap G)$ and $C\cap G\subset F_0\cap G$ we deduce that $\mu^*(C\cap G)=0$. Thus, $C\cap G\in\mathcal{M}$. But $E-G\in\mathcal{A}\subset\mathcal{M}$ so that $C\in\mathcal{M}$ as the union of two elements from $\mathcal{M}$.

5. Given a complete measure space $(X, \overline{\mathcal{A}}, \overline{\mu})$, show that a set $C\subset X$ is measurable if and only if for any $\varepsilon>0$ there exists $A, E\in\overline{\mathcal{A}}$ such that $A\subset C\subset E$ and $\overline{\mu}(E-A)<\varepsilon$.

Hint:

If $C\in\overline{\mathcal{A}}$, then taking $A=B=C$ the necessity follows. Next, assume that the condition holds, then there exist sequences $\{A_k\}_1^\infty$, $\{E_k\}_1^\infty$ from $\overline{\mathcal{A}}$ such that $A_k\subset C\subset E_k$ and $\overline{\mu}(E_k-A_k)< 1/k$. Set

$$A^*=\bigcup_1^\infty A_k, \quad E^*=\bigcap_1^\infty E_k, \quad \text{then}$$

$$A_k \subset A^* \subset C \subset E^* \subset E_k => \overline{\mu}(E^*-A^*) = 0 => \overline{\mu}(C-A^*)=0.$$

Thus, $C-A^*$ is a null set and consequently, measurable. Since $C=A^*\cup(C-A^*)$ the measurability of C follows.

6. Let $(X, \mathcal{A}, \mu)$ be σ-finite and μ^* the outer measure induced by μ. Show that for any increasing sequence $\{A_k\}_1^\infty$ from $\mathcal{P}(X)$

$$\mu^*(\bigcup_1^\infty A_k) = \lim_{n\to\infty} \mu^*(A_n).$$

Hint:

According to proposition 3 every A_k has a measurable cover E_k if $\mu^*(A_k)<\infty$, such that $\mu^*(A_k)=\mu(E_k)$ $k=1, 2, \cdots$. Since

$$\mu^*(\bigcup_1^\infty A_k) \leq \mu(\bigcup_1^\infty E_k) = \lim_{k\to\infty} \mu(E_k) = \lim_{k\to\infty} \mu^*(A_k).$$

On the other hand,

$$\mu^*(A_n) \leq \mu^*(\bigcup_1^\infty A_k)$$

implying that

$$\lim_{n\to\infty} \mu^*(A_n) \leq \mu^*(\bigcup_1^\infty A_k).$$

21. INNER MEASURE

Let $(X, \mathscr{A}, \mu)$ be a σ–finite measure space. Here, as usual, μ^* stands for the outer measure induced by μ and $\bar{\mu} = \mu^* | \mathscr{M}$, where $\mathscr{M} \subset \mathscr{P}(X)$ is the class of all μ^*-measurable sets. According to proposition 6 of section 20, we have that $\overline{\mathscr{A}} = \mathscr{M}$, where $\overline{\mathscr{A}}$ is the "completion" of $\mathscr{A}$.

The "inner measure" μ_* induced by the measure μ is a map $\mu_* : \mathscr{P}(X) \to [0, \infty]$ defined as follows: for any $C \in \mathscr{P}(X)$

$$\mu_*(C) = \sup \{\mu(A); A \subset C \text{ and } A \in \mathscr{A}\}. \tag{1}$$

Clearly, $0 \le \mu_* \le \mu^*$, $\mu_*(C_1) \le \mu_*(C_2)$ if $C_1 \subset C_2$ and $\mu_* = \mu$ on $\mathscr{A}$. Since $\mathscr{A} \subset \mathscr{M}$ we have that

$$\mu_*(C) \le \sup \{\bar{\mu}(A); A \subset C \text{ and } A \in \mathscr{M}\}$$

$$= \sup \{\bar{\mu}(A); A \subset C \text{ and } A \in \overline{\mathscr{A}} \} \tag{2}$$

$$= \sup \{\mu(G); G \subset C \text{ and } G \in \mathscr{A}\} = \mu_*(C),$$

because, according to remark 1 of section 18, given any $A \in \overline{\mathscr{A}}$, there exists $G \in \mathscr{A}$ such that $G \subset A$ and $\bar{\mu}(A) = \mu(G)$.

Given a set $C \subset X$, a measurable subset $K \subset C$ is called a "measurable kernel" of C if any $H \in \mathscr{A}$, such that $H \subset C\text{-}K$ is a μ - null set.

The next proposition is of the existence type.

Proposition 1

Every $C \subset X$ has a measurable kernel.

Proof:

Let $\hat{C}$ be a measurable cover of C and denote by $\overline{M}$ a measurable cover of $\hat{C}\text{-}C$, then the set $K = \hat{C}\text{-}\overline{M}$ is a measurable kernel of C. Indeed,

$$K = \hat{C} - \overline{M} \subset \hat{C} - (\hat{C} - C) = C,$$

and if a measurable set $H \subset C\text{-}K = C\text{-}(\hat{C}-\overline{M}) \subset C \cap \overline{M} \subset \overline{M} - (\hat{C}-C)$, it follows, since $\overline{M}$ is a measurable cover of $\hat{C}\text{-}C$, that $\mu(H) = 0$. ♥

Proposition 2

If $K \subset C$ is a measurable kernel then

$$\mu(K) = \mu_*(C). \tag{3}$$

In addition, if $K_0 \subset C$ is another measurable kernel of C then $\mu(K \Delta K_0) = 0$.

Proof:

From (1) it follows that $\mu(K) \le \mu_*(C)$. If we assume that $\mu(K) < \mu_*(C)$, this will lead to contradiction. Indeed, in this case it follows from the definition of μ_* that there is $K_0 \in \mathcal{A}$ such that $K_0 \subset C$

and $\mu(K) < \mu(K_0)$. Thus, since K is a kernel and

$$K_0 - K \subset C - K \Rightarrow \mu(K_0 - K) = 0 \Rightarrow \mu(K_0) = \mu(K),$$

which is contradiction; this proves (3) and that K_0 is a kernel. Since $K \cup K_0 \subset C$ it follows that

$K \cup K_0 - K_0 \subset C - K_0$ and $K \cup K_0 - K \subset C - K$. This clearly implies that $\mu(K \cup K_0 - K) = \mu(K \cup K_0 - K_0) = 0$,

completing the proof of the proposition. ♥

Corollary 1

Let $\{C_k\}_1^\infty \subset \mathcal{P}(X)$ be a sequence of disjoint sets, then

$$\sum_1^\infty \mu_* (C_k) \leq \mu_* (\bigcup_1^\infty C_k).$$

Indeed, let $K_i \subset C_i$ be a measurable kernel (of C_i), then taking into account (3) we have:

$$\sum_1^\infty \mu_* (C_i) = \sum_1^\infty \mu (K_i) = \mu(\bigcup_1^\infty K_i) \leq \mu_* (\bigcup_1^\infty C_i).$$

Proposition 3

For any disjoint sequence $\{M_k\}_1^\infty$ from $\mathcal{M}$ and any $C \subset X$ we have that

$$\mu_* (C \cap \bigcup_1^\infty M_k) = \sum_1^\infty \mu_* (C \cap M_k).$$

Proof:

Let $K \subset C \cap \bigcup_1^\infty M_k$ be a measurable kernel, then due to the monotonicity property of μ_* and

corollary 1 we have that

$$\mu_* (C \cap \bigcup_1^\infty M_k) = \mu(K) = \sum_1^\infty \overline{\mu} (K \cap M_k) \leq \sum_1^\infty \mu_* (C \cap M_k)$$

$$\leq \mu_* (\bigcup_1^\infty (C \cap M_k)) = \mu_* (C \cap \bigcup_1^\infty M_k),$$

which proves our claim. ♥

Proposition 4

On $\mathcal{M}$, $\mu_* = \mu^* = \overline{\mu}$. In addition, if $\mu_* (G) = \mu^*(G)$ then $G \in \mathcal{M}$.

Proof:

If $C \in \mathcal{M}$, then according to (2) $\mu_* (C) = \overline{\mu} (C)$. But $\overline{\mu} = \mu^* | \mathcal{M}$, which proves the first part of the

proposition. To prove the second part assume that for some $C \subset X$, $\mu_* (C) = \mu^*(C)$. If A and K are measurable

cover and measurable kernel of the set C, respectively, the we have:

$$0 = \mu^*(C) - \mu_*(C) = \mu(A) - \mu(K) = \mu(A\text{-}K).$$

Then according to problem 5 of section 20, $C \in \mathscr{M}$, which proves the proposition. ♥

Proposition 5

For any two disjoint subsets C_1 and C_2 of X we have that

$$\mu_*(C_1 \cup C_2) \le \mu_*(C_1) + \mu^*(C_2) \le \mu^*(C_1 \cup C_2) \tag{4}$$

Proof:

Let $A \supset C_2$ be a measurable cover of C_2 and let $K \subset C_1 \cup C_2$ be a measurable kernel of $C_1 \cup C_2$. Then since

$$\mu(K\text{-}A) = \mu(K) - \mu(K \cap A) \ge \mu(K) - \mu(A),$$

we have that $\mu(A) + \mu(K\text{-}A) \ge \mu(K)$. Since $K\text{-}A = K \cap A^c \subset C_1$, it follows that

$$\mu_*(C_1 \cup C_2) = \mu(K) \le \mu(K\text{-}A) + \mu(A)$$

$$= \mu_*(C_1) + \mu^*(C_2).$$

To prove the second inequality, let $K_1 \subset C_1$ be a measurable kernel of C_1 and let B be a measurable cover of $C_1 \cup C_2$. Then

$$\mu^*(C_1 \cup C_2) = \mu(B) = \mu(K_1 \cup (B\text{-}K_1))$$

$$= \mu(K_1) + \mu(B\text{-}K_1) \ge \mu_*(C_1) + \mu^*(C_2)$$

Because $C_2 \subset B\text{-}K_1$. This completes the proof of the proposition. ♥

Proposition 6

For any $C \subset X$ and $M \in \mathscr{M}$ we have that

$$\overline{\mu}(M) = \mu_*(M \cap C) + \mu^*(M \cap C^c).$$

Proof:

According to propositions 4 and 5

$$\mu_*(M) = \mu_*((M \cap C) \cup (M \cap C^c))$$

$$\le \mu_*(M \cap C) + \mu^*(M \cap C^c) \le \mu^*(M).$$

Since $\mu_* = \mu^*$ on $\mathscr{M}$, the assertion follows. ♥

Proposition 7

A set $C \subset X$ is measurable if and only if

$$\mu_*(C) = \mu^*(C). \tag{5}$$

Proof:

We have seen (proposition 4) if (5) holds, then $C \in \mathscr{M}$. To show that (5) is also necessary, assume that $C \in \mathcal{M}$; then there exists a disjoint sequence $\{A_k\}_1^\infty \subset \mathcal{A}$ such that $C \subset \bigcup_1^\infty A_k$. By proposition 3

$$\mu_*(C) = \sum_1^\infty \mu_*(C \cap A_k) = \sum_1^\infty \bar{\mu}(C \cap A_k)$$

$$= \bar{\mu}(C) = \mu^*(C),$$

which proves our contention. ♥

Remark 1

If $\mu(X) < \infty$ then from proposition 6 we have that

$$\mu(X) = \mu_*(C) + \mu^*(C^c).$$

Thus,

$$\mu_*(C) = \mu(X) - \mu^*(C^c), \tag{6}$$

which is an alternative definition of inner measure. In time past measurability of a set was originally characterized by both μ_* and μ^*. A set $A \subset X$ such that $\mu_*(A) < \infty$ was defined to be measurable if

$\mu_*(A) = \mu^*(A)$.

Problems and complements

1. Show that the infimum and supremum in the formulae defining μ^* and μ_* are attained.

Hint:

We shall demonstrate this in the case of μ_*. To this end let $\mu_*(C)<\infty$ where $C \subset X$. Let also $\{A_i\}_1^\infty$ be a sequence from $\mathscr{A}$ such that $A_i \subset C$ for all $i=1, 2, \cdots$ and such that $\mu(A_n) \to \mu_*(C)$. Then from

$$A_n \subset \bigcup_1^n A_i \subset C$$

it follows that

$$\mu(A_n) \leq \mu(\bigcup_1^n A_i) \leq \mu_*(C), \qquad n=1, 2, \cdots .$$

By letting $n \to \infty$ we have that

$$\mu_*(C) = \mu(\bigcup_1^\infty A_i).$$

2. Show that for each $C \subset X$ such that $\mu^*(C)<\infty$, there exist A and $A' \in \mathscr{A}$ having the following properties: $A \subset C \subset A'$ and

$$\mu_*(C) = \mu(A), \ \mu^*(C) = \mu(A').$$

Hint:

From the previous problem we see that $A=\bigcup_1^\infty A_i$. Now, let $\{A_i'\}_1^\infty$ be a sequence from $\mathscr{A}$ such that $C \subset A_n'$ for every $n=1, 2, \cdots$ and such that $\mu(A_n') \to \mu^*(C)$ as $n \to \infty$. Clearly,

$$C \subset \bigcap_1^n A_i' \subset A_n' \quad \text{for all } n=1, 2, \cdots .$$

Consequently,

$$\mu^*(C) \leq \mu(\bigcap_1^\infty A_i') \leq \mu(A_n');$$

By letting $n \to \infty$ we have that

$$\mu^*(C) = \mu(\bigcap_1^\infty A_i') \quad \text{and} \quad A' = \bigcap_1^\infty A_i'.$$

3. Show that for any $C \subset X$, $\mu_*(C)=\varphi_C(X)$ if $\mu(X)<\infty$, where $\varphi_C(\cdot)$ is the map defined in problem 17 of section 19.

Hint:

For any $C \subset X$ and $A \subset \mathscr{A}_0$ we have

$$\varphi_C(A) = \mu(A)-\mu^*(A-C).$$

Since $\varphi_C(A)$ is monotone in A, it attains its maximum value at X so that

$$\varphi_C(X) = \mu(X) - \mu^*(C^c).$$

This and (6) prove the claim.

4. If $M \in \mathcal{M}$ satisfy $\mu_*(M) > 0$ show that for any $A \subset M$ and $A \notin \mathcal{M}$, $\mu_*(M-A) > 0$.

Hint:

If $\mu_*(M-A) = 0$ then from $A = M - (M-A)$ it would follow that A is measurable.

22. MEASURES ON THE REAL LINE R

Let $(X, \mathscr{A}, \mu)$ be a measure space. Many interesting problems arise when the set X is endowed with some "natural" topological structure. Recall that a "**topology** in X" is a family $\mathscr{T} \subset \mathscr{P}(X)$, closed under finite intersection, arbitrary union and which includes θ and X. Elements of $\mathscr{T}$ are called "open sets." There is no preconceived idea what "open" may mean, except that a subset $O \subset X$ is open if and only if $O \in \mathscr{T}$. A subset $C \subset X$ is called "closed" if $C^c \in \mathscr{T}$. The class $\mathscr{A}$ of measurable sets is then usually the σ-algebra generated by $\mathscr{T}$, i.e.,

$$\mathscr{A} = \sigma\{\mathscr{T}\}$$

(**the Borel algebra**). The elements of $\mathscr{A}$ are called "**Borel sets.**"

The classical example of a measure space is $(R, \mathscr{B}, \mu)$, where R is the real line, $\mathscr{B}$ is the Borel algebra of subsets of R (here $\mathscr{T}$ is the class of all open subsets of R , and $\mathscr{B} = \sigma\{\mathscr{T}\}$) and μ is a measure on $\mathscr{B}$. If μ is finite on bounded Borel sets it is called "**Borel measure.**" It is not difficult to show that $\mathscr{B} = \sigma\{\mathscr{S}\}$, where $\mathscr{S}$ is the "semiring" of the intervals of the form $(a, b]$. Assume for a moment that $\mu(R) < \infty$ and consider the function

$$F(x) = \mu(-\infty, x].$$

The map $F(\cdot) \geq 0$ is often called **the "distribution function" of the Borel measure** μ. It is clearly bounded, non decreasing and if $-\infty < t_1 < t_2 < \infty$

$$\mu(t_1, t_2] = F(t_2) - F(t_1).$$

Since $(t_1, t_2 + 1/n] \downarrow (t_1, t_2]$ as $n \to \infty$, we have that

$$\mu(t_1, t_2] = \lim_{n \to \infty} \mu(t_1, t_2 + \frac{1}{n}] = \lim_{n \to \infty} F(t_2 + \frac{1}{n}) - F(t_1) = F(t_2 + 0) - F(t_1).$$

Consequently, $F(t_2 + 0) = F(t_2)$ which means that $F(\cdot)$ is right continuous on R.

It is clear that every singleton $\{x\} \in \mathscr{B}$. If $\mu\{x\} = 0$ at every $x \in R$, then $F(\cdot)$ is continuous on R and conversely. Indeed,

$$\mu\{t\} = \lim_{n \to \infty} \mu(t - \frac{1}{n}, t] = F(t) - \lim_{n \to \infty} F(t - \frac{1}{n}) = F(t) - F(t - 0).$$

Finally, since $\bigcap_1^\infty (-\infty, -n] = \theta$ we have that

$$\lim_{n \to -\infty} F(n) = 0, \quad \text{and consequently} \quad \lim_{t \to -\infty} F(t) = 0.$$

Next let $F:R \to [0, \infty]$ be an arbitrary non decreasing and right continuous function and define $\mu_F : \mathscr{S} \to [0, \infty]$ by

$$\mu_F (a, b] = F(b) - F(a).$$

We shall prove that μ_F is a measure on the semiring $\mathcal{S}$. Then the general theory of extension of section 19 can be applied to obtain the outer measure μ_F^* and the class $\mathcal{M}$ of all μ_F^*-measurable sets. The restriction

$$\mu_F^* \mid \mathcal{M} = \bar{\mu}_F$$

is called **the "Lebesgue-Stieltjes" measure** on R induced by F. Since $\mathcal{S} \subset \mathcal{M}$ we have that

$$\sigma\{\mathcal{S}\} = \mathcal{B} \subset \mathcal{M}. \tag{1}$$

To prove that μ_F is a measure on the semiring $\mathcal{S}$ we need the next two simple lemmas (see problems 4 and 5).

Lemma 1

If $\{(a_i, b_i]\}_1^N$ are disjoint and $(a, b] \supset \bigcup_1^N (a_i, b_i]$ then

$$\sum_1^N \mu_F (a_i, b_i] \leq \mu_F (a, b].$$

Lemma 2

If $(a, b] \subset \bigcup_1^N (a_i, b_i] \ (\{(a_i, b_i]\}_1^N$ are not necessarily disjoint$)$ then

$$\sum_1^N \mu_F (a_i, b_i] \geq \mu_F (a, b].$$

Proposition 1.

The set function μ_F is a measure on the semiring $\mathcal{S}$.

Proof:

Suppose that $\{(a_i, b_i]\}_1^N$ are disjoint and that

$$(a, b] = \bigcup_1^N (a_i, b_i].$$

Then from the previous two lemmas we have that

$$\mu_F (a, b] = \sum_1^N \mu_F (a_i, b_i].$$

Next, assume that the intervals $\{(a_i, b_i]\}_1^\infty$ are disjoint, such that

$$(a, b] = \bigcup_1^\infty (a_i, b_i],$$

Then, since $\bigcup_{1}^{N}(a_i, b_i] \subset (a, b]$, it follows from lemma 1 that:

$$\sum_{1}^{N} \mu_F(a_i, b_i] \le \mu_F(a, b]$$

for every finite N. Hence

$$\sum_{1}^{\infty} \mu_F(a_i, b_i] \le \mu_F(a, b]. \qquad (2)$$

Next let $\varepsilon > 0$ be such that $a+\varepsilon < b$, then $[a+\varepsilon, b] \subset (a, b]$. By the right continuity of $F(\cdot)$, for each i there is $b_i' > b_i$ such that

$$F(b_i') - F(b_i) < \varepsilon / 2^i,$$

then

$$\mu_F(a_i, b_i'] \le \mu_F(a_i, b_i] + \varepsilon / 2^i.$$

Since $[a+\varepsilon, b] \subset \bigcup_{1}^{\infty}(a_i, b_i')$, by the **"Heine-Borel theorem"** there is a finite sub collection, say $\{(a_i, b_i')\}_1^k$, such that

$$[a+\varepsilon, b] \subset \bigcup_{1}^{k}(a_i, b_i').$$

This and lemma 2 yield

$$\mu_F(a+\varepsilon, b] \le \sum_{1}^{k}\mu_F(a_i, b_i'] \le \sum_{1}^{\infty}\mu_F(a_i, b_i] + \varepsilon.$$

By letting $\varepsilon \to 0$ and taking into account the right continuity of $F(\cdot)$ and inequality (2), the proof is now completed. ♥

Let $c \in R$; the function $f_c : R \to R$ defined by

$$f_c(x) = x + c$$

is a continuous bijection. The map

$$f_c : \mathscr{P}(R) \to \mathscr{P}(R)$$

is then

$$f_c(K) = K + c$$

where $K+c = \{x+c;\ x \in K\}$. For instance, $f_c(a, b) = (a, b) + c = (a+c, b+c)$. For this reason $f_c(\cdot)$ is called a "translation" map.

By means of translation maps we can easily establish the following identities: given $x \in R$ then for any family $\{G_t;\ t \in I\}$ of subsets of R

$$\bigcup_I (G_t + x) = \bigcup_I G_t + x \quad \text{and} \quad \bigcap_I (G_t + x) = \bigcap_I G_t + x.$$

For instance,

$$\bigcup_I (G_t + x) = \bigcup_I f_x(G_t) = f_x(\bigcup_I G_t) = \bigcup_I G_t + x.$$

In addition, for any $B \subset R$, $(f_x(B))^c = f_x(B^c)$. Indeed, since $f_x(R)=R$ and

$$f_x(B) \cap f_x(B^c) = f_x(B \cap B^c) = \theta$$

$$f_x(B) \cup f_x(B^c) = f_x(B \cup B^c) = R,$$

the assertion follows.

Recall that each open set $O \subset R$ is a union of countably many disjoint open intervals, say $O = \bigcup_k I_k$, and since each $f_x(I_k)$ is an open interval, we have that

$$O + x = f_x(O) = f_x(\bigcup_k I_k) = \bigcup_k f_x(I_k)$$

so that $O+x$ is also an open set.

Proposition 2

For every $B \in \mathcal{B} <=> B + x \in \mathcal{B}$.

Proof:

Denote by

$$\mathcal{B}_0 = \{A \in \mathcal{B}, f_x(A) \in \mathcal{B}\}.$$

Since $\mathcal{B}_0$ contains all open subsets of R, the assertion is true if we show that $\mathcal{B}_0$ is a σ-algebra. Let $\{B_k\}_1^\infty$ be a sequence from $\mathcal{B}_0$, then for each k

$$B_k \in \mathcal{B} \text{ and } f_x(B_k) \in \mathcal{B} => \bigcup_1^\infty B_k \in \mathcal{B} \text{ and } f_x(\bigcup_1^\infty B_k) = \bigcup_1^\infty f_x(B_k) \in \mathcal{B}.$$

Next, if $A \in \mathcal{B}_0$ then

$$A \in \mathcal{B} \text{ and } f_x(A) \in \mathcal{B} => A^c \in \mathcal{B} \text{ and } (f_x(A))^c \in \mathcal{B}.$$

But $(f_x(A))^c = f_x(A^c)$, which implies that $A^c \in \mathcal{B}_0$. Consequently, $\mathcal{B}_0 = \mathcal{B}$. ♥

When $F(x) = x$ the Lebesgue-Stieltjes measure $\bar{\mu}_F$ becomes the ordinary Lebesgue measure $\bar{\lambda}$ on $\mathcal{M}$, the class of Lebesgue measurable sets. The Lebesgue measure $\bar{\lambda}$ is "translation invariant", i.e., for any $x \in R$ and $M \in \mathcal{M}$

$$\bar{\lambda}(M+x) = \bar{\lambda}(M).$$

This follows from the fact that the λ^* is translation invariant because for any $D \subset R$ and $x \in R$

$$D \subset \bigcup_1^\infty (a_i, b_i] \text{ if and only if } D+x \subset \bigcup_1^\infty (a_i + x, b_i + x].$$

Recall that $\mu_F = \lambda$ when $F(x) = x$. Proposition 1 has established that λ is a measure on the semiring $\mathcal{S}$. The class of all λ^*-measurable sets is $\mathcal{M}$ (the Lebesgue measurable sets) and the Lebesgue measure $\bar{\lambda} = \lambda^* | \mathcal{M}$. The next proposition shows that $\bar{\lambda}$ is uniquely determined by its values on the class of open sets $\mathcal{T} \subset \mathcal{P}(R)$.

Proposition 3

Let $\lambda_0^* : \mathcal{P}(R) \to [0, \infty]$ be defined as follows: for any $H \subset R$

$$\lambda_0^*(H) = \inf\{\bar{\lambda}(O); H \subset O \in \mathcal{T}\}.$$

We claim that $\lambda_0^* = \lambda^*$.

Proof:

According to the remarks 1 and 2 of section 19, for any $H \subset R$ and $\varepsilon > 0$ there exists a disjoint sequence $\{(a_i, b_i]\}_1^\infty$ from $\mathcal{S}$ such that

$$H \subset \bigcup_1^\infty (a_i, b_i] \text{ and } \sum_1^\infty (b_i - a_i) \le \lambda^*(H) + \varepsilon.$$

Clearly then,

$$H \subset \bigcup_1^\infty (a_i, b_i + \varepsilon / 2^i) = O \in \mathcal{T},$$

so that

$$\lambda^*(H) < \bar{\lambda}(O) < \sum_1^\infty (b_i - a_i) + \varepsilon < \lambda^*(H) + 2\varepsilon.$$

From this we obtain that for every $\varepsilon > 0$

$$\lambda^*(H) \le \lambda_0^*(H) < \lambda^*(H) + 2\varepsilon,$$

which proves our contention. ♥

Proposition 4

A subset $B \subset R$ is Lebesgue measurable, i.e., $B \in \mathcal{M}$, if and only if for each $\varepsilon > 0$ there exists a closed set F_ε such that $F_\varepsilon \subset B$ and $\lambda^*(B - F_\varepsilon) < \varepsilon$.

Proof:

If $B \in \mathcal{M}$ then by proposition 3 there exists an open set O such that $B^c \subset O$ and $\bar{\lambda}(O - B^c) = \bar{\lambda}(O \cap B) < \varepsilon$. Since $O^c \subset B$, $\bar{\lambda}(B - O^c) = \bar{\lambda}(B \cap O) < \varepsilon$, which proves necessity of the condition. To prove its sufficiency, let $\{F_n\}_1^\infty$ be a sequence of closed sets such that

$$F_n \subset B \text{ and } \lambda^*(B - F_n) < 1/n, \text{ for each } n = 1, 2, \cdots.$$

Set $F = \bigcup_1^\infty F_n$, then $F \subset B$ and

$$\lambda^*(B-F) \le \lambda^*(B-F_n) < 1/n, \quad n=1, 2, \cdots ,$$

which clearly implies that $\lambda^*(B-F) = 0 \Rightarrow B-F \in \mathscr{M}$. Since $B = F \cup (B-F)$, the assertion follows. ♥

Remark 1

From the last equality it follows that for any $B \in \mathscr{M}$ there exists a closed set $F \subset B$ such that $\overline{\lambda}(B) = \overline{\lambda}(F)$. In other words, the Lebesgue measure $\overline{\lambda}$ is determined uniquely by its values on the class of closed subsets of R. If it happens that for some subset $A \subset R$ and some closed set $F_0 \subset A$ we have $\lambda^*(A) = \overline{\lambda}(F_0)$, that does not imply that $A - F_0 \in \mathscr{M}$, since $\lambda^*(A-F_0) > 0$ if $A \notin \mathscr{M}$.

Proposition 5

A subset $A \subset R$ is Lebesgue measurable if and only if, given any $\varepsilon > 0$ there exists an open set $O \in \mathscr{T}$ such that $A \subset O$ and $\lambda^*(O-A) < \varepsilon$.

Proof:

If $A \in \mathscr{M}$ and $\overline{\lambda}(A) < \infty$, then according to proposition 3, given any $\varepsilon > 0$ there is an open set O such that $A \subset O$ and $\overline{\lambda}(O) < \overline{\lambda}(A) + \varepsilon$; hence, $\overline{\lambda}(O-A) < \varepsilon$. If $\overline{\lambda}(A) = +\infty$ consider the sequence $\{B_i\}_1^\infty$ where $B_i = [-i, i]$ and sets $E_i = A \cap B_i$, then clearly $E_i \in \mathscr{M}$, $\overline{\lambda}(E_i) < \infty$ for all $i=1, 2, \cdots$ and $A = \bigcup_1^\infty E_i$. For each E_i there is an open set O_i such that $E_i \subset O_i$ and

$$\overline{\lambda}(O_i - E_i) < \varepsilon / 2^i .$$

Set $O = \bigcup_1^\infty O_i$; clearly, O is an open set and $A \subset O$, so that

$$\overline{\lambda}(O-A) \le \sum_1^\infty \overline{\lambda}(O_i - E_i) < \varepsilon,$$

which proves necessity of the condition.

Next, assume that for each $\varepsilon > 0$ there exists an open set O such that $A \subset O$ and $\lambda^*(O-A) < \varepsilon$. Let $\{O_i\}_1^\infty$ be a sequence of open sets such that $A \subset O_i$ and $\lambda^*(O_i - A) < 1/i$. The set $G = \bigcap_1^\infty O_i \in \mathscr{M}$ and since $A \subset G$ we have that

$$\lambda^*(G-A) \le \lambda^*(O_i - A) < 1/i.$$

Consequently $\lambda^*(G-A) = 0 \Rightarrow G-A \in \mathscr{M}$. Since

$$A = G - (G-A) \Rightarrow A \in \mathscr{M},$$

which proves sufficiency and the proposition. ♥

Remark 2

From the last two propositions we conclude that given any $M \in \mathcal{M}$ and $\varepsilon > 0$, there exist an open set O_ε and a closed set F_ε such that

$$F_\varepsilon \subset M \subset O_\varepsilon, \quad \bar{\lambda}(M\text{-}F_\varepsilon) < \varepsilon \ \text{ and } \ \bar{\lambda}(O_\varepsilon\text{-}M) < \varepsilon.$$

Since

$$O_\varepsilon\text{-}F_\varepsilon = (M - F_\varepsilon) \cup (O_\varepsilon - M) \text{ we have that } \bar{\lambda}(O_\varepsilon - F_\varepsilon) < 2\varepsilon.$$

This gives rise to the following:

Definition 1

Let $(R, \mathcal{B}, \mu)$ be a measure space. If for every $B \in \mathcal{B}$ and $\varepsilon > 0$ there exists a closed set $F_\varepsilon \subset R$

and an open set $O_\varepsilon \subset R$ such that

$$\mathbf{F_\varepsilon \subset B \subset O_\varepsilon \ \text{ and } \ \mu(O_\varepsilon\text{-}F_\varepsilon) < \varepsilon,}$$

the measure μ is said to be "regular." The Lebesgue measure $\bar{\lambda}$ is clearly regular.

Remark 3

The inclusion $\mathcal{B} \subset \mathcal{M}$ is proper. This was first proved by M. Suslin in 1917, who discovered a class of subsets of R, presently known as "analytic sets." Each Borel set is an analytic set. Necessary and sufficient condition that an analytic set be Borel is that its complement is also analytic. Later Luzin proved that every analytic set is Lebesgue measurable.

Do Lebesgue nonmeasurable sets exist? It is not at all obvious that for something like the Lebesgue measure there are any such sets. However, if the axiom of choice holds the answer is affirmative. Without this assumption no Lebesgue nonmeasurable set has been constructed up to the present time.

Example 1

Vitali, in his paper "Sul problema della missura dei gruppi di punti di una retta" (1905) established that the set S in example 1 of section 9 is not Lebesgue measurable. Here is a simple version of his proof: Assume to the contrary, that S is Lebesgue measurable and write $[-1, 1] \cap Q = \{r_1, r_2, \cdots\}$, where $Q \subset R$ is the set of rational numbers. Then $\{S_n\}_1^\infty$, where $S_n = S + r_n$, is a sequence of disjoint measurable sets (see problem 5 of section 9). Since the Lebesgue measure $\bar{\lambda}$ is translation invariant $\bar{\lambda}(S_n) = \bar{\lambda}(S)$ for each $n=1, 2,$ $\cdots$. By observing that $\bigcup_1^\infty S_n \subset [-1, 2]$ we have that

$$\bar{\lambda}\left(\bigcup_1^\infty S_n\right) = \lim_{n \to \infty} (n\, \bar{\lambda}(S)) \le 3,$$

which implies that $\bar{\lambda}(S)=0$. On the other hand,

$$[0, 1] \subset \bigcup_1^\infty S_n \ \Rightarrow \ 1 \le \bar{\lambda}\left(\bigcup_1^\infty S_n\right) = 0$$

which is contradiction. Consequently, S cannot be Lebesgue measurable.

The Cantor ternary set $C \subset [0, 1]$ described in section 7 has some remarkable properties. It has the cardinality of continuum, it is closed and thus Lebesgue measurable, and $\bar{\lambda}(C)=0$ (it is obtained from $[0, 1]$ by removing an infinite sequence of disjoint open intervals whose total length is 1 (see (5) in section 7)). In other words,

$$\text{Card } C = c, \qquad \text{Card } \mathscr{P}(C) = 2^c$$

(see (4) of section 7). Since $C \in \mathscr{M}$ and $\bar{\lambda}(C)=0$

$$\mathscr{P}(C) \subset \mathscr{M} \subset \mathscr{P}(R)$$

which implies (see problem 5 of section 6) that

$$\text{Card } \mathscr{M} = 2^c . \tag{3}$$

Denote by $\mathbf{J}_R$ the collection of all open intervals with rational end points. The set $\mathbf{J}_R$ is a base for the class of all open subsets of R (every open set is a finite or countable disjoint union (see problem 2) of elements from $\mathbf{J}_R$). Consequently,

$$\mathscr{B} = \sigma\{\mathbf{J}_R\}.$$

But $\mathbf{J}_R \sim Q \times Q$, where $Q \subset R$ is the set of all rationales, so that $\mathbf{J}_R$ is countable and consequently

$$\text{Card } \mathscr{P}(\mathbf{J}_R) = c.$$

From this we conclude that the Card $\mathscr{B} \leq c$ and since Card $\mathscr{B} > \aleph_0$, it follows from the continuum hypothesis (see remark 1 of section 5) that

$$\text{Card } \mathscr{B} = c. \tag{4}$$

Remark 4

It was while reading a Lebesgue paper M.Suslin, a student of Lusin, discovered an error in the proof of the theorem that the inverse (if there exists) of a Baire function is also a Baire function (recall a map $f: R \to R$ is a Baire function if and only if $f^{-1}((a, b))$ is a Borel set for every $(a, b) \subset R$). Although the theorem is true, Lebesgue employed the following false argument : If $\{H_n\}_1^\infty$ is a decreasing sequence of sets in R^2, then the projection to the x-axis of $\bigcap_1^\infty H_k$ is equal to the intersection of projection of H_k. The statement is false since the projection of G_δ-set in the plane need not be a Borel set. This gives rise to a new class of subsets of R subsequently called "analytic sets", namely the projections of Borel sets. Every analytic set is Lebesgue measurable. Suslin also proved that Borel sets are exactly those sets which are analytic and whose complements are also analytic. Notice that the projection of a Borel plane set is not necessarily a Borel set of reals.

Problems and complements

1. Show that $\mathscr{B}=\sigma\{\mathscr{S}\}$.

    Hint:

    Since $\mathscr{S}\subset\mathscr{B} \Rightarrow \sigma\{\mathscr{S}\}\subset\mathscr{B}$. On the other hand, $(a, b)\in\mathscr{B}$ and

$$(a, b) = \bigcup_1^\infty (a, b - \frac{1}{k}] \in \sigma\{\mathscr{S}\}.$$

2. Show that every open set $O \subset R$ is a countable union of disjoint open intervals.

    Hint:

    Given $x\in O$ let I_x be the largest open interval which contains x and which $I_x\subset O$. For any $x'\in R$ and $x'\neq x$, either $I_{x'}=I_x$ or $I_{x'}\cap I_x=\theta$. Indeed, if

$$I_{x'} \neq I_x \quad\text{and}\quad I_{x'}\cap I_x \neq \theta$$

then $x\in I_x\cup I_{x'}\subset O$, which contradicts the maximality of I_x. Clearly,

$$O = \bigcup_{x\in O} I_x.$$

Finally, there is only countably many distinct I_x since each such I_x contains a different rational number. Since end points of I_x do not belong to I_x they do not belong to O.

3. Show that every closed set F is a limit of a decreasing sequence of open sets.

    Hint:

    If $F=[a, b]$ then $\{(- \frac{1}{k}+a, b+\frac{1}{k})\}_1^\infty$ is such a sequence. In the general case we proceed as follows:

Given $x\in R$ define

$$\rho(x, F) = \inf_{u\in F} |x-u| .$$

Let us show that for any $x, y\in R$

$$|\rho(x, F) - \rho(y, F)| < |x-y|. \tag{*}$$

Indeed, for any $u\in F$

$$\rho(x, F) \leq |x-u| \leq |x-y| + |y-u|.$$

From this we conclude that

$$\rho(x, F) \leq |x-y| + \rho(y, F).$$

In a similar fashion we obtain that

$$\rho(y, F) \leq |x-y| + \rho(x, F),$$

which proves (*). From this inequality it follows that the function

$$h_F(\cdot) = \rho(\cdot, F)$$

is uniformly continuous. Consequently,

$$\{x; h_F(x)< \frac{1}{n} \}, \qquad n = 1, 2, \cdots$$

are open sets. Since

$$F = \bigcap_{1}^{\infty} \{x; \rho(x, F) < \tfrac{1}{n}\},$$

our contention is true.

4. Prove lemma 1.

Proof:

$$\mu_F(a, b] = F(b) - F(a) \geq F(b_N) - F(a_1)$$

$$= \sum_{1}^{N} [F(b_i) - F(a_i)] + \sum_{1}^{N-1} [F(a_{i+1}) - F(b_i)]$$

$$\geq \sum_{1}^{N} [F(b_i) - F(a_i)] = \sum_{1}^{N} \mu_F(a_i, b_i].$$

5. Prove lemma 2

Proof:

Assume that $\{(a_i, b_i)\}_1^N$ is the minimal set of intervals necessary to cover $(a, b]$. Under this condition

$$a_1 \leq a \leq b_1,\ a_N < b \leq b_N \ \text{and}\ a_{i+1} \leq b_i < b_{i+1}$$

for $i=1, 2, \cdots, N-1$. Then

$$\mu_F(a, b] = F(b) - F(a) \leq \sum_{1}^{N-1} [F(a_{i+1}) - F(a_i)] + F(b_N) - F(a_N)$$

$$\leq \sum_{1}^{N} [F(b_i) - F(a_i)] = \sum_{1}^{N} \mu_F(a_i, b_i].$$

6. Show that for each $x \in R$, $A \in \mathcal{M} \Longleftrightarrow A + x \in \mathcal{M}$.

Hint:

If $A \in \mathcal{M}$, then for any $E \subset R$

$$\lambda^*(E \cap A) + \lambda^*(E \cap A^c) = \lambda^*(E) = \lambda^*(E - x),$$

because λ^* is translation invariant. Next,

$$\lambda^*(E-x) = \lambda^*((E-x) \cap A) + \lambda^*((E-x) \cap A^c)$$

$$= \lambda^*(x + (E-x) \cap A) + \lambda^*(x + (E-x) \cap A^c)$$

$$= \lambda^*(f_x((E-x) \cap A) + \lambda^*((f_x(E-x) \cap A^c))$$

$$= \lambda^*(f_x(E-x) \cap f_x(A)) + \lambda^*(f_x(E-x) \cap f_x(A^c))$$

$$= \lambda^*(E \cap (A+x)) + \lambda^*(E \cap (A+x)^c) = \lambda^*(E).$$

7. Show that to every $M \in \mathcal{M}$ there correspond two Borel sets A and B such that

$$A \subset M \subset B \ \text{and}\ \bar{\lambda}(B-A) = 0.$$

Hint:

According to remark 2, there exists a sequence of closed sets $\{F_n\}_1^\infty$ such that $F_n \subset M$ and $\bar{\lambda}(M-F_n)<1/n$. Denote by $A=\bigcup_1^\infty F_i$; clearly $A \subset M$ and from

$$\bar{\lambda}(M-A) \leq \bar{\lambda}(M-F_n) < 1/n \qquad \text{for every } n=1, 2, \cdots$$

we conclude that $\bar{\lambda}(M-A)=0$.

There is also a sequence of open sets $\{O_n\}_1^\infty$ such that $M \subset O_n$ and

$\bar{\lambda}(O_n-M)<1/n$ for each $n=1, 2, \cdots$. Set $B=\bigcap_1^\infty O_n$, then clearly for every $n=1, 2, \cdots$

$$\bar{\lambda}(B-M) \leq \bar{\lambda}(O_n-M) < 1/n \implies \bar{\lambda}(B-M) = 0.$$

The proof of the claim follows from

$$B-A = (B-M) \cup (M-A).$$

8. Let $(X, \mathscr{A}, \mu)$ be a measure space. Recall that a set $A \in \mathscr{A}$ is called an atom (of the measure μ) if and only if $0<\mu(A)<\infty$ and for any measurable set $H \subset A$, either $\mu(H)=0$ or $\mu(H)=\mu(A)$. If no such a set exists, the measure μ is said to be nonatomic. Show that $(R, \mathscr{M}, \bar{\lambda})$ is nonatomic.

Hint:

Given any $A \in \mathscr{M}$ such that $0<\bar{\lambda}(A)<\infty$, there exists an integer k such that the set $M=[k, k+1] \cap A$ satisfies $\bar{\lambda}(M)=\delta>0$. Split $[k, k+1]$ into a finite number of intervals I of the same length, then for at least one of them $\bar{\lambda}(M \cap I) > 0$. Since $0<\bar{\lambda}(M \cap I)<\delta \leq \bar{\lambda}(A)$, it follows that A cannot be an atom, proving that $(X, \mathscr{M}, \bar{\lambda})$ is nonatomic measure space.

9. (Continuation) Given any $A \in \mathscr{M}$ such that $\bar{\lambda}(A) > \delta > 0$, there is a Lebesgue measurable set $B \subset A$ such that $\bar{\lambda}(B)=\delta$.

Hint:

Assume that for some integer n, $A \subset [-n, n]$ and consider the map $f:[-n, n] \to R$ defined by

$$f(t) = \bar{\lambda}(A \cap [-n, t]).$$

Clearly, $f(\cdot)$ is increasing with $f(-n)=0$ and $f(n)=\bar{\lambda}(A)$. It is also uniformly continuous on $[-n, n]$ since for any $-n \leq s < t \leq n$

$$f(t) = \bar{\lambda}(A \cap [-n, t]) = \bar{\lambda}(A \cap [-n, s]) + \bar{\lambda}(A \cap (s, t]) \leq f(s) + (t-s).$$

By the "intermediate" value theorem there exists an $x_0 \in [-n, n]$ such that $f(x_0)=\delta$. The measurable set $B=[-n, x_0] \cap A$ satisfies $\bar{\lambda}(B)=\delta$ (see problem 9 of section 18).

10. (Continuation) Let $0<\delta<1$; given a Lebesgue measurable set $M \subset (-\infty, \infty)$ such that $\bar{\lambda}(M)>0$; do there exist $-n \leq x < y \leq n$ such that

$$\overline{\lambda}(A \cap [x, y]) = \delta \overline{\lambda}(A)?$$

Hint:

The answer is affirmative if and only if $\delta = 1/N$ where $N = 2, 3, \cdots$. To see this let $\{\frac{k}{N}\overline{\lambda}(M)\}_0^N$ be a partition of $[0, \overline{\lambda}(M)]$. Chose $-n \leq s < t \leq n$ such that $f(s) = \frac{k-1}{N}\overline{\lambda}(M)$ and $f(t) = \frac{k}{N}\overline{\lambda}(M)$.

Clearly then

$$f(t) - f(s) = \overline{\lambda}(A \cap [s, t]) = \frac{1}{N}\overline{\lambda}(M).$$

11. Construct a Lebesgue measurable non Borel set.

Hint:

Recall that the Cantor set C is closed, has cardinality of continuum, $\overline{\lambda}(C) = 0$, and obviously $\mathscr{P}(C) \subset \mathscr{M}$. We shall attempt to identify a member $H \in \mathscr{P}(C)$ such that $H \notin \mathscr{B}$. To this end consider a strictly increasing continuous map $h: R \to R$, and define

$$\mathscr{R}_0 = \{A \subset R; h(A) \in \mathscr{B}\}. \tag{*}$$

We have $\mathscr{R}_0 = \mathscr{B}$. Next, define

$$h(x) = \Phi(x) + x$$

where $\Phi(\cdot)$ is the Cantor function (see (6) of section 7). The image of an interval in C^c by h is an interval of the same length. On the other hand,

$$[0, 2] = h([0, 1]) = h([0, 1] \cap C) \cup h([0, 1] \cap C^c).$$

Consequently,

$$2 = \overline{\lambda}(h(C)) + \overline{\lambda}(h([0, 1] \cap C^c)) = \overline{\lambda}(h(C)) + 1,$$

which yields that $\overline{\lambda}(h(C)) = 1$. There exists $\Gamma \subset h(C)$ which is not (Lebesgue) measurable. Let $H = h^{-1}(\Gamma)$. Since $H \subset C$ we have that $\overline{\lambda}(H) = 0$. But $H \notin \mathscr{B}$, for if it does $h(H) = \Gamma \in \mathscr{B}$ according to (*). Thus, $H \in \mathscr{M}$ but $H \notin \mathscr{B}$.

12. Every finite Borel measure μ on R is regular.

Hint:

Let $(R, \mathscr{B}, \mu)$ be a finite Borel measure space. We can assume, without loss of generality, that $\mu(R) = 1$. Let $\mathscr{R} \subset \mathscr{B}$ be the class of μ-regular sets (see definition 1), i.e., for any $G \in \mathscr{R}$ and $\varepsilon > 0$ there exists an open set O_ε and a closed set F_ε such that $O_\varepsilon \subset G \subset F_\varepsilon$ and $\mu(F_\varepsilon - O_\varepsilon) < \varepsilon$. Since θ and R are both closed and open $\theta \subseteq \theta \subseteq \theta$ and $R \subseteq R \subseteq R$ we have that $\theta, R \in \mathscr{R}$.

The class $\mathscr{R}$ is closed under complementation, for if $G \in \mathscr{R}$ and $\varepsilon > 0$, there exists open set O_ε and closed set F_ε, such that

$$O_\varepsilon \subseteq G \subseteq F_\varepsilon \text{ and } \mu(F_\varepsilon - O_\varepsilon) < \varepsilon.$$

From this we have that

$$O_\varepsilon^c \supseteq G^c \supseteq F_\varepsilon^c \quad \text{and since} \quad O_\varepsilon^c - F_\varepsilon^c = F_\varepsilon - O_\varepsilon$$

it follows that $G^c \in \mathscr{R}$.

Next, assume that $\{G_k\}_1^\infty$ from $\mathscr{R}$ and set $G = \bigcup_1^\infty G_k$. Let $\{O_{k\varepsilon}\}_1^\infty$ and $\{F_{k\varepsilon}\}_1^\infty$ be sequences of open and closed sets, respectively, such that

$$F_{k\varepsilon} \subseteq G_k \subseteq O_{k\varepsilon} \quad \text{and} \quad \mu(O_{k\varepsilon} - F_{k\varepsilon}) \le \varepsilon/3^k, \quad k = 1, 2, \cdots.$$

Set

$$O_\varepsilon = \bigcup_1^\infty O_{k\varepsilon}, \ F_\varepsilon = \bigcup_1^\infty F_{k\varepsilon}$$

and chose n so large that $\mu(F_\varepsilon - \bigcup_1^n F_{i\varepsilon}) < \varepsilon/2$. Clearly, O_ε is open, $\bigcup_1^n F_{i\varepsilon}$ is closed and

$$\bigcup_1^n F_{i\varepsilon} \subseteq G \subseteq O_\varepsilon \quad \text{and} \quad \mu(O_\varepsilon - \bigcup_1^n F_{i\varepsilon}) \le \mu(O_\varepsilon - F_\varepsilon) + \mu(F_\varepsilon - \bigcup_1^n F_{i\varepsilon})$$

$$\le \sum_1^\infty \mu(O_{i\varepsilon} - F_{i\varepsilon}) + \varepsilon/2 \le \sum_1^\infty \varepsilon/3^i + \varepsilon/3 = \varepsilon.$$

Consequently, $\mathscr{R}$ is a σ-algebra.

Finally, let $F \subset R$ be a closed set and $\varepsilon > 0$. According to problem 3 there exists a sequence of open sets $\{O_k\}_1^\infty$ such that $O_k \downarrow F$. Since $\mu(O_k) \downarrow \mu(F)$, there is an index k such that $\mu(O_k - F) < \varepsilon$; Thus, $F \subseteq O_k$, which proves that $F \in \mathscr{R}$. Consequently, $\mathscr{R} = \mathscr{B}$.

23 MEASURABLE FUNCTIONS

Let $\{X, \mathscr{A}\}$ be a measurable space fixed throughout of this section . Here X is a nonempty set and $\mathscr{A} \subset \mathscr{P}(X)$ is a σ-algebra. The elements of $\mathscr{A}$ are called "measurable sets." There is no a preconceived idea what measurable may mean, except that a subset $A \subset X$ is measurable if and only if $A \in \mathscr{A}$. Clearly, this is purely a set theoretic concept which has nothing to do with any particular measure.

Let $f:X \to R$ be a function with the range $f(X) \subset R$. The function f is said to be "$\mathscr{A}$-measurable" or just measurable, if

$$f^{-1}(\mathscr{B}) \subset \mathscr{A},$$

where $f^{-1}(\mathscr{B}) = \{f^{-1}(B);\ B \in \mathscr{B}\}$ and $\mathscr{B}$ is the Borel algebra of subsets of R. Clearly, just as the notion of measurable set, the notion of measurable function is purely a set theoretic concept.

In the above definition of measurability of a function $f(\cdot)$ nothing is said about its image $f(A)$ of a measurable set $A \in \mathscr{A}$. It not need be a Borel set. However, if $f(\cdot)$ is an injection, $f^{-1}(f(A)) = A$. The following is one of the most useful propositions in dealing with measurable functions.

Proposition 1

For an arbitrary map $f:X \to R$ and a subclass $\mathscr{C} \subset \mathscr{P}(R)$ we have that

$$\sigma\{f^{-1}(\mathscr{C})\} = f^{-1}(\sigma\{\mathscr{C}\}). \tag{1}$$

Proof:

Since $\mathscr{C} \subset \sigma\{\mathscr{C}\}$, it follows that $f^{-1}(\mathscr{C}) \subset f^{-1}(\sigma\{\mathscr{C}\})$; consequently

$$\sigma\{f^{-1}(\mathscr{C})\} \subseteq f^{-1}(\sigma\{\mathscr{C}\}).$$

To prove the converse inclusion define

$$\mathscr{H} = \{H \subset R;\ f^{-1}(H) \in \sigma\{f^{-1}(\mathscr{C})\}\}.$$

Clearly, $\mathscr{C} \subset \mathscr{H}$; if $A \in \mathscr{H}$ then $f^{-1}(A) \in \sigma\{f^{-1}(\mathscr{C})\}$. Hence,

$$\{f^{-1}(A)\}^{c} = f^{-1}(A^{c}) \in \sigma\{f^{-1}(\mathscr{C})\} \Rightarrow A^{c} \in \mathscr{H}.$$

In addition, for any sequence $\{A_n\}_1^{\infty}$ from $\mathscr{H}$

$$f^{-1}(A_k) \in \sigma\{f^{-1}(\mathscr{C})\} \text{ for each } k=1, 2, \ldots \text{ Consequently,}$$

$$\bigcup_1^{\infty} f^{-1}(A_k) = f^{-1}\left(\bigcup_1^{\infty} A_k\right) \in \sigma\{f^{-1}(\mathscr{C})\} \Rightarrow \bigcup_1^{\infty} A_k \in \mathscr{H}.$$

Thus, $\mathscr{H}$ is a σ-algebra which contains $\mathscr{C}$ and since $\sigma\{\mathscr{C}\}$ is the least σ-algebra which contains $\mathscr{C}$, it follows that $\sigma\{\mathscr{C}\} \subseteq \mathscr{H}$. Hence,

$$f^{-1}(\sigma\{\mathscr{C}\}) \subseteq \sigma\{f^{-1}(\mathscr{C})\},$$

which completes the prove of (1).

Corollary 1

If $\mathscr{C} \subset \mathscr{B}$ is such that $\sigma\{\mathscr{C}\} = \mathscr{B}$ then a map $f:X \to R$ is measurable if and only if

$$f^{-1}(\mathscr{C}) \subset \mathscr{A}.$$

For, if $f(\cdot)$ is measurable $f^{-1}(\mathscr{C}) \subset f^{-1}(\mathscr{B}) \subset \mathscr{A}$. On the other hand, if the condition holds, we have that

$$\sigma\{f^{-1}(\mathscr{C})\} = f^{-1}(\mathscr{B}) \subset \mathscr{A}.$$

Example 1

Recall that the class $\mathscr{S}^*$ of all intervals of the form $(-\infty, a], (\alpha. \beta], (b, \infty)$ is a semialgebra. Since $\sigma\{\mathscr{S}^*\} = \mathscr{B}$ (see problem 1 of section 22) any map $f:X \to R$ is measurable if and only if $f^{-1}(\mathscr{S}^*) \subset \mathscr{A}$.

A mapping $h:R \to R$ such that $h^{-1}(\mathscr{B}) \subset \mathscr{B}$ is called a "Borel function." If $h^{-1}(\mathscr{M}) \subset \mathscr{M}$, where $\mathscr{M} \subset \mathscr{L}(R)$ is the Lebesgue σ-algebra, the function h is said to be Lebesgue measurable. For example, if $L \in \mathscr{M}$ the indicator function $I_L(\cdot)$ is Lebesgue measurable.

Let $f:X \to R$ be $\mathscr{A}$-measurable and let $h:R \to R$ be a Borel function. We claim that the composition $h \circ f$ is $\mathscr{A}$-measurable. Indeed, since $h^{-1}(\mathscr{B}) \subset \mathscr{B}$

$$(h \circ f)^{-1}(\mathscr{B}) = f^{-1}(h^{-1}(\mathscr{B})) \subset f^{-1}(\mathscr{B}) \subset \mathscr{A}.$$

If, however, h is Lebesgue measurable $h \circ f$ is not necessarily $\mathscr{A}$-measurable. For instance, if $h(\cdot) = I_L(\cdot)$ and $L \in \mathscr{M}$, we have:

$$(h \circ f)^{-1}(\{1\}) = f^{-1}(h^{-1}(\{1\})) = f^{-1}(L) \notin \mathscr{A}$$

in general.

It may happen that given a Borel function h and a map $f:X \to R$, not necessarily measurable, that the composition $h \circ f$ is $\mathscr{A}$-measurable, as the next counterexample shows:

Example 2

The function $h(x) = |x|^\alpha$ where $\alpha \geq 0$, is continuous on R and hence Borel measurable. Let $B \in \mathscr{B}$ and assume that a subset $C \subset B$ is not measurable, then

$$f(\cdot) = I_C(\cdot) - I_{B-C}(\cdot)$$

is not $\mathscr{A}$-measurable. For instance, $f^{-1}(\{1\}) = C \notin \mathscr{B}$. However

$$h \circ f = |f|^\alpha = |I_C + I_{B-C}|^\alpha = I_B.$$

Sometimes, when we have to deal with the extended valued functions $f:X \to \overline{R} = [-\infty, \infty]$, the following proposition is useful:

Proposition 2

A map $f: X \to \overline{R}$ is measurable if

$$f^{-1}([-\infty, t]) \in \mathcal{A} \quad \text{for each } t \in \overline{R} .\tag{2}$$

Proof:

Assume that (2) holds then clearly

$$(f^{-1}([-\infty, t]))^c = f^{-1}((t, \infty]) \in \mathcal{A}.$$

Consequently,

$$f^{-1}(\{-\infty\}) = \bigcap_1^{\infty} f^{-1}([-\infty, -n]) \in \mathcal{A}$$

$$f^{-1}(\{+\infty\}) = \bigcap_1^{\infty} f^{-1}((n, +\infty]) \in \mathcal{A}.$$

In addition,

$$f^{-1}([t, +\infty]) = \bigcap_1^{\infty} f^{-1}((t - \frac{1}{n}, +\infty]) \in \mathcal{A}$$

$$f^{-1}(\{t\}) = f^{-1}([-\infty, t]) \cap f^{-1}([t, +\infty]) \in \mathcal{A}$$

and for any s<t

$$f^{-1}((s, t]) = f^{-1}([-\infty, t]) - f^{-1}([-\infty, s]) \in \mathcal{A},$$

which, taking into account proposition 1, implies that

$$f^{-1}(\mathcal{S}^*) \subset \mathcal{A} \Rightarrow f^{-1}(\mathcal{B}) \subset \mathcal{A}$$

because $\sigma\{\mathcal{S}^*\} = \mathcal{B}.$ ♥

Remark 1

Let $f \in \overline{R}^{\,X}$ be measurable. A number K>0 is said to be an "essential bound" for f if $|f| \leq K$ (a.e), or equivalently, if

$$\mu\{|f| \geq K\} = 0.$$

The essential "supremum" of f, denoted by ess sup $|f|$, is defined to be

$$\text{ess sup} |f| = \inf \{K; \ |f| \leq K \ (a.e)\}.$$

If a measurable function f does not have any essential bound, then it is understood that the ess sup $|f| = +\infty$

.

Problems and complements

1. Denote by $\mathscr{T}$ the collection of all open subsets of the real line R. A map h:R→R is continuous if and only if $h^{-1}(\mathscr{T})\subset\mathscr{T}$. Show that every continuous function is Borel measurable.

Hint:

We must show that $h^{-1}(\mathscr{B})\subset\mathscr{B}$. Since $h^{-1}(\mathscr{T})\subset\mathscr{T}$, we have that

$$\sigma\{h^{-1}(\mathscr{T})\}\subset\sigma\{\mathscr{T}\}=\mathscr{B}.$$

From this and proposition 1 the assertion follows.

2. If f:X→R is $\mathscr{A}$-measurable, show that the function

$$\varphi=\frac{|f|}{1+|f|}$$

is also $\mathscr{A}$-measurable.

Hint:

The function h:R→R defined by

$$h(t)=\frac{|t|}{1+|t|}$$

is continuous. Now, $\varphi=h\circ f$.

3. Let $\mathbf{C}(R)$ be the set of all continuous functions h: R→R. Show that the Borel algebra $\mathscr{B}$ is the least σ-algebra with respect to which every member of $\mathbf{C}(R)$ is measurable.

Hint:

The σ-algebra with respect to which all members of $\mathbf{C}(R)$ are measurable is obviously

$$\mathscr{K}=\sigma\{\bigcup_{h\in\mathbf{C}(R)}h^{-1}(\mathscr{B})\}.$$

Since every $h^{-1}(\mathscr{B})\subset\mathscr{B}\Rightarrow\mathscr{K}\subset\mathscr{B}$. On the other hand, there is $h_0\in\mathbf{C}(R)$ such that $h_0^{-1}(\mathscr{B})=\mathscr{B}$, which implies that $\mathscr{B}\subset\mathscr{K}$. Incidentally, the Baire σ-algebra $\mathscr{B}_0$ is the least σ-algebra with respect to which all bounded continuous functions are measurable. It is not difficult to show that $\mathscr{B}_0=\mathscr{B}$. This, however, is not necessarily true for the continuous maps h:X→R, where X is a metric space.

4. Let h_i:R→R, i=1, 2 be continuous functions such that $h_1=h_2$ on an everywhere dense subset D⊂R. Show that $h_1=h_2$ on R.

Hint:

Set $f=h_1-h_2$, then the set C={x; $|f(x)|=0$} is closed because

$$C=\bigcap_1^{\infty}\{x;\ |f(x)|\le\frac{1}{n}\}.$$

Now, $f(\cdot)=0$ on D and since $D\subset C$ and $\bar{D}=R$ it follows that $C=R$.

5. Recall that a Borel set is of "type F_σ" if it is the union of a finite or countable collection of closed sets. A set which is the intersection of a finite or countable collection of open sets is said to be of "type G_δ". Show that the set of discontinuity points of a (not necessarily measurable) function $f:R\to R$ is of type F_σ.

Hint:

Let $B_f(x, h)$ and $b_f(x, h)$ be the "least upper bound" and the "greatest lower bound" respectively, of the function $f(\cdot)$ in $(x-h, x+h)$, then as $h\downarrow 0$

$$B_f(x, h) - b_f(x, h) \downarrow \omega_f(x).$$

The function $\omega_f(x)\geq 0$ is called **the "oscillation" of the function $f(\cdot)$ at x**. It is clear that $f(\cdot)$ is continuous at x if and only if $\omega_f(x)=0$. Thus, the set of discontinuity points of $f(\cdot)$ is

$$\bigcup_1^\infty \{x;\ \omega_f(x) \geq \frac{1}{n}\}.$$

Consider the set $O_t = \{x;\ \omega_f(x) < t\}$. We claim that O_t is open for every $t\geq 0$. Indeed, choose an $x_0 \in O_t$; we must show that $(x_0 - \varepsilon, x_0 + \varepsilon)\subset O_t$ for some $\varepsilon>0$. To this end notice that $\omega_f(x_0) < t$ implies that for sufficiently small $h>0$

$$B_f(x_0, h) - b_f(x_0, h) < t.$$

Thus, if $y\in(x_0 - h, x_0 + h)$ we have the following inequalities

$$\omega_f(y) < B_f(y, h) - b_f(y, h) < t,$$

from which we deduce that $y\in O_t$ for every $y\in(x_0 - h, x_0 + h)$. Hence $(x_0 - h, x_0 + h)\subset O_t$, proving that O_t is an open set. Consequently,

$$\{x;\ \omega_f(x) \geq \frac{1}{n}\}$$

is closed, proving that the set of discontinuity points of any map $f:R\to R$ is F_σ-type and thus a Borel set.

6. Show that there is no function $f:R\to R$ which is continuous on the set $Q\subset R$ of rational numbers and discontinuous on Q^c.

Hint:

This follows from the fact that $\bar{Q}^c$ is not of type F_σ. For if it is we would have that

$$R = Q^c \cup Q = (\bigcup_1^\infty C_k) \cup (\bigcup_1^\infty \{r_k\}), \quad Q = \bigcup_1^\infty \{r_k\}$$

where each C_k is closed and contains no intervals (for if it does, each interval contains rational numbers). Consequently,

$$\theta = (\bigcap_1^\infty C_k^c) \cap (\bigcap_1^\infty \{r_k\}^c).$$

But this is not possible because each C_k^c and each $\{r_k\}^c$ is an open everywhere dense subset of R and the intersection of a countably many dense sets is a dense set.

To prove this recall that a set $D \subset R$ is dense if every interval $(\alpha, \beta) \subset R$ contains points of D. Let $\{D_k\}_1^\infty$ be a sequence of open dense sets, then an interval (α, β) contains points of D_1. There is a closed interval $I_1 \subset (\alpha, \beta) \cap D_1$. The interval I_1 contains some points of D_2 and consequently, there exists a close interval $I_2 \subset I_1 \cap D_2$, and so on. According to **Cantor's theorem for nested closed intervals** $\bigcap_1^\infty I_n = \{x_0\}$.

Clearly, $x_0 \in \bigcap_1^\infty D_k$ and $x_0 \in (\alpha, \beta)$. This proves that every interval $(\alpha, \beta) \subset R$ contains a point of the set $\bigcap_1^\infty D_n$, proving that it is dense in R. Consequently, the assumption that Q^c is type $\boldsymbol{F}_\sigma$ is false.

7. Construct a function $\Psi : R \to R$ which is discontinuous on Q and continuous on Q^c.

Hint:

Define Ψ by

$$\Psi(x) = \begin{cases} 0 & \text{if} \quad x \in Q^c \\ \dfrac{1}{q} & \text{if} \quad x = \dfrac{p}{q} \end{cases}$$

where p and q are integers with $q \neq 0$. The map $\Psi(\cdot)$ is discontinuous on Q because any two arbitrary "close" rationales p'/q' and p/q must have $q' \neq q$ so that $|1/q - 1/q'|$ cannot be made arbitrary small. The function Ψ is clearly measurable.

8. Let $(X, \mathscr{A}, \mu)$ be a complete measure space and let $f : X \to R$ be measurable. If a map $h : X \to R$ is equal (a.e) on X to $f(\cdot)$, show that h is also measurable.

Hint:

Denote by $A = \{f = h\}$; since A^c is a null set it is measurable, so that $A \in \mathscr{A}$. Next, since

$$h = h I_A + h I_{A^c} = f I_A + h \cdot I_{A^c}$$

we have for any open set $O \in \mathscr{T}$ that

$$h^{-1}(O) = (f^{-1}(O) \cap A) \cup (h^{-1}(O) \cap A^c) \in \mathscr{A}$$

because $h^{-1}(O) \cap A^c$ is subset of the null set A^c.

9. Let $(X, \mathscr{A}, \mu)$ be a measure space and let $f : X \to R^n$; it is clear that in such a case we have that

$$f = (f_1, f_2, \ldots, f_n)$$

where $f_i : X \to R$ for each $i = 1, 2, \ldots, n$. Show that $f(\cdot)$ is measurable if and only if each $f_i(\cdot)$ is measurable.

Hint:

(Necessity). If $f(\cdot)$ is measurable, then for any $O \in \mathcal{T} \subset \mathcal{B}(R)$

$$f_1^{-1}(O) = f^{-1}(O \times R \times \cdots \times R) \in \mathcal{A}$$

$$f_2^{-1}(O) = f^{-1}(R \times O \times \cdots \times R) \in \mathcal{A}$$

$$\cdots \cdots \cdots \cdots \cdots$$

$$f_n^{-1}(O) = f^{-1}(R \times R \times \cdots R \times O) \in \mathcal{A}$$

(Sufficiency). If each f_i is measurable then for any $\{O_k\}_1^n \subset \mathcal{T}$ we have that $f^{-1}(O_1 \times O_2 \times \cdots \times O_n) =$

$f^{-1}(O_1 \times R \times \cdots \times R) \cap \cdots \cap f^{-1}(R \times \cdots \times R \times O_n) = f_1^{-1}(O_1) \cap \cdots \cap f_n^{-1}(O_n) \in \mathcal{A}.$

10. Show that for any measurable function $f : X \to R$ we have that

$$\lim_{n \to \infty} \mu\{|f| \geq n\} = 0.$$

Hint:

Since f is real-valued, $\mu\{|f| = +\infty\} = 0$. On the other hand,

$$\mu\{|f| = +\infty\} = \mu\left(\bigcap_1^\infty \{|f| \geq n\}\right) = \lim_{n \to \infty} \mu\{|f| \geq n\}.$$

11. Let A be an atom of a measure space $(X, \mathcal{A}, \mu\}$. Show that any measurable map $f \in R^X$ is constant on A.

Hint:

Clearly, $f^{-1}(-\infty, t) \cap A = A$ if $A \subset f^{-1}(-\infty, t)$ and a null set otherwise. Define

$$t_0 = \inf\{t \in R;\ A \cap f^{-1}(-\infty, t) = A\}, \text{ then since } \{t_0\} = \bigcap_1^\infty \left(t_0 - \frac{1}{n}, t_0 + \frac{1}{n}\right)$$

$$f^{-1}(\{t_0\}) \cap A = \lim_{n \to \infty} f^{-1}\left(-\infty, t + \frac{1}{n}\right) \cap A,$$

which implies that $A \subset f^{-1}(\{t_0\}) \Rightarrow f(A) \subset \{t_0\}$. Note that t_0 depends on f.

24. SPACES OF MEASURABLE FUNCTIONS

Let $(X, \mathscr{A})$ be a measurable space. As usual $\overline{R}^X$ stands for the collection of all maps $f:X\to\overline{R}$. Denote by

$$L_0 = L_0(X, \mathscr{A}) \subset \overline{R}^X$$

the sub collection of all $\mathscr{A}$-measurable functions. Recall that a function $f:X\to R$ is $\mathscr{A}$- measurable if and only if $f^{-1}(\mathscr{B})\subset\mathscr{A}$, and that an extended real valued function $f:X\to\overline{R}$ is $\mathscr{A}$-measurable if and only if $f^{-1}(\overline{\mathscr{B}})\subset\mathscr{A}$, where the σ-algebra $\overline{\mathscr{B}}$ contains $\mathscr{B}$ and all subsets of $\overline{R}$ which are the union of an element of $\mathscr{B}$ and of a subset of $\{-\infty, \infty\}$. For instance, $[-\infty, t] = \{-\infty\}\cup(-\infty, t]\in\overline{\mathscr{B}}$, $\{-\infty\}$ and $\{+\infty\}\in\overline{\mathscr{B}}$. Clearly, every constant function on X is $\mathscr{A}$-measurable.

It is technically compelling to think of the collection L_0 as a "vector space" and of its elements as "vectors." This is justified by the fact that the class L_0 is closed under the basic algebraic and lattice operations. In other words, the usual algebraic and lattice operations on elements of L_0 produce elements of L_0.

Proposition 1

For any two $f_1, f_2 \in L_0$, $f_1 + f_2 \in L_0$.

Proof:

Let $Q\subset R$ be the set of rational numbers. For any $x\in X$ and $t\in R$ the relation $f_1(x) + f_2(x) < t$ holds if and only if there exists an $r\in Q$ such that

$$f_1(x) < r \text{ and } r < - f_2(x) + t.$$

Hence,

$$\{x; f_1(x) + f_2(x) < t\} = \bigcup_{r\in Q} \{x; f_1(x) < r\} \cap \{r < -f_2(x) + t\},$$

which implies that $\{f_1 + f_2 < t\}\in\mathscr{A}$. We also have that $\{f_1 + f_2 \leq t\}\in\mathscr{A}$ and that $\{f_1 + f_2 > t\}\in\mathscr{A}$ because

$$\{f_1 + f_2 \leq t\} = \bigcap_{n=1}^{\infty} \{f_1 + f_2 < t + \tfrac{1}{n}\}\in\mathscr{A}$$

$$\{f_1 + f_2 > t\} = \{f_1 + f_2 \leq t\}^c \in\mathscr{A}.$$

Consequently, $(f_1 + f_2)^{-1}(\mathscr{S}^*)\subset\mathscr{A}$ (see example 1 of section 23). This and proposition 1 of section 23 yield:

$$(f_1 + f_2)^{-1}(\mathscr{B})\subset\mathscr{A},$$

which proves our contention. ♥

Corollary 1

(i) For any $f \in L_0$ and $c \in R$, $f+c \in L_0$.

(ii) If $h:R \to R$ is continuous (and thus a Borel) function then the composition $h \circ f \in L_0$ if $f \in L_0$ (see the previous section). Thus,

$$f_1 \cdot f_2 = \frac{1}{4} \{ (f_1 + f_2)^2 - (f_1 - f_2)^2 \} \in L_0.$$

It is now easy to see that the space L_0 is a (vector) "lattice", since for any two $f_1, f_2 \in L_0$

$$f_1 \wedge f_2 = \min\{f_1, f_2\} = \frac{1}{2} \{ f_1 + f_2 - | f_1 - f_2 | \} \in L_0$$

$$f_1 \vee f_2 = \max\{f_1, f_2\} = \frac{1}{2} \{ f_1 + f_2 + | f_1 - f_2 | \} \in L_0.$$

Clearly, $f_1 \wedge f_2$ is the greatest lower bound of f_1 and f_2, while $f_1 \vee f_2$ is the least upper bound of f_1 and f_2.

For any $f \in L_0$ we defined $f^+ = f \vee 0$ (the "positive" part of f), $f^- = -(f \wedge 0)$ (the "negative" part of f). Since

$$f^+ = \frac{1}{2}(f + | f |), \qquad f^- = \frac{1}{2}(| f | - f),$$

it follows that $f^+, f^- \in L_0$.

Let $\{f_n\}_1^\infty$ be a sequence of L_0. Since $\overline{R} = [-\infty, \infty]$ is a "complete lattice" (i.e., every subset of $\overline{R}$ bounded from above has a least upper bound) we may define

$$\overline{M}(x) = \sup_{1 \leq n < \infty} f_n(x) \quad \text{and} \quad \underline{M}(x) = \inf_{1 \leq n < \infty} f_n(x).$$

Clearly, $\overline{M}$ and $\underline{M} \in L_0$ since for any $t \in R$

$$\{\overline{M} < t\} = \bigcap_{i=1}^\infty \{f_i < t\}, \qquad \{\underline{M} > t\} = \bigcap_{i=1}^\infty \{f_i > t\}.$$

Denote by

$$g_n(x) = \inf_{n \leq k < \infty} f_k(x), \qquad G_n(x) = \sup_{n \leq k < \infty} f_k(x).$$

Obviously, $\{g_n\}_1^\infty$ is an increasing and $\{G_n\}_1^\infty$ a decreasing sequence of measurable functions such that

$$g_n(x) \leq G_n(x) \qquad \text{for each } x \in X.$$

Suppose now that for every $x \in X$, the sequence $\{f_n(x)\}_1^\infty$ is bounded from below, then

$$g_n \uparrow f_*.$$

If $\{f_n(x)\}_1^\infty$ is bounded from above for every $x \in X$ then

$$G_n \downarrow f^*.$$

Both $f_*, f^* \in L_0$ because,

$$f^* = \lim_{n \to \infty} G_n = \lim_{n \to \infty} (\sup_{n \le k < \infty} f_k) = \inf_n (\sup_{n \le k < \infty} f_k)$$

$$f_* = \lim_{n \to \infty} g_n = \lim_{n \to \infty} (\inf_{n \le k < \infty} f_k) = \sup_n (\inf_{n \le k < \infty} f_k),$$

and

$$\{ f^* < t \} = \bigcup_{n=1}^{\infty} \{ \sup_{n \le k < \infty} f_k < t \} = \bigcup_{n=1}^{\infty} \bigcap_{k=n}^{\infty} \{ f_k < t \}$$

$$\{ f_* < t \} = \bigcap_{n=1}^{\infty} \bigcup_{k=n}^{\infty} \{ f_k < t \}.$$

From $g_n(x) \le G_n(x)$ for every $x \in X$ and $n=1, 2, \cdots$ it follows that $f_*(x) \le f^*(x)$ for each $x \in X$.

The measurable set

$$C = \{ x; f_*(x) = f^*(x) \}$$

is called the set of "**pointwise**" convergence of the sequence $\{ f_n \}_1^{\infty}$. In other words, for every $x \in C$

$$\lim_{n \to \infty} f_n(x) = f(x) \quad \text{where} \quad f(x) = f_*(x) = f^*(x).$$

If $C = X$ we say that the sequence $\{ f_n \}_1^{\infty}$ of real valued measurable functions converges pointwise on X to a real valued measurable function $f \in L_0$. Thus, we have just proved the following:

Proposition 2

Let $\{ f_n \}_1^{\infty}$ be a sequence from L_0. If $\{ f_n(x) \}_1^{\infty}$ is bounded and convergent for every $x \in X$, there exists $f \in L_0$ such that on X

$$f_n \to f \quad \text{pointwise.}$$

This result can be generalized as follows:

Proposition 3

If a sequence $\{ f_n \}_1^{\infty}$ from L_0 converges pointwise to a function $f: X \to \overline{R}$, then $f \in L_0$.

Proof:

Let us show that

$$\{ x; f(x) < t \} = \bigcup_{k=1}^{\infty} \bigcup_{n=1}^{\infty} \bigcap_{i=n}^{\infty} \{ x; f_i(x) < t - \frac{1}{k}). \tag{1}$$

If $x \in \{ f < t \} \Rightarrow f(x) < t$; thus, for some integer $k > 0$ $f(x) < t - \frac{1}{2k}$, and since $f_i(x) \to f(x)$ there exists $n > 0$ such that $f_i(x) < 1 - \frac{1}{k}$ for all $i \ge n$. Thus $x \in \{ f_i < t - \frac{1}{k} \}$ for all $i \ge n \Rightarrow x \in \bigcap_{i=n}^{\infty} \{ f_i < t - \frac{1}{k} \}$. This proves that

$$\{f<t\} \subset \bigcup_{n=1}^{\infty} \bigcap_{i=n}^{\infty} \{f_i < t - \frac{1}{k}\}.$$

Conversely, if x belongs to the right-hand side of (1), there is an integer k>0 such that

$$x \in \bigcup_{n=1}^{\infty} \bigcap_{i=n}^{\infty} \{f_i < t - \frac{1}{k}\}.$$

From this we conclude that $x \in \{f_i < t - \frac{1}{k}\}$ for all $i \geq n$, which implies that $f_i(x) < t - \frac{1}{k}$ for all $i \geq n$ and

consequently, that $f(x) < t - \frac{1}{k}$. Thus, $x \in \{f<t\}$, which completes the proof of the proposition.   ♥

Problems and complements

1. Show that: (i) if $c>0$ then $(c\,f)^+ = c\cdot f^+$; $(c\,f)^- = c\cdot f^-$; (ii) if $c<0$ then $(c\,f)^+ = -c\,f^-$; $(c\,f)^- = -c\,f^+$;

(iii) $(f_1+f_2)^+ \le f_1^+ + f_2^+$; $(f_1+f_2)^- \le f_1^- + f_2^{-1}$.

 Hint:

 Trivial.

2. For any $f\in L_0$ show that the following sets are measurable:

 (i) $\{f=+\infty\}$; (ii) $\{f=t\}$; (iii) $\{f=-\infty\}$.

 Hint:

 (i) $\{f=+\infty\} = \bigcap_{n=1}^{\infty} \{f\ge n\}$;   (iii) $\{f=-\infty\} = \bigcap_{n=1}^{\infty} \{f\le -n\}$;

 (ii) For every $t\in R$, $\{f=t\}=\{f\le t\} \cap \{f\ge t\}$.

3. Let $h:R\to R_+ =[0,\infty)$ be continuous function such that $h=0$ on a dense subset $D\subset R$. Show that $h\equiv 0$.

 Hint:

 Clearly,

$$C = \{x\in R;\ h(x)=0\} = \bigcap_{n=1}^{\infty} \{x\in R;\ h(x) \le \tfrac{1}{n}\}$$

is closed subset of R. Since $D\subset C$ we have that $\overline{D}\subseteq C$ ($\overline{D}$ is the closure of D). But $\overline{D}=R$.

4. A function $h\in L_0$ is said to be "simple" if its range $h(X)\subset R$ is a finite set. If $f\in L_0$ is non-negative show that there is a sequence of non-negative simple functions $h_n \in L_0$ such that

$$h_n \uparrow f \text{ pointwise on } X.$$

 Hint:

 Define

$$h_n(x) = \begin{cases} \dfrac{i-1}{2^n} & \text{if} & \dfrac{i-1}{2^n}\le f(x)<\dfrac{i}{2^n} & i=1,2,...,n\,2^n \\ n & \text{if} & f(x)\ge n \end{cases} ,$$

or in a more compact form

$$h_n(x) = \sum_{i=1}^{n\,2^n} \frac{i-1}{2^n} I_{B_{ni}}(x)+n\, I_{\{f\ge n\}}(x),$$

 where

$$B_{ni} = f^{-1}([\tfrac{i-1}{2^n},\tfrac{i}{2^n})).$$

 Now, since

$$B_{ni} = f^{-1}(\tfrac{i-1}{2^n},\tfrac{2i-1}{2^{n+1}})\cup[\tfrac{2i-1}{2^{n+1}},\tfrac{2i}{2^{n+1}})) = B_{n+1,\,2i-1} \cup B_{n+1,\,2i}$$

 we have that

$$\frac{i-1}{2^n}\, I_{B_{ni}}(x) \le \frac{i-1}{2^n}\, I_{B_{n+1,\,2i-1}}(x) + \frac{2i-1}{2^{n+1}}\, I_{B_{n+1,\,2i}}(x).$$

From this we conclude that for each $n=1, 2, \ldots$

$$h_n(x) \le h_{n+1}(x) \text{ for all } x \in X.$$

It follows from the definition of h_n that for every $x \in X$ for which $f(x) < \infty$

$$f(x) - h_n(x) \le \frac{1}{2^n}.$$

5. Let $\{f_n\}_1^\infty$ be a sequence from R^X converging pointwise on X to a function $f{:}X \to \overline{R}$. Show that for any open set $O \subset R$

$$f^{-1}(O) \subset \bigcup_{n=1}^\infty \bigcap_{k=n}^\infty f_k^{-1}(O).$$

Does the converse inclusion hold?

    Hint:

    If $x \in f^{-1}(O) \Rightarrow f(x) \in O$ and since $f_n(x) \to f(x)$ and O is open, there is for that x a positive integer n

such that $f_k(x) \in O$ for all $k \ge n$. Thus, $x \in f_k^{-1}(O)$ for each $k \ge n$ and hence $x \in \bigcap_{k=n}^\infty f_k^{-1}(O)$, proving the

inclusion. The converse inclusion does not hold since from $x \in \bigcap_{k=n}^\infty f_k^{-1}(O)$ it does not follow that

$x \in f^{-1}(O)$.

6. (Continuation) . Show that for each closed set $F \subset R$

$$f^{-1}(F) \supset \bigcup_{n=1}^\infty \bigcap_{k=n}^\infty f_k^{-1}(F).$$

    Hint:

    Let $x \in \bigcup_{n=1}^\infty \bigcap_{k=n}^\infty f_k^{-1}(F)$ then $x \in \bigcap_{k=n}^\infty f_k^{-1}(F)$ for some n. Consequently, $x \in f_k^{-1}(F)$ for all $k \ge n$

$\Rightarrow f_k(x) \in F$, $k \ge n$. Since F is closed, $\lim_{k \to \infty} f_k(x) = f(x) \in F$, proving that $x \in f^{-1}(F)$.

7. (Continuation). For an open set $O \subset R$ define

$$A_n = \{t \in R;\ \rho(t, O^c) \ge 1/n\},\ n=1, 2, \ldots,$$

where $\rho(\cdot, \cdot)$ was introduced in problem 3 of section 22. Show that

$$f^{-1}(A) = \bigcup_{i=1}^\infty \bigcup_{n=1}^\infty \bigcap_{k=n}^\infty f_k^{-1}(G_i)$$

where

$$A = \bigcup_{i=1}^\infty A_i \text{ and } \qquad G_i = \{t \in R;\ \rho(t, O^c) > 1/i\}.$$

Hint:

Notice that $A_k \uparrow A$ and that A is an open set such that

$$A = \bigcup_{i=1}^{\infty} G_i$$

(see problem 3 of section 22). Thus,

$$f^{-1}(A) = \bigcup_{i=1}^{\infty} f^{-1}(A_i) \supset \bigcup_{i=1}^{\infty} \bigcup_{n=1}^{\infty} \bigcap_{k=n}^{\infty} f_k^{-1}(G_i).$$

On the other hand,

$$f^{-1}(A) = \bigcup_{i=1}^{\infty} f^{-1}(G_i) \subset \bigcup_{i=1}^{\infty} \bigcup_{n=1}^{\infty} \bigcap_{k=n}^{\infty} f_k^{-1}(G_i)$$

proving our contention. Note, if $\{f_n\}_1^{\infty}$ is a sequence from L_0 it follows from this that $f \in L_0$.

8. Let $f:[a, b] \to R$ be continuous. Show that:

    (i) If $C \subset [a, b]$ is closed, its image $f(C)$ is also closed set.

    (ii) If the image of every $\bar{\lambda}$–null subset of $[a, b]$ is null set and $B \subset [a, b]$ is measurable, so is $f(B)$.

    Hint:

    (i) Let $\{y_n\}_1^{\infty}$ be a sequence from $f(C)$ and choose $x_n \in f^{-1}(\{y_n\})$. Notice that for any $i \neq j$ either

$f^{-1}(\{y_i\}) = f^{-1}(\{y_j\})$ or $f^{-1}(\{y_i\}) \cap f^{-1}(\{y_j\}) = \theta$. Thus, according to the axiom of choice there

exists $\{x_n\}_1^{\infty} \subset C$. Assume for the sake of simplicity that $x_n \to x_0$, then $x_0 \in C$, and consequently

$f(x_0) \in f(C)$. Hence,

$$\lim_{n \to \infty} y_n = \lim_{n \to \infty} f(x_n) = f(x_0).$$

    (ii) Let $B \subset [a, b]$ be measurable, then since $\bar{\lambda}$ is regular (see remark 2 of section 22), for every n=1, 2, $\cdots$ there exists a closed set $C_n \subset B$ such that

$$0 \leq \bar{\lambda}(B) - \bar{\lambda}(C_n) < \frac{1}{n}.$$

Then the set $D = \bigcup_1^{\infty} C_n$ is of type $\boldsymbol{F}_\sigma$ (see problem 4 of section 24). Thus,

$$\bar{\lambda}(B-D) \leq \bar{\lambda}(B - C_n) \leq \frac{1}{n} \to 0 \text{ as } n \to \infty.$$

Consequently, $B = D \cup N$ where $N \subset [a, b]$ is a null set. Sin $f(B) = f(D) \cup f(N)$, $f(D) = \bigcup_1^{\infty} f(C_n)$ and each

$f(C_n)$ is closed ,the $f(B)$ is of the type $\boldsymbol{F}_\sigma$ (see problem 5 of section 23).

25 ALMOST EVERYWHERE CONVERGENCE

Let $(X, \mathcal{A}, \mu)$ be a measure space. The almost everywhere concept, abbreviated (a.e), was introduced in section 17. Recall that a certain property is said to hold μ-(a.e) on X if it holds at every $x \in X$ outside of a μ-null set $N \subset X$. Since in this section μ is the only measure we deal with, the prefix μ- will be omitted.

Two measurable functions, say $f_1, f_2 \in \overline{R}^X$, are said to be "equivalent" if and only if $f_1 = f_2$ (a.e) on X. One often uses the symbol $\overset{a.e}{=}$ instead of $= $ (a. e). It is not difficult to see that $\overset{a.e}{=}$ is an equivalence relation in L_0. Similarly, $f_1 \leq f_2$ (a.e) on X if $\{f_1 > f_2\}$ is a null set. Clearly $\leq$ (a. e) is not an equivalence relation. It is material to observe that a measurable function can be equivalent to a function which is not measurable, if the measure space $(X, \mathcal{A}, \mu)$ is complete.

Proposition 1

If $\Psi : X \to \overline{R}$ is equivalent to a function $f : X \to \overline{R}$ which is measurable, then Ψ is also measurable (see problem 8 of section 23).

Proof:

By hypothesis $D = \{\Psi \neq f\}$ is a null set. The assertion follows from the fact that for any $t \in R$

$$\{\Psi < t\} = (\{\Psi < t\} \cap D^c) \cup (\{\Psi < t\} \cap D)$$
$$= (\{f < t\} \cap D^c) \cup (\{\Psi < t\} \cap D) \in \mathcal{A}$$

because $\{\Psi < t\} \cap D \subset D$ is a negligible set and thus measurable. ♥

A sequence of measurable functions $\{f_n\}_1^\infty$ from R^X is said to converge (a.e) on the set X if

$$\limsup f_n \overset{a.e}{=} \liminf f_n = f.$$

Then we write $f_n \to f$ (a.e). The limit f is clearly a measurable function uniquely determined up to "equivalence."

A sequence of measurable functions $\{f_n\}_1^\infty$ from R^X is said to be **"fundamental"** (a e) if for every $x \in X$ outside of a null set $N \subset X$, $\{f_n(x)\}_1^\infty$ is a Cauchy sequence.

Proposition 2 (Cauchy criterion)

In order that a sequence of measurable functions $\{f_n\}_1^\infty$ converges (a.e) to an (a.e) finite measurable function f, it is necessary and sufficient to be fundamental (a.e).

Proof:

If $f_n \to f$ (a.e) then from

$$|f_n - f_m| \leq |f_n - f| + |f_m - f|,$$

it follows that $|f_n - f_m| \to 0$ (a.e) if m, n $\to \infty$. On the other hand, if $\{f_n\}_1^\infty$ is fundamental (a.e) then for each x outside of a null set $N \subset X$ $\{f_n(x)\}_1^\infty$ is a Cauchy sequence and consequently converges to a finite limit. Define

$$f(x) = \lim_{n \to \infty} f_n(x) \qquad \text{if } x \in X\text{-}N$$

and $f(x)=0$ if $x \in N$. Clearly then $f_n \to f$ (a.e). ♥

Remark 1

If $\{f_n\}_1^\infty$ and $\{h_n\}_1^\infty$ are sequences of (a.e) finite measurable functions on X such that $f_n \to f$ (a.e) and $h_n \to h$ (a.e) then $f_n + h_n \to f + h$ (a. e) on X. Indeed, since by hypothesis $f_n(x) \to f(x)$ for all x outside of a null set N_0 and $h_n(x) \to h(x)$ for all x outside of a null set N_1,

$$f_n(x) + h_n(x) \to f(x) + h(x)$$

outside of the null set $N_0 \cup N_1$.

Proposition 3

In order that a sequence $\{f_n\}_1^\infty$ of real- valued measurable functions, on a finite measure space $(X, \mathscr{A}, \mu)$, converges (a.e) to a measurable function f:$X \to R$, it is necessary and sufficient that for all $\varepsilon > 0$

$$\lim_{n \to \infty} \mu \left(\bigcup_{k=n}^{\infty} \{|f_k - f| > \varepsilon\} \right) = 0 \tag{1}$$

Proof:

An interpretation of condition (1) is that all but finitely many measurable sets $\{|f_n - f| > \varepsilon\}$ have measure zero.

If $f_n \to f$ (a. e) then $h_n = |f_n - f| \to 0$ (a e). Consequently,

$$h^* = \limsup h_n = 0 \qquad (a\ e).$$

Thus, for any $\varepsilon > 0$

$$0 = \mu\{h^* > \varepsilon\} = \mu \left(\bigcap_{n=1}^{\infty} \bigcup_{k=n}^{\infty} \{h_k > \varepsilon\} \right) = \lim_{n \to \infty} \mu \left(\bigcup_{k=n}^{\infty} \{h_k > \varepsilon\} \right),$$

which proves necessity of (1). ♥

Conversely, if (1) holds then $\mu\{h^* > \varepsilon\} = 0$ for all $\varepsilon > 0$. But

$$\{h^* > 0\} = \bigcup_{n=1}^{\infty} \{h^* > \tfrac{1}{n}\} \Rightarrow \mu\{h^* > 0\} = 0.$$

Thus, $h^* = 0$ (a.e).

Corollary 1

It follows from (1) that

$$\sum_{k=1}^{\infty} \mu\{|f_k - f| > \varepsilon\} < \infty$$

is a sufficient condition for $\{f_n\}_1^{\infty}$ to converge (a.e) to f.

Recall that a sequence of functions $\{f_n\}_1^{\infty}$ from R^X (not necessarily measurable) is said to converge "uniformly" to a function $f{:}X{\to}R$ if, given an $\varepsilon{>}0$, there exists a positive integer $n_0 = n_0(\varepsilon)$ such that for all $x{\in}X$

$$|f_n(x) - f(x)| < \varepsilon \ \text{ if } n \geq n_0(\varepsilon).$$

Clearly, the uniform convergence implies the pointwise convergence. The next proposition due to Egorov connects (a.e) and the uniform convergence.

Proposition 4

Let $\{f_n\}_1^{\infty}$ from R^X be a sequence of measurable functions on a finite measure space $(X, \mathscr{A}, \mu)$.

If $f_n{\to}f$ (a.e) then, given any $\varepsilon{>}0$, there exists $H{\in}\mathscr{A}$ such that $\mu(H){<}\varepsilon$ and $f_n{\to}f$ uniformly on X-H.

Proof:

Without any loss of generality we may assume that $f_n{\to}f$ pointwise on X, then

$$\lim \sup |f_n - f| = 0 \text{ on } X.$$

Denote by

$$H_n^k = \{ \sup_{j \geq n} |f_j - f| > \frac{1}{k} \}, \ k = 1, 2, \ldots .$$

Clearly, for k fixed, $H_n^k \downarrow \theta$ as $n{\to}\infty$. Thus for every k there exists an integer $n_k > 0$ such that

$$\mu(H_{n_k}^k) < \varepsilon/2^k, \qquad (n_1 < n_2 < \cdots).$$

Set $H = \bigcup_{k=1}^{\infty} H_{n_k}^k$, then obviously $\mu(H){<}\varepsilon$; in addition

$$H^c = \bigcap_{k=1}^{\infty} (H_{n_k}^k)^c \subset (H_{n_k}^k)^c = \{ \sup_{j \geq n_k} |f_j - f| \leq \frac{1}{k} \}.$$

For given $\delta{>}0$ choose k so that $\delta{>}1/k$, then clearly

$$|f_j - f| \leq \delta \text{ on } H^c \text{ if } j \geq n_k,$$

which proves the proposition.   ♥

Egorov's theorem gives rise to the concept of almost **"uniform" convergence**. A sequence of real-valued measurable functions $\{f_n\}_1^{\infty}$ defined on a measure space $(X, \mathscr{A}, \mu)$ is said to converge "almost uniformly" (abbreviated (a.u)) to a measurable function $f{:}X{\to}R$ if, given $\varepsilon{>}0$, there exists $H{\in}\mathscr{A}$ with $\mu(H) {<}\varepsilon$ and $f_n{\to}f$ uniformly on X-H.

Remark 2.

Let $\{C_k\}_1^n$ be a sequence of closed disjoint subsets of R, then $C=\bigcup_1^n C_k$ is also a closed set. The

function

$$f= \sum_{i=1}^n \alpha_i I_{C_i}$$

is continuous on C. For, if $\{x_n\}_1^\infty$ is a sequence from C which converges to x_0 , then $x_0 \in C$. Clearly, x_0

belongs to one of the closed sets C_i. Assume $x_0 \in C_k$, then each C_i, $i \neq k$, contains at most finitely many

x_j. Thus, starting from an index m, all $x_i \in C_k$ if $i \geq m$. Consequently,

$$f(x_i) = f(x_0) \quad \text{for all } i \geq m.$$

Remark 3

If $\{f_n\}_1^\infty$ is a sequence of measurable functions such that $f_n \to f$ (a.e) and $f_n \geq 0$ (a.e) for each

n=1, 2, $\cdots$, then $f \geq 0$ (a.e). Indeed, set H={$f_n \nrightarrow f$} and E_n={$f_n<0$}, then $F=(\bigcup_{n=1}^\infty E_n)\cup H$ is a null set, and

on F^c

$$f= \lim_{n\to\infty} f_n \geq 0.$$

If g is measurable and $f_n \leq g$ (a.e) then $f \leq g$ (a.e). To see this notice that $g-f_n \geq 0$ (a.e) for all n=1, 2, $\cdots$.

Problems and complements

1. Show that every monotone sequence of measurable functions $\{f_n\}$ from R^X converges pointwise to a measurable function.

  Hint:

  Assume for the sake of definiteness that $f_1 \le f_2 \le \cdots$, then since

$$f^* = \inf_{n \ge 1} (\ \sup_{n \le k < \infty} f_k) = \inf_{n \ge 1} (\ \sup_{1 \le k < \infty} f_k) = \sup_{1 \le k < \infty} f_k$$

$$f_* = \sup_{n \ge 1} (\ \inf_{n \le k < \infty} f_k) = \sup_{1 \le n < \infty} f_n,$$

we have that $f_* = f^* = f$, which is the desired measurable function.

2. Let $\{f_n\}_1^\infty$ be a sequence of real-valued measurable functions on a measure space $(X, \mathcal{A}, \mu)$. Construct the set on which the sequence converges.

  Hint:

  Let $C \subset X$ be the required set, i.e., for every $x \in C$ the limit $\lim f_n(x)$ exists and is finite. Assume for the sake of simplicity that on C the sequence converges to zero, then

$$\lim \sup |f_n| = 0 \text{ on } C.$$

Hence,

$$C = \{f_n \to 0\} = \{\lim \sup |f_n| = 0\} = \bigcap_{k=1}^\infty \{\lim \sup |f_n| < \tfrac{1}{k}\}$$

$$= \bigcap_{k=1}^\infty \bigcup_{n=1}^\infty \bigcap_{j=n}^\infty \{|f_j| < \tfrac{1}{k}\}.$$

3. Let $\{f_n\}_1^\infty$ be a sequence of measurable functions on a complete measure space $(X, \mathcal{A}, \mu)$ such that $f_n \to f$ (a.e) on X. Show that f is measurable.

  Hint:

  Set $B = \{x;\ f_n(x) \to f(x)\}$, then $\mu(B^c) = 0$. Next, given $\alpha \in R$ we have (see (1) of section 24)

$$B \cap f^{-1}((\alpha, \infty)) = B \cap (\bigcup_{n=1}^\infty \bigcap_{k=n}^\infty f_k^{-1}((\alpha + \tfrac{1}{n}, \infty))).$$

From this we see that $B \cap f^{-1}((\alpha, \infty))$ is measurable. Thus,

$$f^{-1}((\alpha, \infty)) = (B \cap f^{-1}((\alpha, \infty)) \cup B^c \cap f^{-1}((\alpha, \infty)))$$

is measurable.

4. Let $\{f_n\}_1^\infty$ be a sequence of real-valued measurable functions such that $f_n \to f$ (a. e). Except that f is also measurable we can say little about the structure of the limit f. To shade more light on this problem denote by

$$\sigma\{f_n\} = f_n^{-1}\,(\mathscr{B})\ \text{ and by }\ \mathscr{F}_n = \sigma\{\bigcup_{k=n}^{\infty} f_k^{-1}\,(\mathscr{B})\} = \sigma\{f_n,\, f_{n+1},\, \cdots\},$$

where $\mathscr{B}$ is the σ-algebra of Borel subsets of R. Clearly, $\{\mathscr{F}_n\}_1^{\infty}$ is a sequence of decreasing σ-algebras, i.e. , $\mathscr{F}_1 \supset \mathscr{F}_2 \supset \cdots$. Show that

$$f^{-1}(\mathscr{B}) \subset \mathscr{F}_{\infty}\ \text{ where }\ \mathscr{F}_{\infty} = \bigcap_{n=1}^{\infty} \mathscr{F}_n.$$

Hint:

For any $t \in R$

$$\{f < t\} = \{\limsup f_n < t\} = \limsup\{f_n < t\}$$

$$= \bigcup_{n=1}^{\infty} \bigcap_{k=n}^{\infty} \{f_k < t\}.$$

Since

$$\bigcap_{k=n}^{\infty} \{f_k < t\} \in \mathscr{F}_n \text{ for every } n=1, 2, \cdots, \text{ we have that}$$

$$\bigcup_{n=1}^{\infty} \bigcap_{k=n}^{\infty} \{f_k < t\} = \lim_{n \to \infty} \bigcap_{k=n}^{\infty} \{f_k < t\} \in \bigcap_{n=1}^{\infty} \mathscr{F}_n = \mathscr{F}_{\infty}.$$

For instance, if $\mathscr{F}_{\infty}$ is the trivial σ-algebra, i.e., $\mathscr{F}_{\infty} = \{\theta, X\}$, the limit $f = \text{const}$.

5. Let $\{f_n\}_1^{\infty}$ be a sequence of real-valued measurable functions on a finite measure space $(X, \mathscr{A}, \mu)$. In order that the sequence converges (a.e) it suffices that there exists a sequence of positive numbers $\{\alpha_k\}_1^{\infty}$ satisfying

$$\sum_1^{\infty} \alpha_k < \infty \text{ and such that } \sum_{n=1}^{\infty} \mu\{|f_{n+1} - f_n| > \alpha_n\} < \infty . \qquad (*)$$

Hint:

Set $A_k = \{|f_{k+1} - f_k| > \alpha_k\}$, then since $\sum_{k=1}^{\infty} \mu(A_k) < \infty$ it follows that

$$\mu(\limsup A_n) = \mu(\bigcap_{n=1}^{\infty} \bigcup_{k=n}^{\infty} A_k) \le \sum_{k=n}^{\infty} \mu(A_k) \text{ for all } n=1, 2, \cdots .$$

From this and $(*)$ we conclude that

$$\mu(\limsup A_n) = 0.$$

Let $Z : X \to \{0, 1, 2, \cdots\}$ be a map defined by:

$$Z = \sum_{n=1}^{\infty} n\,I_{A_n \cap (\bigcup_{k=n+1}^{\infty} A_k)^c} \quad \text{and } Z = 0 \text{ on } (\bigcup_{k=1}^{\infty} A_k)^c .$$

Clearly, $\{Z=i\}\cap\{Z=j\}=\emptyset$ if $i\neq j$. Thus, any x outside of the null set $(\limsup A_n)$ belongs to one of the sets $\{Z=k\}$, $k=1, 2, \cdots$. If $x\in\{Z=n\}$ then clearly for this x

$$|f_{j+1}(x) - f_j(x)| < \alpha_j \quad \text{for all } j = n, n+1, \cdots .$$

This proves the existence of the limit

$$f(x) = \lim_{n\to\infty} f_n(x) = f_1(x) + \lim_{n\to\infty} \sum_{k=1}^{n} (f_{k+1}(x) - f_k(x)).$$

In other words, for every $x \notin \limsup A_n$ we have a function

$$f(x) = f_1(x) + \sum_{k=1}^{\infty} (f_{k+1}(x) - f_k(x)).$$

6. (Lusin's theorem) Let $f: [a, b]\to R$ be a Borel function. Given $\varepsilon>0$ there exists a closed set $F\subset[a, b]$ such that $\bar\lambda([a, b] - F) < \varepsilon$ and the restriction $\hat f = f\,|\,F$ is continuous on F. Here $\bar\lambda$ is the Lebesgue measure on R.

Hint:

First we shall prove the contention for f when f is a simple function , i.e.,

$$f = \sum_{k=1}^{n} \alpha_k I_{E_k}, \quad E_i\cap E_j = \emptyset \text{ if } i\neq j, \quad \bigcup_{k=1}^{n} E_k = [a, b].$$

Since the Lebesgue measure $\bar\lambda$ is regular, given $\varepsilon>0$ there exists for each $k=1, 2, \cdots, n$, a closed set $F_k\subset E_k$ such that

$$\bar\lambda(E_k - F_k) < \varepsilon/n \quad k = 1, 2, \cdots, n$$

(see definition 1 of section 22). Then the simple function

$$\hat f = \sum_{k=1}^{n} \alpha_k I_{F_k}$$

is continuous on the closed set $F=\bigcup_{1}^{n} F_k$ (see remark 2) and $\bar\lambda(F^c) = \bar\lambda([a, b]) - \sum_{1}^{n}\bar\lambda(F_k) = \sum_{1}^{n}\bar\lambda(E_k - F_k) <$

ε. Clearly, $\hat f = f\,|\,F$ which proves the contention when f is a simple Borel function.

Assume now that $f\geq 0$, then according to problem 4 of the previous section there exists a sequence of simple functions $\{f_k\}_1^{\infty}$ such that $f_n\uparrow f$ pointwise on [a, b]. According to what we have just proved, given an $\varepsilon>0$ there exists a closed set $C_n\subset [a, b]$ such that $\bar\lambda(C_n^c) < \varepsilon/2^n$ and such that the restriction

$$\hat f_n = f_n\,|\,C_n \quad \text{is continuous on } C_n.$$

The set $H=\bigcap_1^{\infty} C_k$ is clearly closed and $\bar\lambda(H^c) = \bar\lambda(\bigcup_1^{\infty} C_k^c) \leq \sum_1^{\infty}\bar\lambda(C_k^c) < \varepsilon$. Since $\hat f_n\uparrow\hat f$ uniformly on H,

$\hat f$ is continuous on H. An interpretation of this result due to Borel is that any bounded measurable function on [a, b] is "almost continuous."

7. Show that on a finite measure space (a.u) convergence implies (a.e) convergence.

Hint:

Suppose that a sequence $\{f_n\}_1^\infty$ from R^X of measurable functions converges (a.u) to a measurable function $f:X\to R$. Then for each $k=1, 2, \cdots$ there is a measurable set H_k such that $\mu(H_k)<1/k$ and $f_n\to f$ uniformly on $X-H_k$, which implies that $f_n\to f$ on $\bigcup_1^\infty (X-H_k)$. Then, since $H=\bigcap_1^\infty H_k$ is obviously a null set, and $f_n\to f$ on $X-H$, the assertion holds.

26 CONVERGENCE IN MEASURE

In this section we discuss a convergence concept that is intimately related to the structure of a measure space $(X, \mathscr{A}, \mu)$.

A sequence $\{f_n\}_1^\infty$ from $\mathbf{L}_0(X, \mathscr{A}, \mu)$ (of real-valued measurable functions) is said **to converge in "measure"** to a measurable function $f: X \to R$ if for any $\varepsilon > 0$

$$\lim_{n \to \infty} \mu\{|f_n - f| > \varepsilon\} = 0. \tag{1}$$

Than we write $f_n \xrightarrow{\mu} f$. If

$$\mu\{|f_m - f_n| > \varepsilon\} \to 0 \quad \text{as } m, n \to \infty,$$

we say that the sequence $\{f_n\}_1^\infty$ is **"fundamental"(or Cauchy) in measure**.

Here we study some properties of this type of convergence and its relation to other modes of convergence. For example, from the inclusion

$$\{|f_m - f_n| > \varepsilon\} \subset \{|f_m - f| > \tfrac{\varepsilon}{2}\} \cup \{|f_n - f| > \tfrac{\varepsilon}{2}\},$$

we deduce that any sequence which converges in measure is fundamental in measure.

For an arbitrary measurable function $f: X \to R$ and $\varepsilon > 0$ we have:

$$\mu\left(\bigcup_{k=n}^{\infty} \{|f_k - f| > \varepsilon\}\right) \geq \mu\{|f_n - f| > \varepsilon\}.$$

From this and proposition 3 of section 25 it follows, if μ is a finite measure, that

$$f_n \xrightarrow{a.e} f \qquad \text{implies} \qquad f_n \xrightarrow{\mu} f. \tag{2}$$

Thus, if the measure μ is finite, (a.e) convergence implies the convergence in measure. If the measure μ is not finite this is not true in general as the following counter example shows:

Example 1

Let $\{f_n\}_1^\infty$ be a sequence from $(R, \mathscr{B}, \bar{\lambda})$ defined by

$$f_n(t) = I_{[n, n+1]}(t).$$

Clearly, $\lim\limits_{n \to \infty} f_n(t) = 0$ for any $t \in R$ fixed, so that $f_n \to 0$ pointwise. On the other hand, for any $\varepsilon \in (0, 1)$

$$\bar{\lambda}\{|f_n - 0| > \varepsilon\} = \bar{\lambda}([n, n+1]) = 1$$

so that f_n does not converge in measure to 0.

The limit function f in (1) is unique up to equivalence. In other words, if f_0 is another measurable function such that $f_n \xrightarrow{\mu} f_0$, then $f \stackrel{a.e}{=} f_0$. Indeed, for any $\varepsilon > 0$

$$\mu\{|f - f_0| > \varepsilon\} \leq \mu\{|f_n - f| > \tfrac{\varepsilon}{2}\} + \mu\{|f_n - f_0| > \tfrac{\varepsilon}{2}\} \to 0$$

as $n \to \infty$. Consequently,

$$\mu\{f_0 \neq f\} = \mu\{|f_0 - f| > 0\} = \mu\left(\bigcup_{k=1}^{\infty} \{|f_0 - f| > \frac{1}{k}\}\right) = 0,$$

since every member of the union is a null set. In addition,

$$\mu\{|f| = +\infty\} = \mu\{|f_n - f| = +\infty\} \leq \mu\{|f_n - f| > \varepsilon\} \to 0$$

as $n \to \infty$.

Remark 1

Recall that a sequence $\{f_n\}_1^{\infty}$ of functions from R^X (not necessarily measurable) is said to be **"uniformly fundamental"** if, given any $\varepsilon > 0$, there exists an integer $k > 0$ such that

$$|f_m - f_n| < \varepsilon \text{ everywhere on } X \text{ if } m, n \geq k.$$

If for every $\varepsilon > 0$ there exists $H \in \mathscr{A}$ such that $\mu(H) < \varepsilon$ and $\{f_n\}_1^{\infty}$ is uniformly fundamental on X-H, we say that the sequence $\{f_n\}_1^{\infty}$ is **"almost uniformly"** (a.u) fundamental. If $f_n \to f$ (a.u) (i.e., given any $\varepsilon > 0$, there exists a set $H \in \mathscr{A}$ such that $\mu(H) \leq \varepsilon$ and $f_n \to f$ uniformly on X-H), then from the inequality

$$|f_m - f_n| \leq |f_m - f| + |f_n - f|$$

it follows that the sequence $\{f_n\}_1^{\infty}$ is (a.u) fundamental.

Proposition 1

If $\{f_n\}_1^{\infty}$ from R^X is (a.u) fundamental, there exists a function $f : X \to R$ such that $f_n \to f$ (a.u).

Proof:

By hypothesis for each integer $m > 0$ there exists a set $H_m \in \mathscr{A}$ such that $\mu(H_m) < 1/m$ and $\{f_n\}_1^{\infty}$ is uniformly fundamental on X-H_m. Clearly, $H = \bigcap_1^{\infty} H_m$ is a null set ($\mu(H) \leq \mu(H_m) < 1/m$ for every $m = 1, 2,$

$\cdots$). Moreover, for each $x \in H^c = \bigcup_1^{\infty}(X$-$H_m)$, $\{f_n(x)\}_1^{\infty}$ is a Cauchy sequence. Thus, the $\lim_{n \to \infty} f_n(x)$ exists. Let $f : X \to R$ be a function defined as follows:

$$f = \lim_{n \to \infty} f_n \text{ on } X\text{-}H \text{ and } f = 0 \text{ on } H.$$

It is clear from this definition that $f_n \to f$ uniformly on X-H_m (for each $m = 1, 2 \cdots$), which proves the proposition. ♥

The next proposition shows that (a.u) convergence is a "stronger" mode of convergence than (a.e) convergence .

Proposition 2

If $\{f_n\}_1^{\infty}$ is a sequence of measurable function which converges (a.u) to a function $f : X \to R$, then

$$f_n \xrightarrow{a.e} f.$$

Proof:

If $x \in \bigcup_{1}^{\infty}(X\text{-}H_m)$, then $x \in (X\text{-}H_m)$ for some $m \geq 1$ and consequently $f_n(x) \to f(x)$. Thus $f_n(x) \to f(x)$

for every $x \notin H$. ♥

Proposition 3

If $\{f_n\}_1^{\infty}$ is a sequence of measurable functions such that $f_n \to f$ (a.u) then $f_n \xrightarrow{\mu} f$.

Proof:

It is clear from proposition 2 that the limit f is measurable. Given $\varepsilon > 0$ there exists an index υ_0 such

that on the set $X\text{-}H_m$, $|f_n\text{-}f| < \varepsilon$ if $n \geq \upsilon_0$. Thus, $|f_n\text{-}f| \geq \varepsilon$ on H_m if $n \geq \upsilon_0$. Hence, $\{|f_n\text{-}f| \geq \varepsilon\} \subset H_m$ if

$n \geq \upsilon_0$ and consequently $\mu\{|f_n\text{-}f| \geq \varepsilon\} \leq \mu(H_m) \leq 1/m$. Therefore,

$$\limsup_{n} \mu\{|f_n - f| \geq \varepsilon\} \leq 1/m \quad \text{for all} \quad m=1, 2, \cdots,$$

which implies that

$$\lim_{n \to \infty} \mu\{|f_n\text{-}f| \geq \varepsilon\} = 0.$$

This proves the proposition. ♥

As we have established, if a sequence $\{f_n\}_1^{\infty}$ of real- valued measurable functions converges in

measure to a measurable function, then it is fundamental in measure. The converse is also true but it is much

deeper. To prove it we need the next proposition which is of an independent interest.

Proposition 4

If $\{f_n\}_1^{\infty}$ is a sequence of measurable functions which is fundamental in measure, there exists a

subsequence $\{f_{n_k}\}_1^{\infty}$ of $\{f_n\}_1^{\infty}$ which is fundamental (a.u).

Proof:

By hypothesis, for any integer $k \geq 1$, $\mu\{|f_m - f_n| > \dfrac{1}{2^k}\} \to 0$ as $m, n \to \infty$. Thus, there exists an index

n_k such that

$$\mu\{|f_m - f_n| > \frac{1}{2^k}\} < \frac{1}{k^2} \quad \text{if } m, n \geq n_k.$$

Now consider a sequence $\{f_{n_k}\}_1^{\infty}$, where $1 \leq n_1 < n_2 < \cdots$, and set

$$E_k = \{|f_{n_k} - f_{n_{k+1}}| > \frac{1}{2^k}\},$$

then if $k \leq i < j$ we have on the set $(\bigcup_{\upsilon=k}^{\infty} E_\upsilon)^c$ that

$$|f_{n_i} - f_{n_j}| \leq |f_{n_i} - f_{n_{i+1}}| + |f_{n_{i+1}} - f_{n_j}| \leq |f_{n_i} - f_{n_{i+1}}| + |f_{n_{i+1}} - f_{n_{i+2}}| + |f_{n_{i+2}} - f_{n_j}| \leq$$

$$\cdots \leq \sum_{\upsilon=i}^{\infty} |f_{n_\upsilon} - f_{n_{\upsilon+1}}| \leq \sum_{\upsilon=i}^{\infty} \frac{1}{2^\upsilon} = \frac{1}{2^{i-1}} .$$

Thus, $\{f_{n_i}\}_1^\infty$ is uniformly fundamental on $(\bigcup_{\upsilon=k}^{\infty} E_\upsilon)^c$. Since

$$\mu\left(\bigcup_{\upsilon=k}^{\infty} E_\upsilon\right) \leq \sum_{\upsilon=k}^{\infty} \frac{1}{\upsilon^2}$$

the assertion is proved. ♥

Proposition 5

If $\{f_n\}_1^\infty$ is a sequence of real-valued measurable functions fundamental in measure, then there exists a measurable function $f{:}X{\to}R$ such that

$$f_n \overset{\mu}{\to} f.$$

Proof:

According to the previous proposition there exists a subsequence $\{f_{n_k}\}_1^\infty$ of $\{f_n\}_1^\infty$ and such that

$$f_{n_k} \to f \quad (a.\, u),$$

and consequently (a.e) (see proposition 2) From this it follows that f is measurable. Now for any $\varepsilon{>}0$ we have:

$$\{|f_m - f| > \varepsilon\} \subset \{|f_m - f_{n_k}| > \tfrac{\varepsilon}{2}\} \cup \{|f_{n_k} - f| > \tfrac{\varepsilon}{2}\}.$$

Since $\{f_m\}_1^\infty$ is fundamental in measure, $\mu\{|f_m - f_{n_k}| > \tfrac{\varepsilon}{2}\}\to 0$ as $m, n_k \to \infty$. On the other hand, $\mu\{|f_{n_k} - f| > \varepsilon\}\to 0$ as $n_k \to \infty$ due to proposition 3. This completes the proof of the proposition. ♥

Proposition 6

Let $\{f_n\}_1^\infty$ and $f{:}X{\to}R$ be real-valued measurable functions on a finite measure space $(X, \mathcal{A}, \mu)$, then $f_n \overset{\mu}{\to} f$ if and only if every subsequence of $\{f_n\}_1^\infty$ contains a subsequence which converges (a.e) to f.

Proof:

If $f_n \overset{\mu}{\to} f$, then every subsequence $\{f_{n_k}\}_1^\infty$ of $\{f_n\}_1^\infty$ converges in measure to f. Then according to proposition 4, there exists a subsequence which converges to f (a.u) and consequently, due to proposition 2, (a.e). To prove sufficiency, assume that $f_n \overset{\mu}{\to} f$ is not true, then

$$\limsup \mu\{|f_n - f| > \varepsilon\} = \delta > 0.$$

There exists a subsequence $\{f_{n_k}\}_1^\infty$ such that

$$\mu\{|f_{n_k} - f| > \varepsilon\} \to \delta$$

so neither $\{f_{n_k}\}_1^\infty$ nor any of its subsequences converges in measure to f. Consequently, they do not

converge (a. e) to f either, which proves the proposition.   ♥

Problems and complements

1. If $\{f_n\}_1^\infty$ and $\{h_n\}_1^\infty$ are sequences of real-valued measurable functions on $(X, \mathcal{A}, \mu)$ such that $f_n \xrightarrow{\mu} f$ and $h_n \xrightarrow{\mu} h$, show that

$$f_n + h_n \xrightarrow{\mu} f + h.$$

Hint:

$$\mu\{|(f_n + h_n) - (f + h)| > \varepsilon\} \leq \mu\{|f_n - f| + |h_n - h| > \varepsilon\}$$

$$\leq \mu\{|f_n - f| > \tfrac{\varepsilon}{2}\} + \mu\{|h_n - h| > \tfrac{\varepsilon}{2}\} \to 0 \text{ as } n \to \infty.$$

2. If $f_n \xrightarrow{\mu} f$ show that $|f_n| \xrightarrow{\mu} |f|$.

Hint:

$$||f_n| - |f|| = ||f_n - f + f| - |f|| \leq |f_n - f|.$$

3. Let $\{f_n\}_1^\infty$ be a sequence of real-valued measurable functions on a complete measure space $(X, \mathcal{A}, \mu)$. If $f_n \xrightarrow{\mu} f$ and $f \overset{a.e}{=} h$, show that $f_n \xrightarrow{\mu} h$.

Hint:

$$\mu\{|f_n - h| > \varepsilon\} \leq \mu\{|f_n - f| > \tfrac{\varepsilon}{2}\} + \mu\{|f - h| > \tfrac{\varepsilon}{2}\}.$$

4. If $f_n \xrightarrow{\mu} f$ and $f_n \geq 0$ (a. e), show that $f \geq 0$ (a. e).

Hint:

Given $\varepsilon > 0$ we have for every $n = 1, 2, \cdots$:

$$\{f < -\varepsilon\} = \{(f - f_n) + f_n < -\varepsilon\} \subset \{f - f_n < -\varepsilon\} \subset \{|f_n - f| > \varepsilon\}.$$

Thus, $\mu\{f < -\varepsilon\} = 0$; since

$$\{f < 0\} = \bigcup_{n=1}^\infty \{f < -\tfrac{1}{n}\} \Rightarrow \mu\{f < 0\} = 0.$$

5. Let $\{f_n\}_1^\infty$ and h be measurable functions on a finite measure space $(X, \mathcal{A}, \mu)$. If $h : X \to R$ and $f_n \xrightarrow{\mu} f$, show that $h f_n \xrightarrow{\mu} h f$.

Hint:

For any $\varepsilon > 0$ and integer $k > 0$

$$\{|h f_n - h f| > \varepsilon\} = \{|h| |f_n - f| > \varepsilon\} = \{|h| |f_n - f| > \varepsilon, \ |h| < k\}$$

$$\cup \{|h| |f_n - f| > \varepsilon, \ |h| \geq k\} \subset \{|f_n - f| > \tfrac{\varepsilon}{k}\} \cup \{|h| \geq k\}.$$

6. Let $f_n \xrightarrow{\mu} f$ and assume that $f_n \leq h$ (a. e); show that $f \leq h$ (a. e).

Hint:

Since $h - f_n \geq 0$ (a. e) and $h - f_n \xrightarrow{\mu} h - f \Rightarrow h - f \geq 0$ (a. e).

7. If $f_n \xrightarrow{\mu} f$ and $h_n \xrightarrow{\mu} h$ show that $f_n \cdot h_n \xrightarrow{\mu} f \cdot h$.

Hint:

Assume first that $f = h = 0$, then

$$\{|f_n \cdot h_n| > \varepsilon^2\} = \{|f_n \cdot h_n| > \varepsilon^2, |h_n| > |f_n|\} \cup \{|f_n \cdot h_n| > \varepsilon^2, |h_n| \leq |f_n|\}$$

$$\subset \{|h_n|^2 > \varepsilon^2\} \cup \{|f_n|^2 > \varepsilon^2\} \Rightarrow h_n \cdot f_n \xrightarrow{\mu} 0.$$

Consequently, $(f_n - f)(h_n - h) \xrightarrow{\mu} 0$. But

$$f_n \cdot h_n - f \cdot h = (f_n - f)(h_n - h) + h \cdot (f_n - f) + f \cdot (h_n - h) \xrightarrow{\mu} 0.$$

As an example, assume that $f_n \xrightarrow{\mu} f$, then $f_n^2 \xrightarrow{\mu} f^2$.

8. Let $\{f_n\}_1^\infty$ be a sequence measurable function such that $f_n \geq 0$ for every $n = 1, 2, \cdots$. If $f_n \xrightarrow{\mu} f$, show that

$$f_n^\alpha \xrightarrow{\mu} f^\alpha \qquad \text{for every } \alpha \in (0, 1].$$

Hint:

Consider the map h on $[0, \infty)$ defined by

$$h(u) = 1 + u^\alpha - (1 + u)^\alpha.$$

It is easy to sea that $h(\cdot)$ is nondecreasing with $h(0) = 0$. Thus, $(1 + u)^\alpha \leq 1 + u^\alpha$. If we put $u = y/x$ where $x > 0$, $y \geq 0$, we obtain (see also the inequality (5) of section 35)

$$(x + y)^\alpha \leq x^\alpha + y^\alpha.$$

Consequently,

$$|x - y|^\alpha = |(x - z) + (z - y)|^\alpha \leq |x - z|^\alpha + |y - z|^\alpha.$$

From this we deduce that

$$|x - y|^\alpha - |z - y|^\alpha \leq |x - z|^\alpha$$

$$|z - y|^\alpha - |x - y|^\alpha \geq - |x - z|^\alpha$$

which implies that

$$\left| |x - y|^\alpha - |z - y|^\alpha \right| \leq |x - z|^\alpha.$$

From this for $y = 0$, it follows that

$$\left| |z|^\alpha - |x|^\alpha \right| \leq |x - z|^\alpha.$$

Hence,

$$| f_n^\alpha - f^\alpha | \le | f_n - f |^\alpha$$

so that for any $\varepsilon > 0$

$$\mu\{ | f_n^\alpha - f^\alpha | > \varepsilon \} \le \mu\{ | f_n - f |^\alpha > \varepsilon \} = \mu\{ | f_n - f | > \varepsilon^{1/\alpha} \}.$$

9. Let $\{ f_n \}_1^\infty$ be a sequence of real-valued measurable functions. If $f : X \to R$ is a map such that for every $\varepsilon > 0$

$$\mu^*\{ | f_n - f | > \varepsilon \} \to 0 \text{ as } n \to \infty$$

prove that f is also measurable.

Hint:

Here of course μ^* is the outer measure induced by μ. There exists an integer k_n such that

$$\mu^*\{ | f_j - f | > \frac{1}{n} \} \le \frac{1}{2^n} \quad \text{if } j \ge k_n.$$

Let $1 < n_1 < n_2 < \cdots$ and denote by

$$E_n = \{ | f_{k_n} - f | \ge \frac{1}{n} \} \text{ and set } E = \limsup E_n,$$

then for any $n = 1, 2, \cdots$

$$\mu^*(E) \le \mu^*(\bigcup_{j=n}^\infty E_j) \le \sum_{j=n}^\infty \mu^*(E_j) \le \sum_{j=n}^\infty \frac{1}{2^j} = \frac{1}{2^{n-1}} .$$

This implies that $\mu^*(E) = 0$; consequently, if $x \in E^c$ then $x \in \bigcup_{j=n}^\infty E_j$ for some n, implying that for every $m \ge n$

$$| f_{m_k} (x) - f(x) | \le \frac{1}{m} .$$

Consequently, $f_{n_k} \to f$ pointwise on E^c so that $f_{n_k} \overset{a.e}{\to} f$. Then according to proposition 2 of section 25, f must be measurable.

Chapter III
INTEGRATION

27 THE LEBESGUE INTEGRAL OF A BOUNDED FUNCTION

Let $f:[a, b] \to R$, where $[a, b] \subset R$. According to the classical definition, the integral

$$\int_a^b f(x)\,dx$$

is the limit, when it exists, of the "**Riemann sums**"

$$\sum_{i=0}^{n} (x_{i+1} - x_i)\, f(\xi_i), \qquad a = x_1 < x_2 < \cdots < x_n = b, \ \ \xi_i \in [x_i, x_{i+1}].$$

The existence of the limit, when $f(\cdot)$ is continuous, was proved by Cauchy. Riemann discovered that the limit exists if the function $f(\cdot)$ is bounded and if the set $D \subset [a, b]$ of its discontinuity points is a null set. Recall that the set D is measurable (see problem 5 of section 23).

The class of Riemann integrable functions is too narrow. In addition, it is incomplete as the next example shows:

Example 1

Let $\{r_k\}_1^{\infty} \subset [a, b]$ be an enumeration of the rational numbers in this interval, and let $\{f_n\}_1^{\infty}$ be a sequence of functions defined on $[a, b]$ by

$$f_n(x) = \begin{cases} 1 & \text{if } x = r_1, r_2, \ldots, r_n \\ 0 & \text{otherwise} \end{cases}.$$

Clearly, each f_n is Riemann integrable and

$$\int_a^b f_n(x)\,dx = 0, \qquad n = 1, 2, \ldots.$$

On the other hand, $f_n \uparrow f$ pointwise, where $f(\cdot)$ is the "Dirishlet" function, i.e.,

$$f(x) = \begin{cases} 1 & \text{if } x \in [a, b] \text{ is rational} \\ 0 & \text{otherwise} \end{cases}.$$

But the function $f(\cdot)$ is not Riemann integrable. This shows that the class of Riemann integrable functions is "incomplete."

Most of the drawbacks of the classical integration theory disappear in modern theory of measure and integration developed in the last century on the foundations laid down by Henry Lebesgue. The original work was the Lebesgue thesis "Intégral, longueur, aire" published in 1902. In this paper, Lebesgue introduced the concept of the Lebesgue measure, measurable function and the concept of the Lebesgue integral. This work represents a single most important example, which has given rise to the whole new theory.

The Lebesgue method for construction of the integral on R is so general that it can be carried over directly to measurable functions defined on an abstract measure space. For all these reasons we shall use this section to outline, in an informal fashion, the basic features of the Lebesgue integral in the way that Lebesgue did it more than one hundred and ten years ago.

It is perhaps not totally out of place to point out that contemporary writings on any mathematical subject tend to leave the subject's classical origin somewhat obscure. In a small way this is also a remainder how this odyssey, that started with Lebesgue, had began.

Let $(R, \mathcal{A}, \bar{\lambda})$ be the usual (Lebesgue) measure space (i.e., $R=(-\infty, \infty)$, $\mathcal{A}$ is the Lebesgue σ-algebra of subsets of R and $\bar{\lambda}$ is the Lebesgue measure on R). Let $f(\cdot)$ be a bounded measurable function defined on a set $B \in \mathcal{A}$ satisfying $\bar{\lambda}(B) < \infty$. Assume also that $f(B) \subset [m, M)$ where $-\infty < m < M < \infty$. To construct the Lebesgue integral of $f(\cdot)$ over B we proceeded as follows: Let $\{y_i\}_0^n$ be a partition of the interval $[m, M)$, i.e.,

$$m = y_0 < y_1 < \cdots < y_n = M,$$

and define

$$E_i = B \cap f^{-1}([y_i, y_{i+1})), \qquad z_n = \sup_{0 \le i < n} (y_{i+1} - y_i). \tag{1}$$

Clearly

$$E_i \cap E_j = \emptyset \quad \text{if } i \ne j \quad \text{and} \quad \bigcup_{i=0}^{n-1} E_i = B.$$

Next, we define the lower and the upper sums

$$\underline{S}_n = \sum_{i=0}^{n-1} y_i \, \bar{\lambda}(E_i), \qquad \bar{S}_n = \sum_{i=0}^{n-1} y_{i+1} \, \bar{\lambda}(E_i), \tag{2}$$

respectively; clearly then

$$0 \le \bar{S}_n - \underline{S}_n \le z_n \, \bar{\lambda}(B). \tag{3}$$

By adding a single point from $[m, M) - \{y_i\}_0^n$ to $\{y_i\}_0^n$, we obtain another partition, say $\{y_i'\}_0^{n+1}$ of $[m, M)$. If $\underline{S}_{n+1}$ and $\bar{S}_{n+1}$ are the corresponding lower and upper sum, respectively, then obviously

$$\underline{S}_n \le \underline{S}_{n+1} \le \bar{S}_{n+1} \le \bar{S}_n.$$

If we continue adding new points in such a way that $z_n \downarrow 0$ as $n \to \infty$, we obtain two monotone sequences $\{\underline{S}_n\}_1^\infty$ and $\{\bar{S}_n\}_1^\infty$. Since the limits

$$\underline{L} = \lim_{n \to \infty} \underline{S}_n, \quad \bar{L} = \lim_{n \to \infty} \bar{S}_n \tag{4}$$

exist, and

$$\underline{S}_n \le \underline{L} \le \bar{L} \le \bar{S}_n \tag{5}$$

for every $n=1, 2, \ldots$, it follows from (3) that

$$0 \le \bar{L} - \underline{L} \le \bar{S}_n - \underline{S}_n \le z_n \, \bar{\lambda}(B).$$

By letting $n \to \infty$ we conclude from this that

$$\underline{L} = \overline{L} = L.$$

The common value L is called the "**Lebesgue integral**" of the bounded measurable function $f(\cdot)$ over the measurable set $B \subset R$, with respect to the Lebesgue measure $\overline{\lambda}$, and we write

$$L = \int_B f \, d\overline{\lambda}.$$

It seems intuitively clear that L can be construed as a map

$$L : \mathcal{B}_0 \times \boldsymbol{B} \to R$$

where $\mathcal{B}_0 \subset \mathcal{A}$ is the class of all Borel sets of finite measure and $\boldsymbol{B}$ is the set of bounded measurable functions defined on the elements of $\mathcal{B}_0$.

The essential difference between the method of construction of the Riemann integral and of the Lebesgue integral rests on the method of partitioning the domain of integration. In the Lebesgue case the partition is determined by the integrand $f(\cdot)$. From the partition $\{E_i\}_0^{n-1} \subset \mathcal{A}$ defined in (1) we can form the sum

$$S_n^* = \sum_{i=0}^{n-1} f(\xi_i) \overline{\lambda}(E_i) \qquad \text{(where } \xi_i \in E_i\text{)}$$

which converges to L since $\underline{S}_n \leq S_n^* \leq \overline{S}_n$.

It is clear from (5) that $\underline{S}_n \leq L \leq \overline{S}_n$. Hence we have that

$$m \, \overline{\lambda}(B) \leq \int_B f \, d\overline{\lambda} \leq M \overline{\lambda}(B). \tag{6}$$

Next, we list some simple properties of the Lebesgue integral which is easy to verify. If $f(\cdot) \equiv C$ on the set B (C is a constant) we deduce from (6) that

$$\int_B C \, d\overline{\lambda} = C \, \overline{\lambda}(B).$$

If the measurable set B is the union of two disjoint measurable sets B_1 and B_2, then

$$\int_B f \, d\overline{\lambda} = \int_{B_1} f \, d\overline{\lambda} + \int_{B_2} f \, d\overline{\lambda}. \tag{7}$$

Indeed, since in this case $E_i = E_i^1 \cup E_i^2$ and $E_i^1 \cap E_i^2 = \emptyset$, where

$$E_i^k = B_k \cap f^{-1}([y_i, y_{i+1})), \qquad k=1, 2,$$

we have that

$$\underline{S}_n = \sum_{i=0}^{n-1} y_i \, \overline{\lambda}(E_i) = \sum_{i=0}^{n-1} y_i \, \overline{\lambda}(E_i^1) + \sum_{i=0}^{n-1} y_i \, \overline{\lambda}(E_i^2).$$

From this, by the usual limit argument, we obtain (7).

Example 2

If $f(\cdot)=I_G$, where $G\subset B$ is measurable, we deduce from (7) that

$$\int_B I_G \, d\bar\lambda = \int_G I_G \, d\bar\lambda + \int_{B-G} I_G \, d\bar\lambda = \int_G 1 \, d\bar\lambda + \int_{B-G} 0 \, d\bar\lambda = \bar\lambda(G).$$

When $f(\cdot)$ is the "Dirishlet" function we have:

$$\int_B f \, d\bar\lambda = \int_{B_1} f \, d\bar\lambda + \int_{B-B_1} f \, d\bar\lambda = \int_{B_1} 1 \, d\bar\lambda + \int_{B-B_1} 0 \, d\bar\lambda = 0$$

where $B_1 = B\cap Q$ ($Q\subset R$ is the set of all rational numbers).

For two bounded measurable functions f_1 and f_2 defined on $B\in\mathcal{A}$ with $\bar\lambda(B)<\infty$ we have:

$$\int_B (f_1 + f_2) \, d\bar\lambda = \int_B f_1 \, d\bar\lambda + \int_B f_2 \, d\bar\lambda. \tag{8}$$

Here is a sketch of the proof: the function f_1+f_2 is bounded and measurable with the range (we may assume) in $[m, M]$. Denote by $E_{ij}=E_i'\cap E_j''$, where

$$E_i' = B \cap f_1^{-1}([y_i, y_{i+1})), \qquad E_j'' = B \cap f_2^{-1}([y_j, y_{j+1})).$$

Clearly, $\{E_{ij}\}$ is a measurable partition of B. Thus, according to (7)

$$\int_B (f_1 + f_2) \, d\bar\lambda = \sum_{i=0}^{n-1} \sum_{j=0}^{n-1} \int_{E_{ij}} (f_1 + f_2) \, d\bar\lambda.$$

Now, from (6) we have

$$(y_i + y_j)\, \bar\lambda(E_{ij}) \le \int_{E_{ij}} (f_1 + f_2) \, d\bar\lambda \le (y_{i+1} + y_{j+1})\, \bar\lambda(E_{ij}),$$

from which we obtain

$$\sum_{i=0}^{n-1} \sum_{j=0}^{n-1} (y_i + y_j)\, \bar\lambda(E_{ij}) \le \int_B (f_1 + f_2) \, d\bar\lambda \le \sum_{i=0}^{n-1} \sum_{j=0}^{n-1} (y_{i+1} + y_{j+1})\, \bar\lambda(E_{ij}).$$

After some simple calculation we have for all $n=1,2,\dots$

$$\underline{S}_n' + \underline{S}_n'' \le \int_B (f_1 + f_2) \, d\bar\lambda \le \bar{S}_n' + \bar{S}_n''$$

where

$$\underline{S}_n' = \sum_{i=0}^{n-1} y_i\, \bar\lambda(E_i'), \qquad\qquad \underline{S}_n'' = \sum_{j=0}^{n-1} y_j\, \bar\lambda(E_j'')$$

$$\bar{S}_n' = \sum_{i=0}^{n-1} y_{i+1}\, \bar\lambda(E_i'), \qquad\qquad \bar{S}_n'' = \sum_{j=0}^{n-1} y_{j+1}\, \bar\lambda(E_j'').$$

By letting $n\to\infty$ we obtain (8).

The Lebesgue measure theoretic approach to the theory of integration on the real line was extended first to the Euclidean spaces, but with more general measure, by Radon, Young, Riesz and Lebesgue himself.

This has lead to the consideration of integration in abstract measure spaces by Fréchet (1915). Another remarkable achievement in this period was the concept of outer measure due to C. Carathéodory (1918) and the innovative, linear functional approach by Daniel (1918).

To motivate the forthcoming presentation observe that we can write $\underline{S}_n$ and $\bar{S}_n$ in (2) as follows:

$$\underline{S}_n = \sum_{i=0}^{n-1} y_i \, \bar{\lambda}(E_i) = \sum_{i=0}^{n-1} y_i \int_B I_{E_i} \, d\bar{\lambda} = \int_B \underline{h}_n \, d\bar{\lambda} \tag{9}$$

$$\bar{S}_n = \int_B \bar{h}_n \, d\bar{\lambda},$$

where

$$\underline{h}_n(x) = \sum_{i=0}^{n-1} y_i I_{E_i}(x), \qquad \bar{h}_n(x) = \sum_{i=0}^{n-1} y_{i+1} I_{E_i}(x).$$

Clearly, $\{\underline{h}_n\}_1^\infty$ is monotone increasing and $\{\bar{h}_n\}_1^\infty$ is monotone decreasing such that on B

$$\underline{h}_n(x) \le f(x) \le \bar{h}_n(x).$$

From this and (5) it follows that

$$\sup_n \int_B \underline{h}_n \, d\bar{\lambda} = \inf_n \int_B \bar{h}_n \, d\bar{\lambda} = \int_B f \, d\bar{\lambda}.$$

Hence, since obviously

$$\sup_n \int_B \underline{h}_n \, d\bar{\lambda} \le \sup_{\varphi \le f} \int_B \varphi \, d\bar{\lambda} \le \inf_{\psi \ge f} \int_B \Psi \, d\bar{\lambda} \le \inf_n \int_B \bar{h}_n \, d\bar{\lambda}, \tag{10}$$

where the infinum and supremum of the middle two terms are taken over all simple functions φ and Ψ defined on B and such that $\varphi \le f \le \Psi$, we conclude that

$$\int_B f \, d\bar{\lambda} = \sup_{\varphi \le f} \int_B \varphi \, d\bar{\lambda} = \inf_{\psi \ge f} \int_B \Psi \, d\bar{\lambda}. \tag{11}$$

This gives rise to the following definition:

Let f be a bounded measurable function defined on a measurable set $B \subset R$ with $\bar{\lambda}(B) < \infty$. The Lebesgue integral of $f(\cdot)$ over B is defined by

$$\int_B f \, d\bar{\lambda} = \sup_{\varphi \le f} \int_B \varphi \, d\bar{\lambda}, \tag{12}$$

where the supremum is taken over all simple functions φ defined on B such that $\varphi \le f$. Here, of course, if

$$\varphi = \sum_{i=1}^n \alpha_i I_{B_i}$$

where $\{B_i\}_1^n$ is a partition of B, we define by analogy with (9),

$$\int_B \varphi \, d\bar{\lambda} = \sum_{i=1}^n \alpha_i \bar{\lambda}(B_i). \tag{13}$$

Next we shall prove that every bounded Riemann integrable function is Lebesgue measurable. For this purpose we need the following simple

Lemma 1

If a measurable function $f \geq 0$ on a measurable set $B \subset R$ and

$$\int_B f \, d\bar{\lambda} = 0, \quad \text{then } f = 0 \quad \text{(a.e) on } B.$$

Proof:

For every $n = 1, 2, \ldots$

$$\frac{1}{n} \bar{\lambda}(B \cap \{f > \frac{1}{n}\}) \leq \int_{B \cap \{f > \frac{1}{n}\}} f \, d\bar{\lambda} \leq \int_B f \, d\bar{\lambda} = 0.$$

Consequently,

$$\bar{\lambda}(B \cap \{f > 0\}) = \bar{\lambda}(B \cap \bigcup_{n=1}^{\infty} \{f > \frac{1}{n}\}) \leq \sum_{n=1}^{\infty} \bar{\lambda}(B \cap \{f > \frac{1}{n}\}) = 0,$$

which proves the lemma. ♥

Corollary 1

If $f > 0$ and

$$\int_B f \, d\bar{\lambda} = 0, \quad \text{then} \quad \bar{\lambda}(B) = 0.$$

Indeed,

$$\int_B f \, d\bar{\lambda} = \int_B f I_B \, d\bar{\lambda} = 0 \Rightarrow f I_B = 0 \ \text{(a.e.)} \Rightarrow I_B = 0 \ \text{(a.e.)} \Rightarrow \bar{\lambda}(B) = 0.$$

The following proposition is essentially due to Lebesgue.

Proposition 1

Let f be a bounded Riemann- integrable function on $[a, b] \subset R$, then f is (a.e) continuous and measurable.

Proof:

Given a partition of $[a, b]$

$$a = x_0 < x_1 < \cdots < x_n = b,$$

define

$$m_i = \inf \{f(x); \ x_i \leq x \leq x_{i+1}\}$$

$$M_i = \sup \{f(x); \ x_i \leq x \leq x_{i+1}\},$$

$i = 0, 1, \ldots, n-1$. Then the lower and the upper sum

$$\underline{S}_n = \sum_{i=0}^{n-1} (x_{i+1} - x_i) \, m_i \qquad \qquad \bar{S}_n = \sum_{i=0}^{n-1} (x_{i+1} - x_i) \, M_i$$

satisfy

$$\bar{S}_n - \underline{S}_n = \sum_{i=0}^{n-1} (x_{i+1} - x_i)(M_i - m_i) = \sum_{i=0}^{n-1} \int_{x_i}^{x_{i+1}} (M_i - m_i) \, d\bar{\lambda}$$

$$\geq \sum_{i=0}^{n-1} \int_{x_i}^{x_{i+1}} \omega_f \, d\bar{\lambda} = \int_a^b \omega_f \, d\bar{\lambda}, \tag{14}$$

where $\omega_f(\cdot) \geq 0$ is the oscillation function of $f(\cdot)$ (see problem 5 of section 23). By letting $n \to \infty$ in (14) we obtain that

$$\int_a^b \omega_f \, d\bar{\lambda} = 0.$$

Now, denote by $D \subset [a, b]$ the set of discontinuity points of f (recall that $\omega_f(\cdot)$ is measurable and that D is of type F_σ), then

$$0 = \int_a^b \omega_f \, d\bar{\lambda} = \int_D \omega_f \, d\bar{\lambda} + \int_{[a,b]-D} \omega_f \, d\bar{\lambda} = \int_D \omega_f \, d\bar{\lambda}.$$

Since $\omega_f > 0$ on D it follows from corollary 1 that $\bar{\lambda}(D) = 0$, proving that the function f is (a.e) continuous on [a, b]. Finally, there is a continuous function on [a, b], say $f_0(\cdot)$, such that $f_0 = f$ (a.e) on [a, b]. Since f_0 is measurable so is f (see proposition 1 of section 26). This proves our contention. ♥

We have just established that every bounded Riemann integrable function on $[a, b] \subset R$ is (a.e) continuous on this interval. Actually, the converse also holds implying that a bounded function $f(\cdot)$ defined on a closed interval [a, b] is Riemann integrable if and only if it is continuous (a.e) . To prove this claim we need the following lemma which is a generalization of the well known result that a continuous function defined on a compact set is uniformly continuous.

Proposition 2

Let $f(\cdot)$ be a function defined on $[a, b] \subset R$ such that for some $\varepsilon > 0$

$$\sup \{\omega_f(x); x \in [a, b]\} < \varepsilon, \tag{15}$$

then there exists $h_0 > 0$ such that for any $u, v \in [a, b]$ with $|u-v| < h_0$ we have that

$$|f(u) - f(v)| < \varepsilon.$$

Proof:

Recall that $B(x, h) - b(x, h) \downarrow \omega_f(x)$ as $h \downarrow 0$, where

$$B(x, h) = l u b \{f(s); s \in [a, b] \cap (x-h, x+h)\}$$

$$b(x, h) = g l b \{f(s); s \in [a, b] \cap (x-h, x+h)\}$$

(see problem 5 of section 24). From this and (15) it follows that there exists $h_x > 0$ such that

$$B(x, h_x) - b(x, h_x) < \varepsilon.$$

Consequently, for any $u, v \in (x - h_x, x + h_x)$

$$|f(u) - f(v)| < \varepsilon. \tag{16}$$

Denote by $I_x(h_x)=(x-\frac{1}{2}h_x, x+\frac{1}{2}h_x)$, then the family $\{I_x; x\in[a, b]\}$ is an open covering of $[a, b]$. Let $\{I_{x_1}, I_{x_2}, \ldots, I_{x_n}\}$ be a subcover of $[a, b]$ which exists (by the Heine-Borel lemma) since $[a, b]$ is compact. Denote by

$$h^* = \inf\{h_{x_1}, h_{x_2}, \ldots, h_{x_n}\}$$

and consider some $y, z\in[a, b]$ with $|y-z|<h^*$. If $y\in I_{x_i} = (x_i-\frac{1}{2}h_{x_i}, x_i+\frac{1}{2}h_{x_i})$, then clearly $z\in (x_i-h_{x_i}, x_i+h_{x_i})$. From this and (16) we conclude that

$$|f(y) - f(z)| < \varepsilon.$$

This proves the proposition.

Assume now that the function $f(\cdot)$ is (a.e) continuous on $[a, b]$ and denote by $D\subset[a, b]$ the set of its discontinuity points. The set D is of type F_σ and $\bar\lambda(D)=0$. Consequently, each closed set

$$D_n = \{x\in[a, b];\ \omega_f(x) \geq \frac{1}{n}\}$$

is a null set since

$$D = \bigcup_{n=1}^{\infty} D_n.$$

Thus, there is a partition $a=x_0<x_1<\cdots<x_n=b$ such that the total length L_1 of all those intervals $[x_i, x_{i+1}]$ which meet D_n is less then $1/n$. Let $K=[a, b]-D_n$, then on the set K, $\omega_f(\cdot)<1/n$. Hence, according to the last proposition, there exists $h>0$ such that for any $x, y\in K$ with $|x-y|<h$

$$|f(x) - f(y)| < \frac{1}{n}$$

Let $a=u_0<u_1<\cdots<u_{n_1}=b$ be a refinement of the initial partition such that $z_{n_1} = \sup_i (u_{i+1}-u_i)< h$, then

$$\bar{S}_{n_1} - \underline{S}_n \leq L\cdot\frac{1}{n} + \frac{b-a}{n} \to 0 \text{ as } n\to\infty.$$

Here $L=\max(M_i-m_i)$ where the maximum is taken over the intervals $[u_i, u_{i+1}]$ that meet D_n.

From this and proposition 1 we conclude that a bounded function $f(\cdot)$ on $[a, b]\subset R$ is Riemann integrable if and only if it is (a.e) continuous. In other words, if $\omega_f(\cdot)=0$ (a.e) on $[a, b]$. According to proposition 1 every bounded Riemann integrable function f defined on $[a, b]\subset R$ is measurable and its Lebesgue integral coincides with its Riemann integral (see (10)), i.e.,

$$\int_a^b f(x)\, dx = \int_a^b f\, d\bar\lambda.$$

♥

Problems and complements

1. Let $f(\cdot)$ be a bounded measurable function defined on a measurable set $B\subset R$ with $\bar\lambda(B)<\infty$. Show that for every $u>0$

$$\bar\lambda(B\cap\{|f|>u\}) \le \frac{1}{u}\int_B |f|\,d\bar\lambda$$

(this is well known Chebyshev's inequality).

Hint:

$$\int_B |f|\,d\bar\lambda \ge \int_{B\cap\{|f|>u\}} |f|\,d\bar\lambda \ge u\bar\lambda(B\cap\{|f|>u\}).$$

2. Two measurable functions $f(\cdot)$ and $h(\cdot)$ defined on a measurable set $B\subset R$ are said to be "equivalent" (written $f\sim h$) if

$$\bar\lambda(B\cap\{f\neq h\})=0.$$

If f and h are bounded and equivalent, show that

$$\int_B |f-h|\,d\bar\lambda = 0.$$

Proof:

Denote by

$$B_1 = \{f=h\}\cap B \qquad B_2 = \{f\neq h\}\cap B$$

then

$$\int_B |f-h|\,d\bar\lambda = \int_{B_1} |f-h|\,d\bar\lambda + \int_{B_2} |f-h|\,d\bar\lambda = 0$$

since $\bar\lambda(B_2)=0$ (see lemma 1).

3. Show that

$$\int_0^{\pi/2} f\,d\bar\lambda = 1$$

where

$$f(x) = \begin{cases} \sin x & \text{if } x \text{ is rational} \\ \cos x & \text{if } x \text{ is irrational} \end{cases}.$$

Hint:

Notice that on $[0,\,\pi/2]\quad f\sim\cos x$.

4. Let $\{\varphi_n\}_1^\infty$ be a sequence of simple functions on a measurable set $B\subset R$ with $\bar\lambda(B)<\infty$. If $\varphi_n\downarrow 0$ on B as $n\to\infty$, show that

$$\int_B \varphi_n\,d\bar\lambda \downarrow 0.$$

Hint

Set

$$B_0 = B \cap \{\varphi_1 > 0\}, \quad M = \sup_B \varphi_1 \quad \text{and} \quad G_n = \{\varphi_n > \varepsilon\} \cap B$$

where $\varepsilon > 0$ is arbitrary. Clearly, $\bar{\lambda}(G_n) \downarrow 0$ as $n \to \infty$. Therefore, there exists a positive integer k such that

$$\bar{\lambda}(G_n) < \varepsilon \quad \text{for all} \ n \geq k.$$

From this we have that

$$\varphi_n = \varphi_n I_{B_0} = \varphi_n (I_{B_0 \cap G_n} + I_{B_0 \cap G_n^c})$$

$$\leq \varphi_n I_{G_n} + \varepsilon I_{B_0} .$$

Hence

$$\int_B \varphi_n d\bar{\lambda} \leq \int_{G_n} \varphi_n d\bar{\lambda} + \varepsilon \bar{\lambda}(B_0) \leq M\bar{\lambda}(G_n) + \varepsilon \bar{\lambda}(B_0).$$

5. If, on the other hand, $\varphi_n \uparrow \varphi$, where φ is also a simple function on B, show that

$$\int_B \varphi_n d\bar{\lambda} \uparrow \int_B \varphi \, d\bar{\lambda} .$$

Hint:

Since $(\varphi - \varphi_n) \downarrow 0$ it follows, from the previous problem that

$$\int_B (\varphi - \varphi_n) \, d\bar{\lambda} \downarrow 0 .$$

6. Let $(R, \mathcal{A}, \bar{\lambda})$ be the Lebesgue measure space, and $f(\cdot)$ a bounded measurable function on a measurable set B. Then, given any $\varepsilon > 0$, there exists a $\delta > 0$ such that

$$\left| \int_E f d\bar{\lambda} \right| < \varepsilon$$

for every $E \in \mathcal{A} \cap B = \{A \cap B; A \in \mathcal{A}\}$ with $\bar{\lambda}(E) < \delta$.

Hint:

Clearly

$$\left| \int_E f d\bar{\lambda} \right| \leq \bar{\lambda}(E) \sup_{x \in E} |f(x)|$$

Next, chose E such that $\bar{\lambda}(E) \sup_{x \in E} |f(x)| \leq \varepsilon$, then $\delta = \varepsilon / \sup_{x \in E} |f(x)|$. This property of the set function

$$\mu(E) = \int_E f d\bar{\lambda}$$

is expressed by saying that μ is "absolutely" continuous with respect to the Lebesgue measure $\bar{\lambda}$.

7. Let $f(\cdot) \geq 0$ be a bounded measurable function. Show that the set function μ on $\mathcal{A}$ defined by

$$\mu(B) = \int_E f d\bar{\lambda}$$

is a measure.

Hint:

Let $\{A_k\}_1^{\infty} \subset \mathcal{A}$ be disjoint and assume that $\bar{\lambda}(A) < \infty$ where $A = \bigcup_1^{\infty} A_k$. Set

$$f I_A = \sum_{k=1}^{\infty} f I_{A_k} \quad \text{and} \quad h_n = \sum_{k=1}^{n} f I_{A_k}$$

then clearly $h_n \uparrow f I_A$. In addition,

$$\int_A h_n \uparrow L .$$

Consequently, since

$$\int_A h_n \, d\mu \le \int_A f \, d\bar{\lambda} \implies L \le \int_A f \, d\bar{\lambda} .$$

Next, given $\delta \in (0, 1)$ and a simple function φ on B such that $0 \le \varphi \le f$; define

$$B_k = \{\delta\varphi \le h_k\} \cap A .$$

Clearly $\{B_k\}_1^{\infty} \subset \mathcal{A}$ and $B_k \uparrow A$. Thus,

$$\int_A h_n \, d\bar{\lambda} \ge \int_{B_n} h_n \, d\bar{\lambda} \ge \delta \int_{B_n} \varphi \, d\bar{\lambda} \ge \int_{B_n} \varphi \, d\bar{\lambda}$$

since $\delta \in (0, 1)$ is arbitrary. Thus when $n \to \infty$ we have

$$\int_A f \, d\bar{\lambda} \ge L \ge \int_A \varphi \, d\bar{\lambda} .$$

8. If h_1 and h_2 defined on [a, b] are Riemann integrable, show that the functions

(i) $h_1 + h_2$; (ii) $(h_1 + h_2)^2$; (iii) $|f_1 + f_2|$; (iv) max (h_1, h_2); are also Riemann integrable.

Hint:

(i) If D_1, D_2 and D are sets of discontinuity points of h_1, h_2 and $h_1 + h_2$ respectively, then clearly $D \subset D_1 \cup D_2$.

(ii) Every continuity point of h is a continuity point of h^2, and vice versa. In general, if $f: R \to R$ is continuous, then each continuity point of h is a continuity point of $f(h)$.

(iv) We Have

$$2\max\{h_1, h_2\} = (h_1 + h_2 + |h_1 - h_2|)$$

In this section we lay down the foundations for the development of integration theory for measurable functions defined on a measure space $(X, \mathcal{A}, \mu)$. Here X is a non empty set, μ is a measure on X and $\mathcal{A}$ is a σ-algebra of measurable subsets of X.

To achieve this objective we follow the oldest idea, which is first to integrate some simple functions and then to take the limits. Since the simple functions are the building blocks of integration theory, we shall first briefly outline some basic properties of this class of functions needed in this section.

Recall that a measurable map $h:X\rightarrow R$ is called "simple" if its range $f(X)\subset R$ is a finite set. Usually a simple function is given as a finite linear combination of measurable indicator functions, say

$$h(x) = \sum_{i=1}^{n} d_i I_{A_i}(x), \quad \text{where } \{A_i\}_1^n \text{ is a sequence from } \mathcal{A}. \tag{1}$$

It is clear from (1) that a simple function can have many different representations in this form. However, if for a simple function $f:X\rightarrow R$ we know its range, say $f(X)=\{\alpha_1, \alpha_2, \cdots, \alpha_n\}$ where each $\alpha_i \neq 0$, then we can write

$$f(x) = \sum_{i=1}^{n} \alpha_i I_{B_i}(x), \qquad \text{where } B_i = \{x;\ f(x)=\alpha_i\} \tag{2}$$

$i=1, 2, \cdots, n$. Clearly, $\{B_i\}_1^n$ is a sequence of measurable and pairwise disjoint sets. The form (2) is called the **"standard representation" of the simple function f**. It is clear that a simple function may have only one standard representation.

To obtain the standard representation of (1) we need the set $h(X)$ which is not always easily identifiable from its representation, although it is obvious that

$$\text{Card } (h(X)) \leq 2^n - 1.$$

To obtain elements of the set $f(X)$ we proceed as follows: Let $S\subset\{1, 2, \cdots, n\}$ be an arbitrary non empty subset and denote by $S'=\{1, 2, \cdots, n\}- S$. Consider the sets

$$H_S = (\bigcap_{i\in S} A_i)\cap(\bigcap_{j\in S'} A_j^c). \tag{3}$$

Clearly, there are at most $2^n - 1$ such sets. If $S_1, S_2\subset\{1, 2, \cdots, n\}$ then obviously

$$H_{S_1}\cap H_{S_2} = \theta.$$

It seems rather obvious from (3) that any A_j is the union of all H_S such that $j\in S$, i.e.,

$$A_j = \bigcup_{S\ni j} H_S, \quad j = 1, 2, \cdots, n. \tag{4}$$

We claim that

$$\sum_{S\subset\{1, 2, ..., n\}} \gamma_S I_{H_S}, \qquad \text{where } \gamma_S = \sum_{i\in S} d_i, \tag{5}$$

is the standard representation of the simple function h. Indeed, if $y \in H_S$, it follows from (3) that $y \in \bigcap_{i \in S} A_i$, so that

$$h(y) = \sum_{i=1}^{n} d_i I_{A_i}(y) = \sum_{i \in S} d_i = \gamma_S.$$

If, on the other hand, $y \in A_j$, then due to (4) $y \in H_S$ for some $S \ni j$, so that

$$\sum_{S \subset \{1, 2, \ldots, n\}} \gamma_S I_{H_S}(y) = \gamma_S = \sum_{j \in S} d_j,$$

which proves that

$$\sum_{i=1}^{n} d_i I_{A_i} = \sum_{S \subset \{1, 2, \ldots, n\}} \gamma_S I_{H_S}.$$

Now, by deleting all those I_{H_S} where $H_S = \emptyset$ or where $\gamma_S = 0$ and joining the I_{H_S} with equal coefficients, we obtain the standard representation of the simple function h.

In the development so far the concept of a measure was not required. From now on it will be in constant use. We begin with the following definition.

A simple function $\varphi : X \to R$, say

$$\varphi(x) = \sum_{i=1}^{n} \alpha_i I_{E_i}(x),$$

which vanishes outside of a measurable subset of a finite measure, is called a "step function." In other words, if each $\mu(E_i) < \infty$, φ is a step function. We are now ready to define the integral of φ.

By analogy with (13) of the previous section the integral of a step function φ over the whole space X, usually denoted by $\int \varphi \, d\mu$, is defined "axiomatically" by

$$\int \varphi \, d\mu = \sum_{i=1}^{n} \alpha_i \mu(E_i) \tag{6}$$

Remark 1

From (6) it follows that the integral of a step function always exists and is finite. This, however, is not necessarily true for simple functions. For instance, the map $\psi : R \to R$, defined by

$$\psi(x) = 3 \, I_{(-\infty, u)}(x) - 2 \, I_{(v, \infty)}$$

is a simple function whose integral is

$$\int_{-\infty}^{\infty} \Psi \, d\bar{\lambda} = \infty - \infty$$

and consequently it does not exist. On the other hand, if $G \in \mathcal{A}$ is a set of finite measure, then the integral of any simple function $h : X \to R$ over G exists because

$$\int_G h \, d\mu = \int h \cdot I_G \, d\mu, \tag{7}$$

and, obviously, $h \cdot I_G$ is a step function.

We must show that the integral (6) is "well defined." By this we mean that its value does not depend on a particular representation of the step function φ. To this end we need the following lemma.

Lemma 1

If φ_0 is a step function such that $\varphi_0(x) = 0$ for every $x \in X$ then

$$\int \varphi_0 \, d\mu = 0$$

Proof:

Assume that

$$\varphi_0(x) = \sum_{i=1}^{k} c_i I_{A_i}(x) = 0.$$

According to (5) its standard representation is

$$\varphi_0 = \sum_{S \subset \{1, 2, \dots, n\}} \gamma_S I_{H_S} \qquad \gamma_S = \sum_{i \in S} c_i \,.$$

Since by assumption $\varphi_0 \equiv 0$ on X, it follows that each $\gamma_S = 0$, whenever $H_S \neq \theta$. Thus

$$\sum_{S \subset \{1, 2, \dots, n\}} \gamma_S \, \mu(H_S) = 0.$$

But

$$\sum_{S \subset \{1, 2, \dots, n\}} \gamma_S \, \mu(H_S) = \sum_{S \subset \{1, 2, \dots, n\}} \left(\sum_{i \in S} c_i \right) \mu(H_S)$$

$$= \sum_{i=1}^{n} c_i \, \mu\left(\bigcup_{S \ni i} H_S \right) = \sum_{i=1}^{n} c_i \, \mu(A_i),$$

which proves the lemma. Obviously the lemma holds if $\varphi_0 = 0$ (a.e) on X.

Proposition 1

The integral (6) is well defined.

Proof:

Assume that

$$\sum_{i=1}^{n} \alpha_i I_{A_i} \quad \text{and} \quad \sum_{j=1}^{m} \beta_j I_{B_j}$$

are two different representation of a step function h_0, then

$$\sum_{i=1}^{n} \alpha_i I_{A_i} + \sum_{j=1}^{m} (-1) \beta_j I_{B_j} = 0$$

is the sum of two step functions and as such it is a step function. But, according to the lemma 1, its integral is zero proving that

$$\sum_{i=1}^{n} \alpha_i \, \mu(A_i) = \sum_{j=1}^{m} \beta_j \, \mu(B_j).$$

We now list some simple properties of the integral of step functions. If φ_1 and φ_2 are two step function and c_1, $c_2 \in R$ then obviously

$$\int (c_1\,\varphi_1 + c_2\,\varphi_2)\,d\mu = c_1 \int \varphi_1 d\mu + c_2 \int \varphi_2\,d\mu. \tag{8}$$

In other words, the integral $\int$ is a linear map. It is also non negative, for if $\varphi_0(\cdot) \geq 0$ is a step function then

$$\int \varphi_0\,d\mu \geq 0.$$

Indeed, let

$$\varphi_0(\cdot) = \sum_{i=1}^{n} d_i\, I_{E_i}(\cdot)$$

be its standard representation, then since $\varphi_0 \geq 0$ it is clear that $\mu(E_i)=0$ if $d_i <0$. Thus,

$$\sum_{i=1}^{n} d_i\,\mu(E_i) \geq 0.$$

From this it follows that for any two step functions h_1 and h_2 on X such that $h_1 \leq h_2$ we have that

$$\int h_1\,d\mu \leq \int h_2\,d\mu. \tag{9}$$

We also have that

$$\left| \int \varphi\,d\mu \right| \leq \int |\varphi|\,d\mu \quad \text{and} \quad \int |\varphi_1 + \varphi_2|\,d\mu \leq \int |\varphi_1|\,d\mu + \int |\varphi_2|\,d\mu. \tag{10}$$

Proof of both inequalities is trivial. A version of the next proposition was proved in the previous section (problem 4).

Proposition 2

Let $\{\varphi_n\}_1^\infty$ be a sequence of step functions on a measure space $(X,\ \mathcal{A},\ \mu)$ such that $\varphi_n \downarrow 0$ (a.e) on X, then (the upper continuity)

$$\int \varphi_n\,d\mu \downarrow 0.$$

Proof:

Without any loss of generality we may assume that $\varphi_n \downarrow 0$ everywhere on X. Denote by

$$B = \{\varphi_1 > 0\}, \quad G_n = \{\varphi_n > \varepsilon\}, \quad M = \sup\,\{\varphi_1(x);\ x \in X\}$$

where $\varepsilon > 0$. Since $\mu(G_n) \downarrow 0$ as $n \to \infty$, there exists a positive integer k such that $\mu(G_n) < \varepsilon$ for all $n \geq k$. Thus, for any $n \geq k$

$$\varphi_n \leq \varphi_k = \varphi_k\,I_B \leq \varphi_k\,(I_{G_k} + I_{B \cap G_k^c}) \leq M\,I_{G_k} + \varepsilon\,I_B.$$

From this and (8) we have that

$$0 \leq \int \varphi_n\,d\mu \leq M\,\mu(G_k) + \varepsilon \cdot \mu(B),$$

which proves the proposition. ♥

Corollary 1

If φ and $\{\varphi_k\}_1^\infty$ are step functions such that $\varphi_n \uparrow \varphi$ then

$$\int \varphi_n \, d\mu \uparrow \int \varphi \, d\mu.$$

Indeed, since $\varphi - \varphi_n \downarrow 0$ we have that $\int (\varphi - \varphi_n) \, d\mu \downarrow 0$.

Proposition 3

Let $\{\varphi_n\}_1^\infty$ and $\{h_n\}_1^\infty$ be non-negative step functions such that

$$h_n \uparrow f \text{ and } \varphi_n \uparrow f.$$

If the sequence $\{\int h_n \, d\mu\}_1^\infty$ is bounded so is the sequence $\{\int \varphi_n \, d\mu\}_1^\infty$ and

$$\lim_{n \to \infty} \int h_n \, d\mu = \lim_{n \to \infty} \int \varphi_n \, d\mu.$$

Proof:

For k fixed, we have, as $n \to \infty$ that

$$\varphi_k \wedge h_n \uparrow \varphi_k \wedge f = \varphi_k$$

(see section 24 for notation). This and corollary 1 yield

$$\lim_{n \to \infty} \int (\varphi_k \vee h_n) \, d\mu. = \int \varphi_k \, d\mu.$$

Since $\varphi_k \wedge h_n \leq h_n$ it follows that

$$\int \varphi_k \, d\mu \leq \lim_{n \to \infty} \int h_n \, d\mu$$

for all $k = 1, 2, \cdots$, proving that $\{\int \varphi_k\}_1^\infty$ is bounded. Interchanging the role of φ_k and h_k we have (keeping in mind that $\{\int \varphi_k\}_1^\infty$ is bounded) that

$$\int h_k \, d\mu \leq \lim_{n \to \infty} \int \varphi_n \, d\mu$$

for all $k = 1, 2, \cdots$, which proves the proposition.   ♥

Proposition 4

If $\{\psi_k\}_1^\infty$ is a sequence of step functions such that $\psi_n \uparrow I_B$, then

$$\lim_{n \to \infty} \int \psi_n \, d\mu = \mu(B).$$

Proof:

Since every monotone sequence of measurable functions converge pointwise to a measurable function (see problem 1, section 25) I_B is measurable and consequently $B \in \mathcal{A}$. On the other hand, it is obvious that $\{I_B - \psi_n\} \downarrow 0$. The assertion now follows from corollary 1.   ♥

Remark 2

If the measure space $(X, \mathcal{A}, \mu)$ is finite, i.e., $\mu(X) < \infty$, then every simple function is a step function.

Problems and complements

1. Show that a step function $\varphi = 0$ (a.e) if and only if

$$\int |\varphi|\, d\mu = 0 \qquad\qquad (*)$$

Hint:

If $\varphi = 0$ (a.e) then $|\varphi| = 0$ (a.e). If

$$\varphi = \sum_{i=1}^{n} \alpha_i\, I_{A_i}$$

is its standard representation then, since

$$\mu\{|\varphi|>0\} = \sum_{i=1}^{n} \mu(A_i) = 0, \quad \text{it follows that } \mu(A_i) = 0$$

for every i=1, 2, $\cdots$, n, and therefore

$$\int |\varphi|\, d\mu = \sum_{i=1}^{n} |\alpha_i|\, \mu(A_i) = 0.$$

Conversely, if $(*)$ holds, then since each $|\alpha_i|>0$ it follows that $\mu(A_i)=0$, i=1, 2, $\cdots$, n, and so $\varphi = 0$ (a.e).

2. If h is a simple function on X such that

$$\int |h|\, d\mu < \infty$$

then it must be a step function. Show this.

Hint:

Let

$$h = \sum_{i=1}^{n} a_i\, I_{A_i}$$

be the standard representation of h, then

$$\infty > \int |h|\, d\mu \geq \min_{1\leq i\leq n} |a_i| \sum_{i=1}^{n} \mu(A_i)$$

implies that

$$\mu\{h \neq 0\} < \infty.$$

3. Let $\{G_i\}_1^n \subset \mathcal{A}$ be disjoint. Show that for any step function φ on X

$$\int_{\bigcup_1^n G_i} \varphi\, d\mu = \sum_{i=1}^{n} \int_{G_i} \varphi\, d\mu .$$

Hint:

According to (7)

$$\int_{\bigcup_1^n G_i} \varphi\, d\mu = \int \varphi\, (\sum_{i=1}^{n} I_{G_i})\, d\mu = \int (\sum_{i=1}^{n} \varphi\, I_{E_i})\, d\mu .$$

Now use (8).

4. Find the standard representation of the simple function ψ defined in remark 1.

Hint:

If $v < u$

$$\psi(x) = 3 \cdot I_{(-\infty, v]} + 1 \cdot I_{(v, u)} - 2 \cdot I_{(u, \infty)} .$$

If $u < v$, $\psi(R) = \{-2, 0, 3\}$.

5. Let $\{\varphi_n\}_1^\infty$ and $\{\psi_n\}_1^\infty$ be Cauchy sequences of step functions on X converging (a.e) to the same limit.

Show that the limits of $\{\int \varphi_n \, d\mu\}_1^\infty$ and $\{\int \psi_n \, d\mu\}_1^\infty$ exist and that

$$\lim_{n \to \infty} \int \varphi_n \, d\mu = \lim_{n \to \infty} \int \psi_n \, d\mu .$$

Hint:

Since $|\varphi_n - \varphi_m| \to 0$ as $m, n \to \infty$

$$\left| \int \varphi_n \, d\mu - \int \varphi_m \, d\mu \right| \leq \int |\varphi_n - \varphi_m| \, d\mu \to 0$$

(see proposition 2). Consequently, $\{\int \varphi_n \, d\mu\}_1^\infty$ is a Cauchy sequence of real numbers and as such it is

bounded and convergent. The same obviously holds for the sequence $\{\int \psi_n \, d\mu\}_1^\infty$.

Next, denote by $h_n = \varphi_n - \psi_n$. Them $\{h_n\}_1^\infty$ is Cauchy and $h_n \to 0$ (a.e). Thus, given $\varepsilon > 0$, there is

an integer k_0 such that if $m, n \geq k_0$

$$\int |f_m - f_n| \, d\mu < \varepsilon .$$

Let the map f_{k_0} vanish outside of a set $A \in \mathcal{A}$ such that $\mu(A) < \infty$ and denote by $\gamma = \int |f_{k_0}| \, d\mu$. Then there

exists a measurable subset $L \subset A$ such that

$$\mu(L) \leq \varepsilon / (1 + \gamma)$$

(see problem 5 of section 18) and such that $\{f_n\}_1^\infty$ converges to 0 uniformly on A-L. Hence we have that

$$\int_{A-L} |f_n| \, d\mu < \varepsilon .$$

Finally, for n large we obtain that

$$\int_L |f_n| \, d\mu \leq \int_L |f_n - f_{k_0}| \, d\mu + \int_L |f_{k_0}| \, d\mu$$

$$\leq \int_L |f_n - f_{k_0}| \, d\mu + \mu(L) \gamma < 2 \varepsilon.$$

Since

$$\int_{A^c} |f_n| \, d\mu \leq \int_{A^c} |f_n - f_{k_0}| \, d\mu \leq \int |f_n - f_{k_0}| \, d\mu < 2 \varepsilon$$

we obtain, by taking the sum of the integrals over A^c, L and A-L, that

$$\int |f_n|\, d\mu \le 5\,\varepsilon$$

proving that the sequence $\{(\varphi_n - \psi_n)\}_1^\infty$ converges in "mean" to zero, i.e., that

$$\int |\varphi_n - \psi_n|\, d\mu \to 0$$

as $n \to \infty$.

29 THE INTEGRAL OF A NON NEGATIVE FUNCTION

Let $(X, \mathcal{A}, \mu)$ be a measure space and let

$$f: X \to [0, \infty)$$

be a measurable function. By analogy with expression (12) of section 27, its integral over X is defined by

$$\int f \, d\mu = \sup_{\psi \leq f} \int \psi \, d\mu, \tag{1}$$

where the supremum is taken over all step functions ψ on X satisfying $0 \leq \psi \leq f$.

We say that that the function $f(\cdot)$ is "integrable" or "summable" (to use Lebesgue's terminology) if the supremum on the right-hand side of (1) is finite. In other words if $\int f \, d\mu < \infty$. Many interesting properties of this integral follow at once from its definition. For instance, given any $c \in (0, \infty)$

$$\int c \, f \, d\mu = c \int f \, d\mu. \tag{2}$$

This follows directly from (1) and definition (6) of the previous section. If $f_1 \geq 0$ and $f_2 \geq 0$, defined on X, are measurable and such that $f_1 \leq f_2$, then if f_1 is integrable,

$$\int f_1 \, d\mu \leq \int f_2 \, d\mu. \tag{3}$$

This is obviously a consequence of the inequality

$$\sup_{\psi \leq f_1} \int \psi \, d\mu \leq \sup_{\psi \leq f_2} \int \psi \, d\mu,$$

(notice that $\{\psi; \psi \leq f_1\} \subset \{\psi; \psi \leq f_2\}$) and (1). The next proposition gives **Chebishev's inequality** (see also problem 1 of section 28).

Proposition 1

Let $f \geq 0$ be integrable, then for any $\varepsilon > 0$

$$\mu\{f \geq \varepsilon\} \leq \frac{1}{\varepsilon} \int f \, d\mu. \tag{4}$$

Proof:

On X we clearly have that

$$f \cdot I_{\{f \geq \varepsilon\}} \leq f.$$

This and (3) yield:

$$\int f \, d\mu \geq \int f I_{\{f \geq \varepsilon\}} \, d\mu \geq \varepsilon \int I_{\{f \geq \varepsilon\}} \, d\mu = \varepsilon \, \mu\{f \geq \varepsilon\}. \qquad \blacktriangledown$$

Corollary 1

Any integrable function $f \geq 0$ on X is finite (a.e.). Indeed, since

$$\{f = +\infty\} = \bigcap_{k=1}^{\infty} \{f \geq k\},$$

we have from (4) that for any $n = 1, 2, \ldots$

$$\mu\{f = +\infty\} \leq \mu\{f \geq n\} \leq \frac{1}{n} \int f \, d\mu.$$

Definition (1) of the integral of a non-negative measurable function is somewhat awkward and not easy to use. For instance, it is not clear from this definition that

$$\int (f_1 + f_2)\, d\mu = \int f_1\, d\mu + \int f_2\, d\mu \ ,$$

where $f_1 \geq 0$, and $f_2 \geq 0$ are integrable. A remedy for this situation is the **"Lebesgue monotone" convergence theorem for the sequences of non-negative increasing measurable functions** (unfortunately, no such result exists for monotone decreasing sequences measurable functions). This important result makes it possible to take the limit under the integral sign, which is so often necessary.

Proposition 2 (Lebesgue)

Let $\{f_n\}_1^\infty$ be a sequence of non-negative measurable functions on $(X, \mathcal{A}, \mu)$ such that $f_n \uparrow f$ pointwise. Then

$$\lim_{n \to \infty} \int f_n\, d\mu = \int f\, d\mu . \tag{5}$$

Proof:

The limit f is measurable (proposition 3 of section 24). Since $\{\int f_n\, d\mu\}_1^\infty$ is monotone, the limit

$$\lim_{n \to \infty} \int f_n\, d\mu = L \tag{6}$$

exists in $[0, \infty]$. Since for any $n = 1, 2, \ldots$

$$\int f_n\, d\mu \leq \int f\, d\mu \quad \text{it follows that } L \leq \int f\, d\mu . \tag{7}$$

Given any $\delta \in (0, 1)$ and a step function ψ such that $0 \leq \psi \leq f$, we define

$$B_n = \{\delta\,\psi \leq f_n\}.$$

Clearly, $B_n \subset B_{n+1}$ and since $f_n \uparrow f$, $\bigcup_1^\infty B_n = \{\sup_{1 \leq n < \infty} f_n \geq \delta\psi\} = \{f \geq \delta\psi\} = X$. Now,

$$\int f_n\, d\mu \geq \int_{B_n} f_n\, d\mu \geq \delta \int_{B_n} \psi\, d\mu ,$$

which yields

$$L \geq \delta \int \psi\, d\mu .$$

Since $\delta \in (0, 1)$ is arbitrary, this and (7) prove the proposition. ♥

Corollary 2

Recall that given any measurable function $f \geq 0$ on X, there exists a sequence of non-negative step function $\{\psi_n\}_1^\infty$ such that $\psi_n \uparrow f$ (problem 4 of section 24). Then according to proposition 3 of the previous section, if $\{\varphi_n\}$ is another sequence of non-negative step function such that $\varphi_n \uparrow f$,

$$\lim_{n \to \infty} \int \varphi_n\, d\mu = \lim_{n \to \infty} \int \psi_n\, d\mu$$

Thus, the value of the integral $\int f\, d\mu$ does not depend on a particular sequence of step functions which converges to f on X. As a consequence, we can replace the sign "sup" in (1) with the "lim" sign. In other words, we have that

$$\int f \, d\mu = \lim_{n \to \infty} \int \varphi_n \, d\mu , \tag{8}$$

for any sequence of step functions $\{\varphi_n\}_1^\infty$ such that $0 \le \varphi_n \uparrow f$.

Proposition 3

If f_1 and f_2 are non-negative integrable functions on X then

$$\int (f_1 + f_2) \, d\mu = \int f_1 \, d\mu + \int f_2 \, d\mu . \tag{9}$$

Proof:

If $\{\varphi_n\}_1^\infty$ and $\{\psi_n\}_1^\infty$ are step functions on X such that $\varphi_n \uparrow f_1$ and $\psi_n \uparrow f_2$, then

$$(\varphi_n + \psi_n) \uparrow (f_1 + f_2) \quad \text{on } X.$$

Consequently, according to (8) and the linearity of the integral on the collection of step functions (see the previous section) we have:

$$\int (f_1 + f_2) \, d\mu = \lim_{n \to \infty} \int (\varphi_n + \psi_n) \, d\mu$$

$$= \lim_{n \to \infty} \int \varphi_n \, d\mu + \lim_{n \to \infty} \int \psi_n \, d\mu = \int f_1 \, d\mu + \int f_2 \, d\mu$$

as claimed. ♥

Proposition 4

For a measurable function $f \ge 0$ on X we have

$$\int f \, d\mu = 0 \iff f(\cdot) = 0 \quad \text{(a.e)}.$$

Proof:

By Chebishev's inequality

$$0 = \int f \, d\mu \ge n \, \mu\{f \ge \tfrac{1}{n}\} \qquad \text{for every } n=1, 2, \ldots .$$

Thus,

$$\mu\{f > 0\} = \mu\left(\bigcup_{n=1}^{\infty} \{f \ge \tfrac{1}{n}\} \right) \le \sum_{n=1}^{\infty} \mu\{f \ge \tfrac{1}{n}\} = 0.$$

Conversely, if $f(\cdot) = 0$ (a.e.), then

$$\int f \, d\mu = \int_{\{f > 0\}} f \, d\mu = \int f I_{\{f > 0\}} \, d\mu$$

$$= \int f \cdot \sum_{i=1}^{\infty} I_{\{i-1 < f \le i\}} \, d\mu = \int f \cdot \left(\lim_{n \to \infty} \sum_{i=1}^{n} I_{\{i-1 < f \le i\}} \right) d\mu .$$

Now, invoking the monotone convergence theorem we obtain that

$$\int f \, d\mu = \lim_{n \to \infty} \sum_{i=1}^{n} \int_{\{i-1 < f \le i\}} f \, d\mu \le \sum_{i=1}^{\infty} i \, \mu \{i-1 < f \le i\}$$

$$= \sum_{k=1}^{\infty} \sum_{i=k}^{\infty} \mu\{i-1 < f \le i\} = 0 ,$$

since each $\mu\{i-1 < f \le i\}=0$. This completes the proof of our proposition.

Corollary 3

If $N \in \mathcal{A}$ is a null set and $f(\cdot) \ge 0$ measurable then

$$\int_N f\, d\mu = 0,$$

because $f\, I_N = 0$ (a.e) on X.

Proposition 5

If $f(\cdot) > 0$ on X is integrable and for some $B \in \mathcal{A}$

$$\int_B f\, d\mu = 0,$$

then necessarily $\mu(B)=0$.

Proof:

Since

$$\int_B f\, d\mu = \int f\, I_B\, d\mu = 0 \quad \text{it follows that} \quad f\, I_B = 0 \text{ (a.e) on X,}$$

which implies that $\mu(B)=0$.

Remark 1

As indicated earlier in this section, there is no result corresponding to the monotone convergence theorem for decreasing sequences of non-negative measurable functions. Here is a simple example which proves our contention. Let $\{f_n\}_1^\infty$ be a sequence of measurable functions on $(R, \mathcal{B}, \bar{\lambda})$ defined by

$$f_n = \frac{1}{n}\, I_{[n, \infty)}.$$

Clearly, $f_n \downarrow f = 0$. Thus, $\int f\, d\bar{\lambda} = 0$, while, on the other hand for every $n=1, 2, \cdots$

$$\int f_n\, d\bar{\lambda} = +\infty.$$

Example 1

Let $\bar{\mu}_F$ be the Lebesgue-Stieltjes measure on the interval $[a, b)$ induced by a nondecreasing right continuous function $F(\cdot) \ge 0$ (see section 22). Let $f:[a, b) \to [0, \infty)$ be a $\bar{\mu}_F$-integrable on $[a, b)$. Then by the Lebesgue-Stieltjes integral of $f(\cdot)$ with respect to F, denoted by

$$\int_a^b f(x)\, dF(x),$$

we simply mean the integral

$$\int_{[a, b)} f\, d\bar{\mu}_F.$$

As an illustration let us calculate the Lebesgue-Stieltjes integral

$$\gamma_n = \int_0^1 t^n\, d\Phi(t), \qquad n=0, 1, \cdots$$

where Φ is the Cantor function (see equation (6) of section 7), i.e.,

$$\Phi(t) = \sum_{i=1}^{\eta(t)-1} d_i / 2^{i+1} + 2^{-\eta(t)}.$$

Recall that $\Phi: [0, 1] \to [0, 1]$ is continuous surjection (see Fig. 1). Clearly, for $n = 1, 2, \ldots$

$$\gamma_n = \int_0^{1/3} t^n \, d\Phi(t) + \int_{2/3}^1 t^n \, d\Phi(t), \quad \text{and } \gamma_0 = 1.$$

The substitutions $t = x / 3$ in the first integral and $t = (x+2) / 3$ in the second one yield

$$\gamma_n = \frac{1}{3^n} \int_0^1 x^n \, d\Phi(\frac{x}{3}) + \frac{1}{3^n} \int_0^1 (x + 2)^n \, d\Phi(\frac{x+2}{3}).$$

It is not difficult to establish that

$$\Phi(\frac{x}{3}) = \frac{1}{2} \Phi(t) \quad \text{and that} \quad \Phi(\frac{x+2}{3}) = \frac{1}{2}(\Phi(t) + 1).$$

For instance, if the ternary expansion of x is

$$x = \sum_{i=1}^{\infty} \frac{d_i}{3^i}, \quad \frac{1}{3} x = .d_1' \, d_2' \, d_3' \cdots, \text{ where } d_1' = 0 \text{ and } d_i' = d_{i-1} \text{ for } i = 2, 3, \cdots. \text{Consequently,}$$

$$\eta(\frac{1}{3} x) = \eta(x) + 1$$

so that

$$\Phi(\frac{x}{3}) = \sum_{i=1}^{\eta(\frac{x}{3})-1} \frac{d_i'}{2^{i+1}} = \sum_{i=2}^{\eta(x)} \frac{d_{i-1}}{2^{i+1}} = \sum_{j=1}^{\eta(x)-1} \frac{d_j}{2^{j+2}}$$

$$= \frac{1}{2} \sum_{j=1}^{\eta(x)-1} \frac{d_j}{2^{j+1}} = \frac{1}{2} \Phi(x),$$

and similarly for $\Phi(\frac{x+2}{3})$. Hence, after some straightforward calculations we obtain the recursion

$$\gamma_n = \frac{1}{2 \cdot 3^n} \int_0^1 (x^n + (2+x)^n) \, d\Phi(x)$$

$$= \frac{1}{3^n} \gamma_n + \frac{1}{2 \cdot 3^n} \sum_{i=1}^{n} \binom{n}{i} 2^i \gamma_{n-1},$$

from which we deduce that

$$\gamma_n = \frac{1}{2(3^n - 1)} \sum_{i=1}^{n} \binom{n}{i} 2^i \gamma_{n-1}.$$

For instance, since $\gamma_0 = 1$, we have that $\gamma_1 = 1/2$, $\gamma_2 = 3/8$, $\gamma_3 = 5/16$, $\gamma_4 = 87/320$, $\cdots$.

Remark 2

Let $\{f_n\}_1^\infty$ be a sequence of integrable functions on $(R, \mathcal{A}, \mu)$ such that $f_n \downarrow 0$. If f_1 is bounded and

$\mu\{f_1 > 0\} < \infty$ then

$$\int f_n \, d\mu \downarrow 0.$$

The proof is identical to one used to prove proposition 2 of section 28. Indeed, set $M = \sup_{x \in X} f_1(x)$, then for every $n = 1, 2, \cdots, 0 \leq f_n \leq M \, I_{\{f_1 > 0\}}$.

Problems and complements

1. Let $f \geq 0$, defined on $(X, \mathcal{A}, \mu)$, be measurable. If $f \leq M$ (a.e) on a set $A \in \mathcal{A}$ where $M \in (0, \infty)$ and $\mu(A) < \infty$, show that

$$\int_A f \, d\mu \leq M \, \mu(A).$$

Hint:

This follows from (3),

$$\int_A f \, d\mu \leq \int_A M \, d\mu .$$

2. Let f_1 and f_2 be non-negative integrable functions defined on $(X, \mathcal{A}, \mu)$ such that

$$\int_A f_1 \, d\mu \leq \int_A f_2 \, d\mu \text{ for every } A \in \mathcal{A} .$$

Show that $f_1 \leq f_2$ (a.e) on X.

Hint:

By assumption,

$$0 \leq \int_A (f_2 - f_1) \, d\mu \text{ for every } A \in \mathcal{A} .$$

If

$$A = \{f_1 - f_2 > 0\}, \text{ we have that}$$

$$\int_{\{f_1 - f_2 > 0\}} (f_2 - f_1) \, d\mu = - \int_{\{f_1 - f_2 > 0\}} (f_1 - f_2) \, d\mu \leq 0 ,$$

which is a contradiction unless $\mu\{f_1 - f_2 > 0\} = 0$. Thus, $f_1 \leq f_2$ (a.e). Incidentally, if

$$\int_A f_1 \, d\mu = \int_A f_2 \, d\mu , \text{ for each } A \in \mathcal{A}$$

it follows from the previous that $f_1 = f_2$ (a.e) on X.

3. Let $f \geq 0$ defined on $(X, \mathcal{A}, \mu)$ be integrable, and denote by μ_1 the set function defined by

$$\mu_1(A) = \int_A f \, d\mu, \ A \in \mathcal{A}.$$

Show that μ_1 is finite measure on X.

Hint:

Clearly, $\mu_1(X) = \int f \, d\mu < \infty$. Let $\{A_k\}_1^\infty$ be a sequence of disjoint sets from $\mathcal{A}$ and denote by

$$A = \bigcup_{k=1}^\infty A_k \text{ and by } h_n = \sum_{k=1}^n f I_{A_k} \uparrow f I_A = \sum_{k=1}^\infty f I_{A_k} ,$$

then, according to proposition 2,

$$\int f I_A \, d\mu = \lim_{n \to \infty} \int h_n \, d\mu = \lim_{n \to \infty} \sum_{k=1}^n \int f I_{A_k} \, d\mu = \sum_{k=1}^\infty \int_{A_k} f \, d\mu,$$

proving that

$$\mu_1\left(\bigcup_{k=1}^{\infty} A_k\right) = \sum_{k=1}^{\infty} \mu_1(A_k).$$

4. Let $f \geq 0$ defined on $(X, \mathcal{A}, \mu)$ be integrable. Show that

$$h_n = n\, I_{\{f \geq n\}} \to 0 \ (a.e) \text{ as } n \to \infty.$$

Hint:

Denote by $B_i = \{i \leq f < i+1\}$, then from

$$\int f\, d\mu = \sum_{i=0}^{\infty} \int_{B_i} f\, d\mu \geq \sum_{i=1}^{\infty} i\, \mu(B_i) = \sum_{n=1}^{\infty} \sum_{i=n}^{\infty} \mu(B_i) = \sum_{n=1}^{\infty} \mu\{f \geq n\},$$

we deduce that

$$\sum_{n=1}^{\infty} \mu\{f \geq n\} < \infty.$$

Now, for any $\varepsilon > 0$

$$\mu\{\limsup_n h_n > \varepsilon\} = \lim_{n \to \infty} \mu\left(\bigcup_{k=n}^{\infty} \{h_k > \varepsilon\}\right)$$

$$\leq \lim_{n \to \infty} \sum_{k=n}^{\infty} \mu\{h_k > \varepsilon\} = \lim_{n \to \infty} \sum_{k=n}^{\infty} \mu\{f \geq k\} = 0.$$

Consequently, the $\limsup_n h_n = 0$ (a.e).

5. Let $f \geq 0$ be an integrable function on $(X, \mathcal{A}, \mu)$. Show that the measure μ_1 defined by

$$\mu_1(A) = \int_A f\, d\mu$$

satisfies $\mu_1(N) = 0$ if $\mu(N) = 0$. In such a case the measure μ_1 is said to be "**absolutely continuous**" with respect to μ.

Hint:

We must show that for every $N \in \mathcal{A}$ such that $\mu(N) = 0$ it follows that $\mu_1(N) = 0$. To this end define

$$f_n = f\, I_{\{f \leq n\}} + n\, I_{\{f > n\}} \qquad n = 1, 2, \cdots .$$

Clearly, $f_n \uparrow f$ so that according to the monotone convergence theorem

$$\int f_n\, d\mu \uparrow \int f\, d\mu .$$

Thus, given $\varepsilon > 0$, there is $n_0 = n_0(\varepsilon)$ such that

$$0 \leq \int f\, d\mu - \int f_n\, d\mu < \varepsilon / 2n_0 \ \text{ if } n \geq n_0 .$$

Hence, for any $A \in \mathcal{A}$ satisfying $\mu(A) < \varepsilon / 2$, we have:

$$\mu_1(A) = \int_A f\, d\mu = \int_A f_{n_0}\, d\mu + \int_A (f - f_{n_0})\, d\mu \leq \frac{\varepsilon}{2} + \int_A (f - f_{n_0})\, d\mu < \varepsilon .$$

6. Let $f \geq 0$ defined on $(X, \mathcal{A}, \mu)$ be integrable, then $\mu\{f = +\infty\} = 0$. Given an arbitrary $\varepsilon > 0$ show that there exists $A \in \mathcal{A}$ such that

$$\mu(A) < \infty \text{ and } \int_{A^c} f \, d\mu < \varepsilon.$$

Hint:

Write

$$A_0 = \{f = 0\} \quad \text{and} \quad A_n = \{\tfrac{1}{n} \leq f < n\}, \qquad n = 1, 2, \cdots.$$

Clearly $A_1 \subset A_2 \subset \cdots$ and

$$A = \bigcup_{n=1}^{\infty} A_n = \{0 < f < \infty\}. \text{ Hence}$$

$$A^c = \bigcap_{n=1}^{\infty} A_n^c = A_0 \cup \{f = +\infty\}.$$

Now, since

$$\lim_{n \to \infty} \int_{A_n^c} f \, d\mu = \int_{A_0} f \, d\mu + \int_{\{f = +\infty\}} f \, d\mu = 0,$$

it follows that for any $\varepsilon > 0$ there exists an integer $n_0 > 0$ such that for all $k \geq n_0$, $\int_{A_k^c} f \, d\mu \leq \varepsilon$. On the other hand, since

$$0 \leq \int f \, d\mu = \int_{A_n} f \, d\mu + \int_{A_n^c} f \, d\mu \leq \int_{A_n} f \, d\mu + \varepsilon < \infty,$$

we have that

$$\infty > \int_{A_n} f \, d\mu > \frac{1}{n} \mu(A_n) \Rightarrow \mu(A_n) < \infty.$$

7. (Continuation) Show that for every $\varepsilon > 0$ there exists $\delta > 0$ such that for any $G \in \mathcal{A}$ with $\mu(G) < \delta$

$$\int_G f \, d\mu < \varepsilon.$$

Hint:

On A_n, $f < n$; Since

$$\int_G f \, d\mu = \int_{G \cap A_n} f \, d\mu + \int_{G \cap A_n^c} f \, d\mu \leq n \, \mu(G) + \varepsilon < 2\varepsilon$$

if $\mu(G) < \varepsilon / n$.

8. Let $\{h_n\}_1^{\infty}$ be a sequence of non-negative measurable functions on $(X, \mathcal{A}, \mu)$. Show that

$$\int \left(\sum_{k=1}^{\infty} h_k \right) d\mu = \sum_{k=1}^{\infty} \int h_k \, d\mu.$$

Hint:

Denote by

$$f_n = \sum_{k=1}^{n} h_k.$$

Since $f_n \uparrow \sum_{k=1}^{\infty} h_n$, it follows from the monotone convergence theorem that

$$\lim_{n \to \infty} \int f_n \, d\mu = \sum_{k=1}^{\infty} \int h_k \, d\mu = \int \lim_{n \to \infty} f_n \, d\mu = \int \left(\sum_{k=1}^{\infty} h_k \right) d\mu.$$

9. Determine the integral

$$F(\alpha) = \int_0^1 e^{\alpha x} \, d\Phi(x)$$

where Φ is the Cantor function.

Hint:

We have (see example 1) that

$$F(\alpha) = \int_0^{1/3} e^{\alpha x} \, d\Phi(x) + \int_{1/3}^1 e^{\alpha x} \, d\Phi(x)$$

$$= \frac{1}{2} \left(\int_0^1 e^{\alpha y/3} \, d\Phi(y) + \int_0^1 (e^{2/3\, \alpha + \alpha y/3}) \, d\Phi(y) \right)$$

$$= \frac{1}{2} (1 + e^{2/3\, \alpha}) \, F(\frac{\alpha}{3}).$$

From this recursion we deduce that

$$F(\alpha) = e^{\alpha/2} \prod_{k=1}^{\infty} c\,h\left(\frac{\alpha}{3^k}\right)$$

where

$$c\,h\,x = (e^x + e^{-x})/2 .$$

10. Let f and g be measurable on $(X, \mathcal{A}, \mu)$. If g is integrable and $0 \le a \le f \le b < \infty$ (a.e), show that there exists $\alpha \in [a, b]$ such that

$$\int g \cdot f \, d\mu = \alpha \int g \, d\mu$$

Hint:

This is the "mean- value theorem" for integrals. Since

$$a \int g d\mu \le \int g \cdot f \, d\mu \le b \int g d\mu,$$

we have that

$$a \le \int f \cdot g \, d\mu / \int g \, d\mu \le b; \ \text{Set } \alpha = \int f \cdot g \, d\mu \, / \int g \, d\mu.$$

30 SOME LIMIT THEOREMS

This section is concerned with some limit theorems involving sequences of non-negative measurable functions on $(X, \mathcal{A}, \mu)$. The **Fatou lemma** is a simple consequence of the monotone convergence theorem and will be discussed here first. Its usefulness stems from the fact that it enables us to deal with some asymptotic problems involving sequences of non-negative measurable functions which are not monotone.

Proposition 1 (Fatou)

For any sequence $\{f_n\}_1^\infty$ of non-negative measurable functions on a measure space $(X, \mathcal{A}, \mu)$ we have that

$$\int \liminf f_n \, d\mu \le \liminf \int f_n \, d\mu , \tag{1}$$

where the limit on the right hand side exists in $[0, \infty]$.

Proof:

Set $h_n = \inf_{k \ge n} f_k$, then clearly $h_1 \le h_2 \le \cdots$ and consequently

$$\int h_m \, d\mu \le \int f_n \, d\mu \qquad \text{if } m \le n ,$$

implying that for all $m = 1, 2, \ldots ,$

$$\int h_m \, d\mu \le \liminf \int f_n \, d\mu.$$

Since $\{h_n\}_1^\infty$ is an increasing sequence of nonnegative measurable functions, which converges pointwise to the $\liminf f_n$, the monotone convergence theorem yields:

$$\lim_{m \to \infty} \int h_m \, d\mu = \int \lim_{m \to \infty} h_m \, d\mu = \int \liminf f_n \, d\mu .$$

This and the previous inequality prove (1).   ♥

Remark 1

The Fatou lemma may not hold for sequences of measurable functions which are not positive, as the next example shows: Let $\{f_n\}_1^\infty$ be a sequence of maps on $(R, \mathcal{B}, \bar{\lambda})$ defined by

$$f_n = -\frac{1}{n} \, I_{[0, n]} .$$

Clearly, $f_n \to f = 0$ as $n \to \infty$. Thus,

$$\int f \, d\bar{\lambda} = 0, \quad \text{while} \quad \int f_n \, d\bar{\lambda} = -1 \qquad \text{for all } n = 1, 2, \cdots ,$$

and consequently, $\liminf \int f_n \, d\bar{\lambda} = -1$.

Proposition 2

Let $\{f_n\}_1^\infty$ be a sequence of integrable functions such that $0 \le f_1 \le f_2 \le \cdots$. If $\{\int f_n \, d\mu\}_1^\infty$ is bounded (say, by a constant $M \in (0, \infty)$), then the limit

$$f = \lim_{n \to \infty} f_n = \sup_{1 \le n < \infty} f_n$$

229

is (a.e.) finite, integrable and

$$\lim_{n\to\infty} \int f_n \, d\mu = \int f \, d\mu .$$

Proof:

For any $i = 1, 2, \ldots$

$$\{f \geq i\} = \{ \sup_{1 \leq n < \infty} f_n \geq i\} = \bigcup_{n=1}^{\infty} \{f_n \geq i\}.$$

By the Chebishev inequality

$$i \, \mu\{f_n \geq i\} \leq \int f_n \, d\mu \leq M \;\; \Rightarrow \;\; \mu\{f_n \geq i\} \leq M / i.$$

Hence, by letting $n \to \infty$ we obtain that $\mu\{f \geq i\} \leq M/i$. Now, since

$$\{f = +\infty\} = \bigcap_{i=1}^{\infty} \{f \geq i\} \subset \{f \geq k\} \text{ for all } k = 1, 2, \cdots$$

it follows that

$$\mu\{f = +\infty\} \leq \mu\{f \geq k\} \leq M / k \to 0 \text{ as } k \to \infty,$$

which proves that f is (a.e) finite. To show that f is integrable, we apply the Fatou lemma to obtain

$$\int f \, d\mu = \int \liminf f_n \, d\mu \leq \liminf \int f_n \, d\mu \leq M . \tag{$*$}$$

Finally, the Fatou lemma applied to the sequence $\{f - f_n\}_1^{\infty}$ yields

$$\int \liminf (f - f_n) \, d\mu \leq \liminf \int (f - f_n) \, d\mu = \int f \, d\mu - \limsup \int f_n \, d\mu .$$

Hence, we have that

$$\int \limsup f_n \, d\mu \geq \limsup \int f_n \, d\mu .$$

From this and from $(*)$ we deduce that

$$\int f \, d\mu \leq \liminf \int f_n \, d\mu \leq \limsup \int f_n \, d\mu$$

$$\leq \int \limsup f_n \, d\mu = \int f \, d\mu,$$

which completes the proof of the proposition. ♥

Problems and Complements

1. Let $\{f_n\}_1^\infty$ be a sequence of integrable functions such that for every $n = 1, 2, \cdots$

$$0 \le f_n \le g \qquad \text{(a.e.)},$$

where g is also integrable. Show that

$$\lim \sup \int f_n \, d\mu \le \int \lim \sup f_n \, d\mu .$$

This is, so called, the second Fatou lemma.

 Hint:

 Since $h_n = g - f_n \ge 0$ (a.e.) for each $n = 1, 2, \cdots$, the Fatou lemma applies and we have:

$$\int \lim \inf h_n \, d\mu \le \lim \inf \int h_n \, d\mu .$$

But, $\lim \inf h_n = g - \lim \sup f_n$ (a.e.) so that the last inequality yields

$$\int g \, d\mu - \int \lim \sup f_n \, d\mu \le \int g \, d\mu - \lim \sup \int f_n \, d\mu .$$

2. (Continuation) If $f_n \to f$ (a.e.) show that f is integrable function and that

$$\lim \int f_n \, d\mu = \int \lim f_n \, d\mu = \int f \, d\mu.$$

 Hint:

 Clearly $f \le g$ (a. e.), which proves integrability of f. Next, the sequence

$\{g - f_n\}_1^\infty$ satisfies the requirements of Fatou's lemma and

$$\lim \inf (g - f_n) = g - \lim \sup f_n = g - f \,(\text{a.e.}) .$$

Thus,

$$\int g \, d\mu - \int f \, d\mu = \int (g - f) \, d\mu = \int \lim \inf (g - f_n) \, d\mu.$$

Hence

$$\lim \sup \int f_n \, d\mu \le \int f \, d\mu .$$

Similarly, Fatou's lemma applied to the sequence $\{g + f_n\}_1^\infty$ gives

$$\int g \, d\mu + \int f \, d\mu = \int (g + f_n) \, d\mu = \int \lim \inf (g + f_n) \, d\mu$$
$$\le \lim \inf \int (g + f_n) \, d\mu = \int g \, d\mu + \lim \inf \int f_n \, d\mu,$$

and also

$$\int f \, d\mu \le \lim \inf \int f_n \, d\mu .$$

Consequently, $\lim \int f_n \, d\mu$ exists in R, and

$$\lim \int f_n \, d\mu = \int f \, d\mu$$

holds. This is a version of **the Lebesgue dominated convergence theorem**.

3. If the sequence $\{f_n\}_1^\infty$ in the last problem converges in measure to f, show that in such a case f is

integrable and that

$$\lim_{n \to \infty} \int |f_n - f| \, d\mu = 0.$$

Hint:

By propositions 4 and 5 of section 26, the sequence $\{f_n\}_1^\infty$ has a subsequence that converges (a.e) to a measurable function f, where $f \leq g$ (a.e), which implies integrability of f. We prove the second part of the claim by contradiction. Assume that for some $\varepsilon > 0$ there is a subsequence $\{\bar{f}_k\}_1^\infty$ of $\{f_n\}_1^\infty$ such that

$$\int |\bar{f}_k - f| \, d\mu \geq \varepsilon .$$

By proposition 4 of section 26 there exists a subsequence $\{\bar{f}'_k\}_1^\infty$ of $\{\bar{f}_k\}_1^\infty$ such that $\bar{f}'_k \to f$ (a.e).

Now, the Lebesgue dominated convergence theorem implies that

$$0 < \varepsilon \leq \int |\bar{f}'_k - f| \, d\mu \to 0 \text{ as } k \to \infty$$

which is contradiction.

4. Let $\{f_n\}_1^\infty$ be a sequence of measurable functions on $(R, \mathcal{B}, \bar{\lambda})$ defined by

$$f_n = \begin{cases} I_{[0, \frac{1}{4}]} & \text{if } \quad n \text{ is even} \\ I_{(\frac{1}{4}, 1]} & \text{if } \quad n \text{ is odd} \end{cases} .$$

Show that both of Fatou's inequalities hold.

Hint:

Clearly,

$$\lim \inf f_n = 0, \quad \lim \sup f_n = I_{[0, 1]} .$$

In addition,

$$\int f_n \, d\mu = \begin{cases} \dfrac{1}{4} & \text{if } \quad n \text{ is even} \\ \dfrac{3}{4} & \text{if } \quad n \text{ is odd} \end{cases} .$$

Consequently,

$$0 = \int \lim \inf f_n \, d\bar{\lambda} \leq \lim \inf \int f_n \, d\bar{\lambda} = \frac{1}{4}$$

$$1 = \int \lim \sup f_n \, d\bar{\lambda} \geq \lim \sup \int f_n \, d\bar{\lambda} = \frac{3}{4} .$$

Here the function $g = I_{[0, 1]}$.

5. Let $\{f_n\}_1^\infty$ be defined on $(R, \mathcal{B}, \bar{\lambda})$ by

$$f_n(x) = \begin{cases} 0 & \text{if} \quad x < 0 \\ nx^2 & \text{if} \quad x \in [0, \frac{1}{n}] \\ -n^2 x^2 (x - \frac{2}{n}) & \text{if} \quad x \in [\frac{1}{n}, \frac{2}{n}] \\ 0 & \text{if} \quad x \geq \frac{2}{n} \end{cases}.$$

Verify if Fatou's inequalities hold.

Hint:

Notice that a dominating function g does not exist. It is clear that each $f_n \geq 0$ and measurable. It is easy to verify that

$$\int f_n \, d\overline{\lambda} = 1 \text{ for every } n = 1, 2, \cdots .$$

Thus,

$$\limsup \int f_n \, d\overline{\lambda} = \liminf \int f_n \, d\overline{\lambda} = 1 .$$

On the other hand, for each $x \in R$ $f_n(x) \to 0$ as $n \to \infty$, so that

$$\liminf f_n = \limsup f_n = 0.$$

Consequently,

$$0 = \int \limsup f_n \, d\overline{\lambda} \leq \liminf \int f_n \, d\overline{\lambda} = 1.$$

However,

$$0 = \int \limsup f_n \, d\overline{\lambda} = 0 \quad \text{while } \limsup \int f_n \, d\overline{\lambda} = 1$$

so that the second inequality does not hold (there is no a dominating integrable function g).

Given a measure space $(X, \mathcal{A}, \mu)$, we denoted by $L_0 = L_0 (X, \mathcal{A}, \mu)$ the collection of all measurable functions $f:X \to R$. More precisely, L_0 is the collection of all equivalence classes under (a.e) equality, of measurable real-valued functions on the measure space $(X, \mathcal{A}, \mu)$. It is clear that L_0 is a vector space.

For every $f \in L_0$ the corresponding f^-, $f^+ \in L_0$ (see section 24) and since these two functions are non-negative the integrals

$$\int f^- \, d\mu \quad \text{and} \quad \int f^+ \, d\mu$$

exist in $[0, \infty]$. If at least one of them is finite, we define the integral of the function f over X to be

$$\int f \, d\mu = \int f^+ \, d\mu - \int f^- \, d\mu. \tag{1}$$

If both f^- and f^+ are integrable, the function $f = f^+ - f^-$ is integrable. Since $|f| = f^+ + f^-$, an equivalent definition of integrability of a member $f \in L_0$ is that

$$\int |f| \, d\mu < \infty. \tag{2}$$

The subclass of L_0, which consists of integrable functions, is usually denoted by $L_1 = L_1 (X, \mathcal{A}, \mu)$. From proposition 2 of section 29 it follows that every $f \in L_1$ is (a. e) finite. The collection L_1 is a vector space since for any $f_1, f_2 \in L_1$ the inequality $|f_1 + f_2| \leq |f_1| + |f_2|$ implies that $f_1 + f_2 \in L_1$. If $f \in L_1$ and if $h:X \to R$ differs from f only on a set of measure zero, then $h \in L_1$ and the integral of f and h coincide. Thus, any $f \in L_1$ can always be redefined on a set of measure zero say, by giving it a constant value on such a set, so that the new map is also integrable.

The properties of the integral which we obtained for positive integrable functions are now extended to the integral on L_1. We shall go through these properties systematically once more. We start by repeating that the map

$$\int : L_1 \to R$$

is linear.

Proposition 1

For any $f_1, f_2 \in L_1$

$$\int (f_1 + f_2) \, d\mu = \int f_1 \, d\mu + \int f_2 \, d\mu. \tag{3}$$

Proof:

From

$$(f_1 + f_2)^+ - (f_1 + f_2)^- = (f_1 + f_2) = (f_1^+ - f_1^-) + (f_2^+ - f_2^-)$$

we obtain that

$$(f_1 + f_2)^+ + (f_1^- + f_2^-) = (f_1 + f_2)^- + (f_1^+ + f_2^+).$$

Since both sides of this equation are finite sums of positive integrable functions we have that

$$\int (f_1 + f_2)^+ \, d\mu - \int (f_1 + f_2)^- \, d\mu = \int (f_1 + f_2) \, d\mu = \int f_1^+ \, d\mu - \int f_1^- \, d\mu$$

$$+ \int f_2^+ \, d\mu - \int f_2^- \, d\mu = \int f_1 \, d\mu + \int f_2 \, d\mu.$$

This proves the proposition. ♥

Corollary 1

It follows from (2) and (3) that

$$\sup(f_1, f_2) = f_1 \vee f_2 = (f_1 + f_2 + |f_1 - f_2|) / 2 \in L_1$$

$$\inf(f_1, f_2) = f_1 \wedge f_2 = (f_1 + f_2 - |f_1 - f_2|) / 2 \in L_1$$

if $f_1, f_2 \in L_1$.

Proposition 2

For any $f \in L_1$ and $c \in R$

$$\int c\, f \, d\mu = c \int f \, d\mu \qquad (4)$$

Proof:

If $c > 0$ then $c\, f = (c\, f)^+ - (c\, f)^- = c\, f^+ - c\, f^-$ so (4) obviously holds. If $c < 0$, $(c\, f) = (c\, f)^+ - (c\, f)^- = -c\, f^- + c\, f^+$ and hence

$$\int c\, f \, d\mu = -\int c\, f^- \, d\mu + c \int f^+ \, d\mu = c \int f \, d\mu. \qquad ♥$$

Corollary 2

If $f_1, f_2 \in L_1$ and $c_1, c_2 \in R$ then

$$\int (c_1 f_1 + c_2 f_2) \, d\mu = c_1 \int f_1 \, d\mu + c_2 \int f_2 \, d\mu.$$

This clearly follows from (3) and (4).

Corollary 3

For any $f \in L_1$

$$\left| \int f \, d\mu \right| = \left| \int (f^+ - f^-) \, d\mu \right| \leq \int (f^+ + f^-) \, d\mu = \int |f| \, d\mu.$$

Remark 1

The integral of a function $f \in L_1$ over a measurable set $B \in \mathcal{A}$ is defined to be

$$\int_B f \, d\mu = \int f I_B \, d\mu$$

From this and (3) we have for any two disjoint $B_1, B_2 \in \mathcal{A}$ that

$$\int_{B_1 \cup B_2} f \, d\mu = \int f (I_{B_1} + I_{B_2}) \, d\mu = \int_{B_1} f \, d\mu + \int_{B_2} f \, d\mu.$$

Proposition 3

For any two $h_1, h_2 \in L_1$ such that $h_1 \leq h_2$ (a.e) on X

$$\int h_1\, d\mu \le \int h_2\, d\mu .$$

Proof:

Since $h_2 = h_1 + (h_2 - h_1)$ we have that

$$\int h_2\, d\mu = \int h_1\, d\mu + \int (h_2 - h_1)\, d\mu \ge \int h_1\, d\mu .$$

$\heartsuit$

Proposition 4

If $f \in L_1$ is such that

$$\int_G f\, d\mu = 0 \qquad \text{for all } G \in \mathcal{A}$$

then $f=0$ (a.e.) on X.

Proof:

Denote by $A=\{f>0\}$ and by $B=\{f<0\}$ then since by assumption

$$0 = \int_A f\, d\mu = \int f I_A\, d\mu \qquad \text{and } f>0 \ \text{ on } A \Rightarrow I_A = 0 \quad (a.e)$$

(see proposition 4 of section 29). Thus $\mu(A)=0$. Similarly we conclude that $\mu(B)=0$ which implies that $f = 0$ (a.e).

A version of the next proposition was discussed previously (see problem 5 of section 29).

Proposition 5

Given any $f \in L_1$ and $\varepsilon > 0$, there exists a $\delta > 0$ such that

$$\left| \int_B f\, d\mu \right| \le \varepsilon \ \text{ for every } B \in \mathcal{A} \ \text{ such that } \mu(B) < \delta.$$

Proof:

Existence of such a $\delta>0$ is trivial if f is bounded, say $|f| \le M < \infty$, since for any $B \in \mathcal{A}$

$$\left| \int_B f\, d\mu \right| \le \int_B |f|\, d\mu \le M\, \mu(B).$$

From this we conclude that $\delta = \varepsilon/M$ is the required number. To prove the proposition in general, consider a sequence $\{B_n\}_1^\infty$ of disjoint sets from $\mathcal{A}$ where

$$B_n = \{n - 1 \le |f| < n\} \cap B, \quad n = 1, 2, \cdots$$

and write

$$D_n = \bigcup_{k=1}^n B_k , \quad C_n = B - D_n .$$

Then we have that

$$\int_B |f|\, d\mu = \sum_{i=1}^\infty \int_{B_i} |f|\, d\mu = \int_{D_n} |f|\, d\mu + \int_{C_n} |f|\, d\mu .$$

Since $C_n \downarrow \theta$ there exists an integer $n > 0$ such that

$$\int_{C_n} |f|\, d\mu \le \varepsilon/2 .$$

On the other hand, $|f| \le n$ on D_n so that

$$\int\limits_B |f|\, d\mu \le n\, \mu(B) + \varepsilon/2 ,$$

from which we deduce a $\delta = \varepsilon/2n$. ♥

Next, we formulate and prove the **"Lebesgue dominated" convergence theorem,** which is one of the most useful results in the theory of integration.

Proposition 6

Let $\{f_n\}_1^\infty$ be a sequence from $L_1(X, \mathcal{A}, \mu)$ satisfying $|f_n| \le g$ (a.e) on X for all $n=1, 2, \cdots$, where $g \in L_1$. If $f_n \to f$ (a.e) then $f \in L_1$ and

$$\lim_{n \to \infty} \int f_n\, d\mu = \int f\, d\mu.$$

Proof:

Clearly, $|f| \le g$ (a.e) on X so that $f \in L_1$. Since $g - f_n \ge 0$ (a.e) the sequence $\{g - f_n\}_1^\infty$ is subject to Fatou's lemma. Bearing in mind that

$$\lim \inf(g - f_n) = \lim_{n \to \infty}(g - f_n) = g - f \quad (a.e)$$

we have that

$$\int (g - f)\, d\mu = \int \lim \inf(g - f_n) \le \lim \inf \int (g - f_n)\, d\mu$$

$$= \int g\, d\mu - \lim \sup \int f_n\, d\mu .$$

Hence,

$$\lim \sup \int f_n\, d\mu. \le \int f\, d\mu. \tag{$*$}$$

Now, by applying the Fatou lemma to the sequence $\{g + f_n\}_1^\infty$ we obtain:

$$\int (g + f)\, d\mu = \int \lim \inf(g + f_n)\, d\mu \le \lim \inf \int (g + f_n)\, d\mu$$

$$= \int g\, d\mu + \lim \inf \int f_n\, d\mu,$$

yielding

$$\int f\, d\mu \le \lim \inf \int f_n\, d\mu .$$

Combining this and $(*)$ we have

$$\int f\, d\mu \le \lim \inf \int f_n\, d\mu \le \lim \sup \int f_n\, d\mu \le \int f\, d\mu,$$

which proves the proposition. ♥

Remark 2 (L_p spaces)

In the following we make the convention of setting $1/\infty = 0$, $0 \cdot \infty = 0$ and for any $0 < p < \infty$, we denote by $L_p = L_p(X, \mathcal{A}, \mu)$ the collection of all functions $f \in L_0$ for which

$$\|f\|_p = (\int |f|^p \, d\mu)^{1/p} < \infty. \tag{5}$$

The function space $\mathbf{L}_\infty = \mathbf{L}_\infty(X, \mathcal{A}, \mu)$ consists of all essentially bounded maps from $\mathbf{L}_0$. The mapping

$$\|\cdot\|_p: \mathbf{L}_p \to [0, \infty)$$

is a norm for all $p \geq 1$; in other words,

    (i) $\|f\|_p \geq 0$ and $=0$ if and only if $f = 0$ (a.e)

    (ii) if $c \in R$, $\|cf\| = |c| \, \|f\|$ (6)

    (iii) if $f_1, f_2 \in \mathbf{L}_p$, then $\|f_1 + f_2\|_p \leq \|f_1\|_p + \|f_2\|_p$.

Properties (i) and (ii) follow directly from (5). For the proof of (iii) see problem 5. In the case of $\mathbf{L}_\infty$ we have

$$\|f\|_\infty = \operatorname{ess\,sup} |f| = \inf \{c; \ |f| \leq c \ (\text{a.e})\}. \tag{7}$$

Strictly speaking $\|\cdot\|_p$ is a "**seminorm**" because from $\|f\|_p = 0$ it does not follow that $f \equiv 0$ on X, only that $f = 0$ (a.e). In the following we shall identify two functions from $\mathbf{L}_0$ which are equal (a.e). With this convention $\|\cdot\|_p$ is a norm on $\mathbf{L}_p$ for all $1 \leq p \leq \infty$. Let us point out that certain authors use the symbol $\mathcal{L}_p$ for what we use $\mathbf{L}_p$. For us, $\mathbf{L}_p$ is a normed vector space for each $1 \leq p \leq \infty$.

Example 1

The following counter example shows that the map $\|\cdot\|_p$ is not a norm if $0 < p < 1$. Set $X = [0, 1]$ and let $\mathcal{A}$ be the σ-algebra of Borel subsets of $[0, 1]$. Assume that $\mu = \bar{\lambda}$, then if $f_1 = I_{(0, \frac{1}{2})}$, $f_2 = I_{(\frac{1}{2}, 1)}$, we clearly have that

$$\|f_1 + f_2\|_p = (\int (I_{(0, \frac{1}{2})} + I_{(\frac{1}{2}, 1)})^p \, d\bar{\lambda})^{1/p} = 1 > 2^{1 - \frac{1}{p}}$$

$$= (\tfrac{1}{2})^{1/p} + (\tfrac{1}{2})^{1/p} = \|f_1\|_p + \|f_2\|_p.$$

Problems and complements

1. Show that

$$\sum_{n=1}^{\infty} \frac{(-1)^{n+1}}{n} = \log 2.$$

Hint:

Clearly

$$\sum_{n=1}^{\infty} \frac{(-1)^{n+1}}{n} = \sum_{n=0}^{\infty} \left(\frac{1}{2n+1} - \frac{1}{2n+2}\right).$$

Since for any $0 \leq x < 1$

$$\frac{1}{2} \log \frac{1+x}{1-x} = \sum_{n=0}^{\infty} \frac{x^{2n+1}}{(2n+1)},$$

we have that (see problem 8 of section 29)

$$\frac{1}{2} \int_0^1 \log \frac{1+x}{1-x} \, dx = \sum_{n=0}^{\infty} \int_0^1 \frac{x^{2n+1}}{2n+1} \, dx = \sum_{n=0}^{\infty} \left(\frac{1}{2n+1} - \frac{1}{2n+2}\right).$$

On the other hand,

$$\int_0^1 \log \frac{1+x}{1-x} \, dx = [(1+x) \log (1+x) - (1-x) \log (1-x)] \Big|_0^1$$

$$= 2 \log 2 .$$

2. Let $\{f_n\}_1^{\infty}$ be a sequence from L_0. If there exists a map $g \in L_0$ such that $g^- \in L_1$ and $g \leq f_n$ (a.e) for all $n=1, 2, \cdots$, then

$$\int \lim \inf f_n \, d\mu \leq \lim \inf \int f_n \, d\mu.$$

Hint:

Here is a sequence of not necessarily positive measurable function for which Fatou's theorem holds.

Since $g^- \in L_1$

$$-\infty < \int g \, d\mu \leq \int f_n \, d\mu \quad \text{for all} \quad n=1, 2, \cdots .$$

If $\int g \, d\mu = +\infty$ the inequality is trivial. Assume that $g \in L_1$. Since $f_n - g \geq 0$ the Fatou lemma applies to $\{f_n - g\}_1^{\infty}$. Hence

$$\int \lim \inf (f_n - g) \, d\mu = \int \lim \inf f_n \, d\mu - \int g \, d\mu$$

$$\leq \lim \inf \int f_n \, d\mu - \int g \, d\mu.$$

On the other hand, if $f_n \leq g$ (a.e.) for every $n=1, 2, \cdots$ and $g_- \in L_1$

$$\int \lim \inf (g - f_n) \, d\mu \leq \lim \inf \int (g - f_n) \, d\mu ,$$

which yields that

$$\lim \sup \int f_n \, d\mu \leq \int \lim \sup f_n \, d\mu.$$

3. Show that a map $f \in \mathbf{L}_0$ is integrable if and only

$$\sum_{n=-\infty}^{\infty} 2^n \mu\{2^{n-1} < |f| \le 2^n\} < \infty. \tag{$*$}$$

Hint:

The sets $A_n = \{2^{n-1} < |f| \le 2^n\}$, $B = \{f = 0\}$ and $C = \{|f| = +\infty\}$ form a measurable partition of X. Thus,

$$\int |f|\, d\mu = \sum_{n=-\infty}^{\infty} \int_{A_n} |f|\, d\mu + \int_C |f|\, d\mu.$$

Now, if $f \in \mathbf{L}_1$, $\mu(C) = 0$ and since

$$\int_{A_n} |f|\, d\mu \ge \frac{1}{2} 2^n \mu(A^n),$$

the condition $(*)$ clearly holds. To prove its sufficiency, assume $(*)$ holds, then since

$$\{|f| = +\infty\} = \bigcap_{i=1}^{\infty} \{|f| \ge i\} \subset \{|f| \ge n\} = \bigcup_{k=n}^{\infty} A_k$$

$$\mu\{|f| = +\infty\} \le \sum_{k=n}^{\infty} \mu(A_k) \to 0 \quad \text{as } n \to \infty.$$

Hence,

$$\int |f|\, d\mu = \sum_{n=-\infty}^{\infty} \int_{A_n} |f|\, d\mu \ge \frac{1}{2} \sum_{n=-\infty}^{\infty} 2^n \mu(A^n).$$

4. Let $f_1 \in \mathbf{L}_p$ and $f_2 \in \mathbf{L}_q$, where $p > 1$ and

$$\frac{1}{p} + \frac{1}{q} = 1.$$

Show that $f_1 \cdot f_2 \in \mathbf{L}_1$ and that (Hölder's inequality)

$$\| f_1 \cdot f_2 \|_1 \le \| f_1 \|_p \cdot \| f_2 \|_q$$

Hint:

Consider the function $\varphi(t) = t^\alpha - \alpha t$, where $t \ge 0$ and $\alpha \in (0,1)$. Cleary, $\varphi'(t) > 0$ in $(0, 1)$ and $\varphi'(t) < 0$ if $t > 1$. In addition, $\varphi(1) \ge \varphi(t)$ for all $t \ge 0$, so that

$$t^\alpha \le \alpha t - (1 - \alpha).$$

For $t = a/b$, where $a, b \in (0, \infty)$, we obtain, after a simple transformation, the following inequality

$$a^\alpha b^{1-\alpha} \le \alpha a + (1 - \alpha) b.$$

Put $\alpha = 1/p$, $a = (|f_1| / \| f_1 \|_p)^p$ and $b = (|f_2| / \| f_2 \|_q)^q$ to obtain

$$\frac{|f_1 \cdot f_2|}{\|f_1\|_p \cdot \|f_2\|_q} \leq \frac{1}{p} \cdot \frac{|f_1|^p}{(\|f_1\|_p)^p} + \frac{1}{q} \cdot \frac{|f_2|^q}{(\|f_2\|_q)^q} \ ,$$

which shows that $|f_1 \cdot f_2| \in L_1$. Hence,

$$\frac{\int |f_1 \cdot f_2| \, d\mu}{\|f_1\|_p \cdot \|f_2\|_q} \leq \frac{1}{p} + \frac{1}{q} = 1.$$

5 Show that for any $1 \leq p < \infty$ and $f_1, f_2 \in L_p$ we have:

$$\|f_1 + f_2\|_p \leq \|f_1\|_p + \|f_2\|_p.$$

Hint:

This is the "Minkovski inequality". It clearly holds for p=1. To prove for it p>1 notice that

$$|f_1 + f_2|^p \leq (2 \max\{|f_1|, |f_2|\})^p \leq 2^p(|f_1|^p + |f_2|^p) \tag{**}$$

which implies that $|f_1 + f_2| \in L_p$. Next,

$$|f_1 + f_2|^p \leq |f_1 + f_2|^{p-1} |f_1 + f_2| \leq |f_1 + f_2|^{p-1} (|f_1| + |f_2|).$$

Hence, we have:

$$\int |f_1 + f_2|^p \, d\mu \leq \int |f_1 + f_2|^{p-1} |f_1| \, d\mu + \int |f_1 + f_2|^{p-1} |f_2| \, d\mu.$$

By Hölder's inequality,

$$\int |f_1 + f_2|^{p-1} |f_i| \, d\mu \leq \|f_i\|_p (\| |f_1 + f_2|^{p-1} \|_q).$$

Since $(1 - p)q = p$ we have

$$\| |f_1 + f_2|^{p-1} \|_q) = (\int |f_1 + f_2|^{(p-1)q})^{1/q} = (\|f_1 + f_2\|_p)^{p/q}.$$

which yields

$$(\|f_1 + f_2\|_p)^p \leq (\|f_1\|_p + \|f_2\|_p)(\|f_1 + f_2\|_p)^{p/q},$$

proving that $\|\cdot\|_p$ is a norm on L_p for each $p \geq 1$.

6. Show that for any $0 < p < 1$

$$\|f_1 + f_2\|_p \leq 2^{1/p}(\|f_1\|_p + \|f_2\|_p).$$

Hint:

The function $h(t) = 1 + t^p - (1 + t)^p$ is non-negative (and non-decreasing) for all $t \geq 0$. Hence $1 + t^p \geq (1 + t)^p$ which for $t = x/y$ yields

$$(x^p + y^p) \geq (x + y)^p. \tag{***}$$

Using this and (**) we obtain that

$$\|f_1 + f_2\|_p = (\int |f_1 + f_2|^p \, d\mu)^{1/p} \leq (\int |f_1|^p \, d\mu + \int |f_2|^p \, d\mu)^{1/p}$$

$$\leq 2^{1/p}\left(\,\|f_1\|_p + \|f_2\|_p\,\right).$$

The inequality $(***)$ can be obtained for $0 < p < 1$ more directly as follows: let $a, b \in (0, \infty)$ then

$$(a+b)^p = (a+b)(a+b)^{p-1} = \frac{a}{(a+b)^{1-p}} + \frac{b}{(a+b)^{1-p}} \leq \frac{a}{a^{1-p}} + \frac{b}{b^{1-p}} = a^p + b^p.$$

7. In general the L_p-spaces are not comparable. However, if $(X, \mathcal{A}, \mu)$ is a finite measure space, show that

$L_t \subset L_s$ for every $0 \leq s < t$.

Hint:

Denote by $A=\{|f|\leq 1\}$, $B=\{|f|\geq 1\}$ where $f \in L_t$, then

$$\int |f|^s \, d\mu = \int_A |f|^s \, d\mu + \int_B |f|^s \, d\mu \leq \mu(A) + \int_B |f|^t \, d\mu,$$

implying that $f \in L_s$.

8. (Continuation) If $f \in L_t$ show that

$$\|f\|_s \leq \|f\|_t \qquad \text{if } \mu(X) = 1.$$

Hint:

Clearly, $f \in L_s$. Put $p = t/s$ and choose q so that

$$\frac{1}{p} + \frac{1}{q} = 1.$$

Since $|f|^s \subset L_p$, it follows from Hölder inequality that

$$(\|f\|_s)^s = \int |f|^s \, d\mu \leq \left(\int |f|^{ps} \, d\mu\right)^{1/p} \left(\int 1^q \, d\mu\right)^{1/q}$$

$$= \left(\int |f|^t \, d\mu\right)^{s/t} = (\|f\|_t)^s.$$

9. Given a finite measure space $(X, \mathcal{A}, \mu)$, show that

$$\lim_{p \to \infty} \|f\|_p = \|f\|_\infty \qquad \text{for every } f \in L_\infty.$$

Hint:

Let $M < \|f\|_\infty = \text{ess sup} |f|$, then $A=\{|f|>M\}$ has positive measure. Since

$$\|f\|_p \geq \left(\int_A |f|^p \, d\mu\right)^{1/p} \geq M(\mu(A))^{1/p}$$

and $(\mu(A))^{1/p} \to 1$ as $p \to \infty$ we have that

$$\liminf \|f\|_p \geq M.$$

On the other hand,

$$\|f\|_p \leq \left(\int \|f\|_\infty^p \, d\mu\right)^{1/p} = \|f\|_\infty (\mu(X))^{1/p}$$

yielding

$$\limsup \|f\|_p \leq \|f\|_\infty.$$

10. If $f \in L_p$, where $p \geq 1$ and $\dfrac{1}{p} + \dfrac{1}{q} = 1$, we have that

$$\|f\|_p = \sup \{|\textstyle\int f h \, d\mu|; \ \|h\|_q \leq 1\}$$

Hint:

We shall restrict ourselves to the case $1 < p < \infty$. By Hölder's inequality

$$|\textstyle\int f h \, d\mu| \leq \int |f h| \, d\mu \leq \|f\|_p \|h\|_q \leq \|f\|_p.$$

Hence

$$\sup \{|\textstyle\int f h \, d\mu|; \ \|h\|_q \leq 1\} \leq \|f\|_p.$$

To prove the converse inequality, write

$$h_0 = f |f|^{p-2} \implies |h_0|^q = |f|^p \implies \|h_0\|_q = \|f\|_p^{\,p/q}.$$

Next, put

$$h = (\|f\|_p)^{-p/q} \cdot h_0 \implies \|h\|_q = 1.$$

Consequently,

$$\textstyle\int f h \, d\mu = (\|f\|_p)^{-p/q} \int f h_0 \, d\mu = \|f\|_p,$$

so that

$$\|f\|_p = \inf \{|\textstyle\int f h \, d\mu|; \ \|h\|_q = 1\}$$

$$\geq \inf \{|\textstyle\int f h \, d\mu|; \ \|h\|_q \leq 1\}$$

$$\leq \sup \{|\textstyle\int f h \, d\mu|; \ \|h\|_q \leq 1\} = \|f\|_p.$$

32. THE HAHN DECOMPOSITION

A σ-additive set function μ on a measurable space $(X, \mathscr{A})$ is said to be a "signed measure" if

$$\mu(\theta) = 0 \ \text{ and if } \ \mu : \mathscr{A} \to (-\infty, \infty].$$

A set $H \in \mathscr{A}$ is said to be "μ-positive" if every measurable subset of it has a non-negative measure, i.e., if

$$\mu : \mathscr{A} \cap H \to [0, \infty].$$

If $\mu : \mathscr{A} \cap H \to (-\infty, 0]$ the set H is said to be "μ-negative." Recall that $\mathscr{A} \cap H = \{A \cap H;\ A \in \mathscr{A}\}$.

Denote by $\mathscr{N}_+ \subset \mathscr{A}$ the collection of all μ- positive sets and by $\mathscr{N}_- \subset \mathscr{A}$ the collection of all μ-negative sets. Clearly $\theta \in \mathscr{N}_+ \cap \mathscr{N}_-$, implying that the intersection is not empty. Since any $G \in \mathscr{N}_+ \cap \mathscr{N}_-$ satisfies $0 \leq \mu(G) \leq 0$, we have that $\mu(G) = 0$. Because every measurable subset of G has μ-measure 0, we call G a μ-null set. If $M \in \mathscr{A}$ and $\mu(M) = 0$, M is not necessarily a μ-null set. Later we shall show that every $B \in \mathscr{A}$ is the union of a μ- positive and a μ- negative set. Clearly, if $E \in \mathscr{N}_+ - \mathscr{N}_-$ and $F \in \mathscr{N}_- - \mathscr{N}_+$ then $E \cap F = \theta$. Since every measurable subset of a μ- positive set is a μ- positive set and every measurable subset of a μ- negative set is a μ- negative set, both classes $\mathscr{N}_+$ and $\mathscr{N}_-$ are closed under intersection and difference. In the rest of this section we shall omit the prefix μ and say simply a "positive" set and a "negative" set.

Lemma 1

Both classes $\mathscr{N}_+$ and $\mathscr{N}_-$ are closed under countable union.

Proof:

Let $\{H_k\}_1^\infty$ be a sequence of positive sets and write

$$H = \bigcup_{k=1}^{\infty} H_k = H_1 \cup (H_2 \cap H_1^c) \cup (H_3 \cap H_1^c \cap H_2^c) \cup \cdots.$$

If $E \subset H$ is measurable then each

$$E_n = E \cap H_n \cap H_1^c \cap \cdots \cap H_{n-1}^c$$

is a positive set. Consequently,

$$\mu(E) = \sum_{k=1}^{\infty} \mu(E_k) \geq 0,$$

which proves that $H \in \mathscr{N}_+$. In a similar fashion one can show that $\mathscr{N}_-$ is closed under countable union. $\heartsuit$

Lemma 2

For any $A \in \mathscr{A} - \mathscr{N}_-$ satisfying $\mu(A) < 0$ there exists a measurable subset $A_0 \subset A$ such that

$0 < \mu(A_0) < \infty.$

Proof:

For any $A_0 \in \mathcal{A} \cap A$, we have

$$\mu(A) = \mu(A_0) + \mu(A - A_0),$$

from which we deduce that $\mu(A_0) < \infty$. If $\mu(A_0) \leq 0$ for each $A_0 \in \mathcal{A} \cap A$ it would mean that $A \in \mathcal{N}_-$, which contradicts its definition. Thus, there exists $A_0 \in \mathcal{A} \cap A$ such that $\mu(A_0) > 0$. This proves the lemma. ♥

Similarly, if $A \in \mathcal{A} - \mathcal{N}_+$ satisfies $\mu(A) > 0$ there exists a measurable subset $A_0 \subset A$ such that $\mu(A_0) < 0$.

Any signed measure μ on $(X, \mathcal{A})$ splits the space X into two measurable parts, one which is μ-positive and another which is μ-negative. This was proved by Hahn and represents the principal result of this section. Below is an "elementary" proof of this remarkable result, known as **the "Hahn decomposition."** Another proof, based on the Hausdorff maximal principle, is given as the problem 7.

Proposition 1

Let μ be a signed measure on a measurable space $(X, \mathcal{A})$. There exists a set $A \in \mathcal{N}_+$ such that

$A^C = B \in \mathcal{N}_-.$

Proof:

We shall prove existence of the set B. To this end let

$$\alpha = \inf \{\mu(C); \, C \in \mathcal{N}_-\}.$$

There exists a sequence $\{B_k\}_1^\infty$ of sets from $\mathcal{N}_-$ such that $\mu(B_k) \to \alpha$. Since by lemma 1

$$B = \bigcup_{k=1}^\infty B_k \in \mathcal{N}_-,$$

it is clear that $\alpha \leq \mu(B) \leq \mu(B_k) \to \alpha$ as $k \to \infty$. Thus $\alpha = \mu(B)$. Let us show that the set B is the required set, i.e. that $B^C = A \in \mathcal{N}_+$. Our proof is by contradiction.

Assume that $A \notin \mathcal{N}_+$, then by lemma 2 there exists $A_0 \in \mathcal{A} \cap A$ such that $\mu(A_0) < 0$. The set $A_0 \notin \mathcal{N}_-$, because if it does $B \cup A_0 \in \mathcal{N}_-$, which implies that $\alpha \leq \mu(B \cup A_0) = \alpha + \mu(A_0) < \alpha$, and which is a contradiction. According to lemma 2 there is a measurable subset $A_1 \subset A_0$ such that $0 < \mu(A_1) < \infty$. Let $n_1 > 0$ be the smallest integer such that

$$\mu(A_1) \geq \frac{1}{n_1}.$$

Clearly, $A_0 - A_1 \notin \mathcal{N}_-$, and

$$\mu(A_0 - A_1) \le \mu(A_0) - \frac{1}{n_1} < 0,$$

so we can apply the same argument to $A_0 - A_1$. There is a measurable subset $A_2 \subset A_0 - A_1$ with $\mu(A_2) > 0$. Let $n_2 > 0$ be the smallest integer such that

$$\mu(A_2) \ge \frac{1}{n_2},$$

then $A_0 - (A_1 \cup A_2) \notin \mathcal{N}_-$, and

$$\mu(A_0 - (A_1 \cup A_2)) \le \mu(A_0 - A_1) - \frac{1}{n_2} < \mu(A_0) - \frac{1}{n_1} - \frac{1}{n_2}$$

and so on. The sequence $\{A_n\}_1^\infty$ so obtained consists of disjoint measurable subsets of A_0. Since $\mu(A_0) < 0$ it follows that

$$\mu\left(\bigcup_1^\infty A_k\right) = \sum_1^\infty \mu(A_k) < \infty,$$

which implies that

$$\sum_{i=1}^\infty \frac{1}{n_i} < \infty \quad \text{so that} \quad \frac{1}{n_i} \to 0 \quad \text{as} \quad i \to \infty.$$

We claim that

$$D = A_0 - \bigcup_1^\infty A_k \in \mathcal{N}_-.$$

For if not, there is $H \in \mathcal{A} \cap D$ such that $\mu(H) > 0$. Clearly,

$$H \subset A_0 - \bigcup_{k=1}^n A_k \quad \text{for every} \quad n = 1, 2, \cdots.$$

Let $m_n > 0$ be the smallest integer for which

$$\mu(H) \ge \frac{1}{m_n}.$$

Since then

$$\mu(H) < \frac{1}{m_n - 1} \quad \text{for all} \quad n = 1, 2, \cdots,$$

we have that $\mu(H) \le 0$ and consequently the set D does not contain a measurable subset of positive measure so that $D \in \mathcal{N}_-$. Finally, since $B \cap D = \theta$ we have that

$$\alpha \le \mu(B \cup D) = \mu(B) + \mu(D) < \alpha,$$

which is obviously a contradiction, so that $\mu(A_0) < 0$ is not true. Thus, $B^c \in \mathcal{N}_+$, which proves the proposition. $\qquad\qquad\blacktriangledown$

We say that $A\in\mathcal{N}_+$ and $B\in\mathcal{N}_-$ form a Hahn decomposition of X for μ. Proposition 1 states the existence of a Hahn decomposition for every signed measure on $(X, \mathcal{A})$. Such a decomposition is not unique. However, if (A_1, B_1) is another Hahn's decomposition of X with respect to the same signed measure μ, we have the following:

Proposition 2

$$\mu(A \triangle A_1) = \mu(B \triangle B_1) = 0$$

Proof:

Since $A-A_1 \subset A$ it is clear that $\mu(A-A_1)\geq 0$. On the other hand,

$$A - A_1 = A \cap A_1^c = A \cap B_1 \subset B_1$$

which implies that $\mu(A-A_1)\leq 0$. Thus

$$\mu(A - A_1) = 0.$$

By symmetry we obtain that $\mu(A_1-A)=0$ which shows that $\mu(A \triangle A_1)=0$. In the same way we get that $\mu(B \triangle B_1)=0$. This completes our proof. ♥

Remark 1

Let μ be a signed measure on $(X, \mathcal{A})$ and let (A, B) be a Hahn decomposition for μ, then for any $M\in\mathcal{A}$ we have that

$$M = (A \cap M) \cup (B \cap M),$$

which implies that every measurable set is the union of its positive and its negative part. Consequently, since

$$\mu(M) = \mu(A \cap M) + \mu(B \cap M),$$

the Hahn decomposition (A, B) of X with respect to μ gives rise to two uniquely defined measures μ^+ and μ^- on $(X, \mathcal{A})$ specified by

$$\mu^+ (M) = \mu(A \cap M) \quad \text{and} \quad \mu^- (M) = - \mu(B \cap M).$$

Clearly,

$$\mu = \mu^+ - \mu^- \tag{1}$$

The measures μ^+ and μ^- are called the positive and negative parts of μ. Since by assumption $\mu >- \infty$, μ^- is a finite measure. If μ^+ and μ^- are both finite, μ is a finite signed measure.

The decomposition (1) of μ is called the **"Jordan decomposition."** According to (1) each signed measure is the difference of two measures, at least one of which is finite. The converse is not necessarily true, i.e. the difference of two measures $\mu_1 - \mu_2$, both defined on a same measure space, is not a signed measure unless one of them is finite.

As we have observed, μ^+ and μ^- are uniquely defined because they are independent of the particular Hahn decomposition. For instance, if (A, B) and (A_1, B_1) are two Hahn decompositions of X for μ then according to proposition 2 for any $M \in \mathcal{A}$

$$\mu(A \cap M) = \mu(A_1 \cap M) \qquad \mu(B \cap M) = \mu(B_1 \cap M).$$

Remark 2

Let μ be a signed measure on $(X, \mathcal{A})$. The set function $|\mu|$ defined by

$$|\mu| = \mu^+ + \mu^-$$

is called the "total variation" of μ. According to the definition of μ^+ and μ^-

$$\mu^+(B) = \mu(A \cap B) = 0 \quad \text{and} \quad \mu^-(A) = \mu(B \cap A) = 0.$$

Two measures μ_1 and μ_2 on $(X, \mathcal{A})$ are said to be "mutually singular" or "orthogonal" (in symbol $\mu_1 \perp \mu_2$) if there is set $D \in \mathcal{A}$ such that $\mu_1(D) = \mu_2(D^c) = 0$. The measures μ^+ and μ^- are clearly mutually singular.

The next proposition shows that μ^+ and μ^- can be characterized as the smallest (positive) measures which majorises μ and $-\mu$, respectively.

Proposition 3

Let μ be a signed measure on $(X, \mathcal{A})$, then for any $G \in \mathcal{A}$ we have:

(i) $\qquad \mu^+(G) = \sup \{\mu(H); H \in \mathcal{A} \cap G\}$

(ii) $\qquad \mu^-(G) = -\inf \{\mu(H); H \in \mathcal{A} \cap G\}$

Proof:

Let (A, B) be a Hahn decomposition of X for μ, then

$$\mu^+(G) = \mu(A \cap G) \leq \sup \{\mu(H); H \in \mathcal{A} \cap G\}.$$

On the other hand, for any $H \in \mathcal{A} \cap G$

$$\mu(H) = \mu(A \cap H) + \mu(B \cap H) \leq \mu(A \cap G) = \mu^+(G).$$

Hence, $\sup \{\mu(H); H \in \mathcal{A} \cap G\} \leq \mu^+(G)$ which proves (i). Since $\mu^- = (-\mu)^+$ the assertion follows. $\qquad \heartsuit$

Problems and Complements

1. Let μ be a signed measure on $(X, \mathcal{A})$ and let $A\in\mathcal{A}$. If $\mu(A_0)=+\infty$ for some $A_0\in\mathcal{A}\cap A$ show that $\mu(A)=+\infty$.

Hint:

Since $A=A_0\cup(A-A_0)$, $\mu(A)=\mu(A_0)+\mu(A-A_0)$. The assertion now follows from the fact that $-\infty<\mu(A-A_0)$.

2. Let μ be a signed measure on $(X, \mathcal{A})$. If (A, B) and (A_1, B_1) are two Hahn's decompositions for μ, show that for any $H\in\mathcal{A}$

$$\mu(H\cap A) = \mu(H\cap A_1) \text{ and } \mu(H\cap B) = \mu(H\cap B_1).$$

In other words, if $\mu^+(H)=\mu(A\cap H)$ and $\mu_1^+(H)=\mu(A_1\cap H)$, then $\mu^+(H)=\mu_1^+(H)$ and similarly for $\mu^-(H)=\mu_1^-(H)$, so μ^+ and μ^- are uniquely defined.

Hint:

$A-A_1\in\mathcal{N}_+$ and since $A-A_1=A\cap B_1\in\mathcal{N}_-$ it follows that $(A-A_1)\in\mathcal{N}_+\cap\mathcal{N}_-$. Hence, for any $H\in\mathcal{A}$

$$\mu(H\cap(A-A_1)) = \mu(H\cap A) - \mu(H\cap A\cap A_1) = 0.$$

Thus, $\mu^+(H)=\mu(H\cap A\cap A_1)$, and since $A-A_1\in\mathcal{N}_+\cap\mathcal{N}_-$

$$\mu(H\cap(A_1-A)) = \mu(H\cap A_1) - \mu(H\cap A\cap A_1) = 0,$$

yielding that $\mu_1^+(H) = \mu(H\cap A\cap A_1)$. Consequently, $\mu^+=\mu_1^+$ on $\mathcal{A}$.

3. Show that (i) $-\mu^-\le\mu\le\mu^+$; (ii) $|\mu(\cdot)|\le|\mu|(\cdot)$.

Hint:

(i) follows trivially from $\mu^+-\mu^-$.

(ii) $|\mu^+-\mu^-|\le\mu^++\mu^-$.

4. Let μ be a finite signed measure on $(X, \mathcal{A})$ and let $\{E_t; t\in I\}\subset\mathcal{A}$ be a collection of disjoint sets such that $\mu(E_t)>0$ for all $t\in I$. Show that the index set I is at most countable.

Hint:

For an each integer $k>0$ define

$$I_k = \{t\in I; \mu(E_t) > \frac{1}{k}\}.$$

Clearly, $I_k\uparrow I$ as $k\to\infty$. But each I_k is a finite set for every $k<\infty$, for if it is not, choose $\{t_j\}_1^\infty\subset I_k$, then bearing in mind that $\{E_{t_j}\}_1^\infty$ are disjoint and that μ is finite

$$\sum_{j=1}^{\infty} \mu(E_{t_j}) < \infty.$$

This, however contradicts assumption that each $\mu(E_{t_j})>1/k$ for all $t_j \in I_k$.

5. If μ is a finite signed measure on $(X, \mathcal{A})$, show that there exists $W \in \mathcal{A}$ such that for all $A \in \mathcal{A}$

$$\mu(W \cap A) = \mu(A).$$

Hint:

Let $\mathcal{E} = \{E_t \in \mathcal{A}; t \in I\}$ be a class of mutually disjoint sets such that $\mu(E_t)>0$. The class $\mathcal{E}$ is at most countable. By the usual sort of transfinite argument we (see problem 8) deduce that there is a maximal such class, say $\mathcal{M}=\{M_t ; t \in I_0\}$. Since $\mathcal{M}$ is countable $W = \cup\{M_t ; t \in I_0\} \in \mathcal{A}$. Hence, for $H \in \mathcal{A}$ arbitrary we have that $\mu(H-W) = 0$, so that

$$\mu(H) = \mu(H \cap W) + \mu(H \cap W^c) = \mu(H \cap W).$$

6. Let μ be a finite signed measure on $(X, \mathcal{A})$ and let $E \in \mathcal{A}$ be such that $0<\mu(E)$. Show that $\mathcal{N}_+ \cap E \neq \theta$.

Hint:

Let $\{F_t ; t \in I\} \subset \mathcal{A} \cap E$ be the maximal family of disjoint sets such that $\mu(F_t)<0$ for each $t \in I$. This family is clearly countable so that

$$F = \bigcup_{t \in I} F_t \in \mathcal{A} \cap E \text{ and } \mu(F) < 0.$$

We claim that

$$E_0 = E - F \in \mathcal{N}_+ .$$

Clearly, $\mu(E_0)>0$; if $M \in \mathcal{A} \cap E_0$ and $\mu(M)<0$, then, since $M \cap F = \theta => M \cap F_t = \theta$ for every $t \in I$, which contradicts the maximality of the collection $\{F_t ; t \in I\}$. Thus, $\mu(M) \geq 0$, proving that $E_0 \in \mathcal{N}_+ \cap E$.

7. Prove Hahn's decomposition theorem using the previous problem.

Hint:

Without loss of generality we may assume that $\mu: \mathcal{A} \to (-\infty, \infty)$. We shall first prove the existence of a positive set $A \in \mathcal{A}$. To this end consider

$$\beta = \sup \{\mu(C); C \in \mathcal{N}_+ \},$$

and let $\{C_k\}_1^\infty$ be a sequence from $\mathcal{N}_+$ such that

$$\beta = \lim_{k \to \infty} \mu(C_k).$$

Clearly,

$$A = \bigcup_{k=1}^{\infty} C_k \in \mathcal{N}_+$$

and $\mu(A) \le \beta$. Since $A - C_i \subset A \Rightarrow \mu(A - C_i) \ge 0$ for each $i = 1, 2, \cdots$. Hence, since

$$\mu(A) = \mu(C_i) + \mu(A - C_i) \ge \mu(C_i),$$

we conclude that $\mu(A) \ge \beta$. From this we deduce that

$$\mu(A) = \beta < \infty \ (\mu \text{ is finite}).$$

Let us prove that $B = A^c \in \mathcal{N}_-$. According to the previous problem it suffices for this purpose to show that $\mu(E) = 0$ for any $E \in \mathcal{N}_+ \cap B$. Since $A \cap E = \theta$ and

$$\beta \ge \mu(A \cup E) = \mu(A) + \mu(E) = \mu(E) + \beta$$

it follows that $\mu(E) = 0$. Thus the set B does not have a positive subset which proves that $B \in \mathcal{N}_-$.

8. Let μ be a finite signed measure on $(X, \mathcal{A})$. Denote by $\mathcal{E} = \{E_t \,; t \in I\} \subset \mathcal{A}$ the disjoint family such that each $\mu(E_t) > 0$. As we have established in problem 4, the class $\mathcal{E}$ is countable. If $\widetilde{\mathcal{H}}$ is the collection of all such classes, show that $\widetilde{\mathcal{H}}$ contains a maximal element.

Hint:

We shall prove the contention by constructing the maximal class. Under inclusion $\subset$ $(\widetilde{\mathcal{H}}, \subset)$ is a partially ordered set which, according to the Hausdorff maximal principle, contains a maximal chain, say $\widetilde{C} \subset \widetilde{\mathcal{H}}$. We claim that

$$\mathcal{M} = \bigcup_{\mathcal{E} \in \widetilde{C}} \mathcal{E}$$

is the maximal class. First, let us show that $\mathcal{M} \in \widetilde{\mathcal{H}}$. If $E_1, E_2 \in \mathcal{M}$, then $E_1 \in \mathcal{E}_1$ and $E_2 \in \mathcal{E}_2$. Since $\mathcal{E}_1, \mathcal{E}_2 \in \widetilde{C}$ which is chain, either $\mathcal{E}_1 \subset \mathcal{E}_2$ or $\mathcal{E}_2 \subset \mathcal{E}_1$. Assume that $\mathcal{E}_1 \subset \mathcal{E}_2$, then both $E_1, E_2 \in \mathcal{E}_2$ and hence $E_1 \cap E_2 = \theta$ and $\mu(E_i) > 0$, $i = 1, 2$. This proves that $\mathcal{M} \in \widetilde{\mathcal{H}}$. To show that $\mathcal{M}$ is maximal suppose that there exists $\mathcal{E}' \supset \mathcal{M}$ such that $\mathcal{E}' \in \widetilde{\mathcal{H}}$. But then

$$\widetilde{C} \cup \{\mathcal{E}'\} \supset \widetilde{C}$$

which contradicts the maximality property of $\widetilde{C}$. This proves that $\mathcal{M}$ is maximal.

9. Let μ_1 and μ_2 be finite signed measures on $(X, \mathcal{A})$. Show that $\mu_1 \vee \mu_2$ and $\mu_1 \wedge \mu_2$ defined by

$$\mu_1 \vee \mu_2 = \mu_1 + (\mu_2 - \mu_1)^+$$

$$\mu_1 \wedge \mu_2 = \mu_1 - (\mu_2 - \mu_1)^-$$

are the least upper bound and the greatest lower bound, respectively, of μ_1 and μ_2.

Hint:

Clearly, $\mu_1 \vee \mu_2$ and $\mu_1 \wedge \mu_2$ are signed measures. If (A_0, B_0) is a Hahn's decomposition for $(\mu_1 - \mu_2)$ then for any $G \in \mathscr{A}$

$$(\mu_1 \vee \mu_2)(G) = \mu_1(G) + (\mu_2 - \mu_1)(G \cap A_0)$$

$$= \mu_1(G \cap A_0) + \mu_1(G \cap B_0) + \mu_2(G \cap A_0) - \mu_1(G \cap A_0)$$

$$= \mu_1(G \cap B_0) + \mu_2(G \cap A_0).$$

To show that $\mu_1 \vee \mu_2$ is the least upper bound of μ_1 and μ_2, suppose that ψ is a measure on $(X, \mathscr{A})$ larger than μ_1 and μ_2, then

$$(\mu_1 \vee \mu_2)(G) \le \psi(G \cap B_0) + \psi(G \cap A_0) = \psi(G).$$

Finally, let us show that $\mu_1 \vee \mu_2 = \mu_2 \vee \mu_1$. The identities

$$\mu^- = (-\mu)^+ \qquad \text{and} \qquad \mu^+ = (-\mu)^-$$

yield

$$\mu_2 \wedge \mu_1 = \mu_2 - (\mu_1 - \mu_2)^- = \mu_2 - (-(\mu_1 - \mu_2))^-$$

$$= \mu_2 - (\mu_2 - \mu_1)^+ = \mu_1 + (\mu_2 - \mu_1) - (\mu_2 - \mu_1)^+$$

$$= \mu_1 - (\mu_2 - \mu_1)^-.$$

10. Show that

$$(\mu_1 \vee \mu_2) = \frac{1}{2}\,(\mu_1 + \mu_2 + |\mu_1 - \mu_2|).$$

Hint:

$$\mu_1 \vee \mu_2 = \mu_1 + (\mu_2 - \mu_1)^+$$

$$\mu_2 \vee \mu_1 = \mu_2 + (\mu_1 - \mu_2)^+.$$

Hence

$$2(\mu_1 \vee \mu_2) = \mu_1 + \mu_2 + (\mu_2 - \mu_1)^+ + (\mu_1 - \mu_2)^+$$

$$= \mu_1 + \mu_2 + (\mu_2 - \mu_1)^+ + (-(\mu_2 - \mu_1))^+$$

$$= \mu_1 + \mu_2 + (\mu_2 - \mu_1)^+ + (\mu_2 - \mu_1)^-$$

$$= \mu_1 + \mu_2 + |\mu_1 - \mu_2|.$$

33. THE LEBESGUE DECOMPOSITION

Recall **that two measures μ and ν on a measurable space $(X, \mathscr{A})$ are (mutually) singular ($\mu \perp \nu$) if there exists a set $D \in \mathscr{A}$ such that $\mu(D)=\nu(D^C)=0$.** In other words, μ is supported by D^C and ν is supported by D (or D^C is a set of μ-full measure and D is a set of ν-full measure, see section 17). Note that "supported by" should not be confused by the term "support."

If every μ-null set is necessarily a ν-null set we say that the measure μ is "absolutely continuous" with respect to ν and we write $\nu \ll \mu$. In other words, if

$$\mathscr{N}_\mu = \{H \in \mathscr{A}, \mu(H)=0\} \quad \text{and} \quad \mathscr{N}_\nu = \{H \in \mathscr{A}, \nu(H)=0\},$$

then $\nu \ll \mu \iff \mathscr{N}_\nu \supset \mathscr{N}_\mu$. If $\mathscr{N}_\mu = \mathscr{N}_\nu$ we say that μ and ν are "equivalent" and write $\mu \equiv \nu$. Note that $\nu \leq \mu$ implies that $\nu \ll \mu$ but not conversely. The "absolute continuity concept" has been already encountered in the problem 5 of section 30. If the measure ν is finite, a useful characterization of this notion is provided by the following:

Proposition 1

Let $(X, \mathscr{A}, \mu)$ be a measure space. If ν is a finite measure on $\mathscr{A}$, then $\nu \ll \mu$ if and only if:

(i) Given $\varepsilon > 0$ there exists $\delta = \delta(\varepsilon)$ such that for any $H \in \mathscr{A}$ with $\mu(H) < \delta$, necessarily $\nu(H) < \varepsilon$.

Proof:

For each integer $n > 0$ let $\delta_n > 0$ be such that for any $H \in \mathscr{A}$ with $\mu(H) < \delta_n$ we have that $\nu(H) \leq 1/n$. Now, if $\mu(H)=0$ then $\mu(H) < \delta_n$ for all n, hence $\nu(H) \leq 1/n$ for all n implying that $\nu(H)=0$. Thus $\nu \ll \mu$.

Assume now that $\nu \ll \mu$ and that (i) is false. If for some $\varepsilon > 0$ $\{H_n\}_1^\infty$ is a sequence from $\mathscr{A}$ such that $\mu(H_n) \leq 1/2^n$ and $\nu(H_n) > \varepsilon$, $n=1, 2, \ldots$, then since by assumption ν is finite, we have that

$$\nu\left(\bigcap_{n=1}^\infty \bigcup_{k=n}^\infty H_k\right) = \lim_{n \to \infty} \nu\left(\bigcup_{k=n}^\infty H_k\right) \geq \varepsilon.$$

On the other hand,

$$\mu\left(\bigcap_{n=1}^\infty \bigcup_{k=n}^\infty H_k\right) = \lim_{n \to \infty} \mu\left(\bigcup_{k=n}^\infty H_k\right) \leq \lim_{n \to \infty} \frac{1}{2^{n-1}} = 0.$$

But this contradicts the assumption that $\nu \ll \mu$, so (i) is not false. This proves the proposition. ♥

If ν and μ are signed measures defined on a measurable space $(X, \mathscr{A})$, we say that ν is absolutely continuous with respect to μ if $|\nu| \ll |\mu|$. In other words, if $\nu^+ \ll \mu^+$ and $\nu^- \ll \mu^-$ (see problem 2 below). In general, two measures, say ν and μ, on the same measurable space are neither singular nor absolutely continuous. However, according to the classical result of Henri Lebesgue, one of them, say μ, can be decomposed into two components, say μ_1 and μ_2 such that

$$\mu = \mu_1 + \mu_2, \qquad\qquad\qquad (1)$$

where the measures μ_1 and μ_2 are uniquely determined by the measure ν so that

$$\mu_1 << \nu \quad \text{and} \quad \mu_2 \perp \nu.$$

The purpose of this section is to prove the Lebesgue result. To this end the following lemma is needed:

Lemma 1

Let ν and μ be two finite measures on $(X, \mathcal{A})$. Let $\mathbf{F} \subset L_1(X, \mathcal{A}, \nu)$ be the subclass of all non-negative functions such that for every fixed $h \in \mathbf{F}$

$$\int_A h \, d\nu \leq \mu(A) \quad \text{for all } A \in \mathcal{A}. \qquad\qquad (2)$$

Then the class $\mathbf{F}$ has a maximal element.

Proof:

This contention can be proved by the usual sort of transfinite argument. The proof given here is somewhat more direct. First, we prove the next two properties of $\mathbf{F}$:

(i) If $h_1, h_2 \in \mathbf{F}$ then $h_1 \vee h_2 \in \mathbf{F}$. Indeed, set $B = \{ h_1 \leq h_2 \}$, then for any $A \in \mathcal{A}$

$$\int_A (h_1 \vee h_2) \, d\nu = \int_{A \cap B} h_2 \, d\nu + \int_{A \cap B^c} h_1 \, d\nu \leq \mu(A \cap B) + \mu(A \cap B^c) = \mu(A).$$

(ii) If $\{h_n\}_1^\infty$ is a sequence from $\mathbf{F}$ such that $h_n \uparrow h$ then $h \in \mathbf{F}$. Indeed, for any $A \in \mathcal{A}$ we have that

$$\int_A h_n \, d\nu \leq \mu(A) \;\; => \;\; \lim_{n \to \infty} \int_A h_n \, d\nu = \int_A h \, d\nu \leq \mu(A).$$

Denote by

$$\alpha = \sup \left\{ \int h \, d\nu; \; h \in \mathbf{F} \right\},$$

then there exists a sequence $\{f_n\}_1^\infty$ from $\mathbf{F}$ such

$$\lim_{n \to \infty} \int f_n \, d\nu = \alpha \qquad\qquad (3)$$

(see problem 8 below). It obviously follows from property (i) that

$$\tilde{f}_n = \sup_{1 \leq k \leq n} f_k \in \mathbf{F} \quad \text{for every } n = 1, 2, \cdots .$$

Clearly then $\int f_n \, d\nu \leq \int \tilde{f}_n \, d\nu$ and if $\tilde{f} = \lim_{n \to \infty} \tilde{f}_n$, then from (3) and the monotone convergence theorem we infer that

$$\alpha \leq \lim_{n \to \infty} \int \tilde{f}_n \, dv = \int \tilde{f} \, dv.$$

But, according to property (ii) $\tilde{f} \in \mathbf{F}$ implying that $\int \tilde{f} \, dv \leq \alpha$. Consequently,

$$\alpha = \int \tilde{f} \, dv,$$

which proves that the function $\tilde{f}$ is the largest (up to equivalence) member from $\mathbf{F}$ for which (2) holds.   ♥

Proposition 2 (Lebesgue)

Let $(X, \mathcal{A}, v)$ be a σ-finite measure space. If μ is another σ-finite measure on $(X, \mathcal{A})$, there exists a positive function $f \in L_0(X, \mathcal{A}, v)$, uniquely determined up to equivalence, such that for every $A \in \mathcal{A}$

$$\mu(A) = \int_A f \, dv + \mu(A \cap N), \tag{4}$$

where $N \in \mathcal{A}$ is a fixed v-null set.

Proof:

We shall prove the proposition in the case when the measures v and μ are finite. The extension from finite to σ-finite case is not difficult. Consider the finite measure μ' on $\mathcal{A}$ defined by

$$\mu'(A) = \mu(A) - \int_A \tilde{f} \, dv, \quad \text{where} \quad \tilde{f} \in \mathbf{F}, \tag{5}$$

and let (D_n, D_n^c) be a Hahn's decomposition for the signed measure

$$\mu' - \frac{1}{n} v, \qquad n = 1, 2, \cdots .$$

Then for any $H \in \mathcal{A}$,

$$\mu'(H \cap D_n^c) - \frac{1}{n} v(H \cap D_n^c) \leq 0, \qquad \mu'(H \cap D_n) - \frac{1}{n} v(H \cap D_n) \geq 0. \tag{6}$$

From this (5) and (6) we have:

$$\int_H (\tilde{f} + \frac{1}{n} I_{D_n}) \, dv = \int_H \tilde{f} \, dv + \frac{1}{n} v(H \cap D_n) \leq \int_H \tilde{f} \, dv + \mu'(H \cap D_n) = \int_H \tilde{f} \, dv + \mu(H \cap D_n) -$$

$$\int_{H \cap D_n} \tilde{f} \, d\mu = \mu(H \cap D_n) + \int_{H \cap D_n^c} \tilde{f} \, d\mu = \mu(H \cap D_n) + \mu(H \cap D_n^c) - \mu'(H \cap D_n^c) \leq \mu(H)$$

so that $(\tilde{f} + \frac{1}{n} I_{D_n}) \in \mathbf{F}$. But this contradicts the maximality property of $\tilde{f}$ so that $v(D_n) = 0$. The set

$$N = \bigcup_1^\infty D_n \text{ is thus } v\text{-null. On the other hand, for every } n = 1, 2, \ldots$$

$$\mu'(N^C) \leq \mu'(D_n^C) \leq \frac{1}{n}\, \nu(D_n^C) \to 0 \quad \text{as} \quad n \to \infty$$

since ν is finite. Consequently, $\mu'(N^C)=0$ which yields:

$$\mu'(H) = \mu'(H \cap N) + \mu'(H \cap N^C) = \mu'(H \cap N) \ .$$

From this and (5) we have that

$$\mu(H) = \mu'(H) + \int_H \tilde{f}\ d\nu = \mu'(H \cap N) + \int_H \tilde{f}\ d\nu,$$

which proves existence of f in (4) when μ and ν are finite. To prove that f is unique up to equivalence, assume that

$$\mu(H) = \int_H h\, d\nu + \mu_1(H \cap N)$$

is another decomposition of μ, then

$$0 = \int_H (f-h)d\nu + \mu(H \cap N) - \mu_1(H \cap N) \ \Rightarrow\ \mu_1(H \cap N) \geq \int_H (f-h)d\nu.$$

Since $\nu \perp \mu_1(\cdot \cap N)$, there exists $D \in \mathscr{A}$ such that $\mu_1(D \cap N)=0$ and $\nu(D^C)=0$. Therefore,

$$0 \leq \int_H (f-h)d\nu = \int_{H \cap D} (f-h)d\nu \leq \mu_1(D \cap N) = 0.$$

This proves that $f = h$, $\nu-$(a.e).

To show that the proposition holds when ν and μ are σ-finite write (4) as $\mu = \mu_1 + \mu_2$, where

obviously $\mu_1 << \nu$ and $\mu_2 \perp \nu$. There exists a measurable partition $\{X_n\}_1^\infty$ of X such that $\nu(X_n) < \infty$ and $\mu(X_n) < \infty$ for all $n = 1, 2, \cdots$. Define

$$\nu^n(\cdot) = \nu(\cdot \cap X_n), \qquad \mu^n(\cdot) = \mu(\cdot \cap X_n);$$

Since ν^n and μ^n are finite measures on $\mathscr{A} \cap X_n$ we can write $\mu^n = \mu_1^n + \mu_2^n$ where $\mu_1^n << \nu^n$ and

$\mu_2^n \perp \nu^n$, $n = 1, 2, \cdots$. Set

$$\mu = \sum_{n=1}^\infty \mu^n, \quad \mu_1 = \sum_{n=1}^\infty \mu_1^n, \quad \mu_2 = \sum_{n=1}^\infty \mu_2^n,$$

then we clearly have that $\mu_1 << \nu$, $\mu_2 \perp \nu$ and that $\mu = \mu_1 + \mu_2$, which completes the proof of the proposition. $\qquad \blacktriangledown$

Corollary 1

The Lebesgue proposition holds even if ν and μ are σ-finite signed measures, because by the Jordan decomposition each signed measure is the difference of two measures (of which at least one is finite).

Problems and complements

1. Let v and μ be signed measures on a measurable space $(X, \mathcal{A})$. Show that

$$\mu \perp v <=> |\mu| \wedge |v| = 0.$$

Hint:

Assume first that v and μ are measures and that $\mu \perp v$, then there exists a set $D \in \mathcal{A}$ such that

$v(D) = \mu(D^C) = 0$. Consequently, since $v \wedge \mu$ is a measure, we have:

$$(v \wedge \mu)(X) = (v \wedge \mu)(D) + (v \wedge \mu)(D^C)$$

$$\leq v(D) + \mu(D^C) = 0.$$

On the other hand, if $\mu \wedge v = 0$ on $\mathcal{A}$ then for any $A \in \mathcal{A}$, $\mu(A) \wedge v(A) = 0$ so that at least one $\mu(A)$ or $v(A)$ is

equal to zero. Assume that $\mu(A) = 0$ and that $v(A) > 0$, then $\mu(A^C) > 0$ and $v(A^C) = 0$. Hence

$$\mu(A) = v(A^C) = 0 \text{ so that } \mu \perp v = 0.$$

2. Let v and μ be signed measures on $(X, \mathcal{A})$. Show that

(i) $|v| << |\mu|$ <=> (ii) $v << |\mu|$.

Hint:

(i) => (ii); if for some $H \in \mathcal{A}$, $|\mu|(H) = 0 => |v|(H) = 0$, and since $|v(H)| \leq |v|(H) = 0 => v^+(H) = 0$ and

$v^-(H) = 0$, => $v(H) = 0$.

(ii) => (i). Let (A, B) be a Hahn decomposition of X for v. If $|\mu|(H) = 0$ then

$$0 \leq |\mu|(H \cap A) \leq |\mu|(H) = 0 \quad 0 \leq |\mu|(H \cap B) \leq |\mu|(H) = 0.$$

Hence,

$$v^+(H) = v(H \cap A) = 0, \; v^-(H) = v(H \cap B) = 0,$$

implying that $v(H) = 0$.

3. Let μ be a signed measure on $(X, \mathcal{A})$, and let for some $H \in \mathcal{A}$, $\mu(H) < \infty$. Show that any $G \in \mathcal{A} \cap H$ satisfies

$\mu(G) < \infty$ and

$$\mu(H-G) = \mu(H) - \mu(G).$$

Hint:

Since $H = G \cup (H-G)$ we have that

$$\mu(H) = \mu(G) + \mu(H-G) => \mu(G) < \infty.$$

If on the other hand, any $A \in \mathcal{A} \cap H$ satisfies $\mu(A) = +\infty$, then obviously $\mu(H) = +\infty$.

4. Given two measures v and μ on $(X, \mathcal{A})$ such that $v << \mu$. If v is finite and μ σ-finite, show that there exists

an $\varepsilon > 0$ and a set $A \in \mathcal{A}$ with $0 < \mu(A) < \infty$ such that

$$\varepsilon \cdot \mu(\,\cdot\,) \le \nu(\,\cdot\,) \text{ on } \mathscr{A} \cap A.$$

Hint:

Since μ is σ-finite there exists a measurable partition $\{X_n\}_1^\infty$ of X such that $\mu(X_n)<\infty$, n=1, 2, $\ldots$. Let $\nu(X_n)>0$ and choose $\varepsilon>0$ so that

$$\nu(X_n)-\varepsilon\,\mu(X_n) = (\nu-\varepsilon\,\mu)(X_k) > 0.$$

Let $A\in\mathscr{A}\cap X_k$ be a positive subset for the signed measure $(\nu-\varepsilon\mu)$. Clearly $\mu(A)<\infty$; in addition, $\mu(A)>0$ (for if $\mu(A)=0$ we would have

$$(\nu-\varepsilon\,\mu)(A) > 0 => \nu(A) > 0$$

which contradicts $\nu<<\mu$). Since $(\nu-\varepsilon\mu)(B)\ge0$ for every measurable $B\subset A$, it implies that $\nu(B) \ge \varepsilon\mu(B)$ as claimed.

5. If μ_1 and μ_2 are two σ-finite measures on $(X, \mathscr{A})$ such that $\mu_1<<\mu_2$, show that for any $C\in\mathscr{P}(X)$ such that $\mu_2^*(C)=0$ we have that $\mu_1^*(C)=0$.

Hint:

Assume that $\mu_2(C)=0$, then by proposition 4 of section 20, there exists a measurable cover, say B of C. Since by assumption $\mu_2^*(C)=0$ we have that $\mu_2(B)=0 =>\mu_1(B)=0 =>\mu_1^*(C)=0$.

6. (Continuation) Denote by $\mathscr{M}_{\mu_i}$ the class of μ_i-measurable sets (see definition 1 of section 19). Show that $\mu_1<<\mu_2$ implies $\mathscr{M}_{\mu_2} \subset \mathscr{M}_{\mu_1}$.

Hint:

Notice that $\mathscr{A}\subset\mathscr{M}_{\mu_1}\cap\mathscr{M}_{\mu_2}$. Given $M\in\mathscr{M}_{\mu_2}$ with $\mu_2^*(M)<\infty$, there exists a measurable cover, say B of M. Then according to proposition 5 of section 20 $\mu_2^*(B-M)=0$ which, according to the previous problem, implies that $\mu_1^*(B-M)=0$. Thus, $B-M\in\mathscr{M}_{\mu_1}$ and since

$$M = B-(B-M) => M \in \mathscr{M}_{\mu_1}$$

7. Let μ_1 and μ_2 be finite measures on $(X, \mathscr{A})$. Let (A_n, B_n) be Hahn's decomposition of X for the signed measure $\mu_2-n\mu_1$, n=1, 2, $\cdots$. Show that

(i) A_m is a positive set for $\mu_2-k\mu_1$ if $m\ge k$ and that B_m is negative set if $m\le k$.

(ii) If $\mu_1(E) > 0$, then $\mu_2(E)=+\infty$

Hint:

(i) If $G\in\mathscr{A}\cap A_m$ then

$$(\mu_2-k\,\mu_1)(G) = (\mu_2-m\,\mu_1)(G) + (m-k)\,\mu_1(G) \ge 0$$

and similarly for $G \in \mathcal{A} \cap B_m$.

 (ii) Set $A = \bigcap_{k=1}^{\infty} A_k$, $B = \bigcup_{k=1}^{\infty} B_k$; if $E \in \mathcal{A} \cap A$ then $(\mu_2 - m\mu_1)(E) = 0$ for every $n=1, 2, \cdots$.

Consequently,

$$\mu_2(E) = n\,\mu_1(E) \qquad n=1, 2, \cdots$$

implying that $\mu_2(E) = +\infty$ if $\mu_1(E) = 0$.

8. Show that there exists a sequence $\{f_n\}_1^{\infty}$ from $\mathbf{F}$ such that (3) holds.

 Hint:

 There exists $\{h_n\}_1^{\infty}$ from $\mathbf{F}$ such that

$$\alpha - \int h_n \, dv \le \frac{1}{n} \qquad n=1, 2, \cdots. \tag{*}$$

Set

$$f_n = \sup_{1 \le k \le n} h_k \qquad \text{and let} \quad A_k = \{x \in X;\, h_k(x) = f_n(x)\},$$

then clearly $\bigcup_1^{\infty} A_k = X$. Denote by $B_1 = A_1$, $B_2 = A_2 \cap A_1^c$, $\cdots$, $B_n = A_n \cap A_1^c \cap \cdots \cap A_{n-1}^c$, then for any

$E \in \mathcal{A}$

$$\int_E f_n \, dv = \sum_{k=1}^{n} \int_{E \cap B_k} h_k \, dv \le \sum_{k=1}^{n} v(E \cap B_k) \le v(E),$$

which proves that $f_n \in \mathbf{F}$. Since $f_n \uparrow f$ it follows that $f \in \mathbf{F}$ (property ii). Now from $(*)$ we have that

$$\left(\alpha - \frac{1}{n}\right) \le \int h_n \, dv \le \int f_n \, dv \le \alpha, \quad n=1, 2, \cdots$$

which proves that $\{f_n\}_1^{\infty}$ is the required sequence.

9. $f \in L_1(X, \mathcal{A}, v)$ and set for any $A \in \mathcal{A}$

$$\mu(A) = \int_A f \, dv.$$

Show that

$$\mu^+(A) = \int_A f^+ \, dv \qquad \text{and} \qquad \mu^-(A) = \int_A f^- \, dv.$$

 Hint:

 We must show that the sets

$$A = \{f \ge 0\} \text{ and } B = \{f \le 0\}$$

form a Hahn decomposition of X for μ. Indeed, given any $G \in \mathscr{A}$

$$\mu(A \cap G) = \int_G f I_A \, dv = \int_G f^+ \, dv \geq 0$$

$$\mu(B \cap G) = \int_G f I_B \, dv = - \int_G f^- \, dv \leq 0.$$

But $\mu(A \cap G) = \mu^+(G)$ and $\mu(B \cap G) = \mu^-(G)$.

10. Let μ be a signed measure on a measure space $(X, \mathscr{A}, v)$. If $\mathscr{N}_+$ and $\mathscr{N}_-$ are the collection of μ-positive and negative sets, respectively, show that $\mu(E) = 0 \Longleftrightarrow E \in \mathscr{N}_+ \cap \mathscr{N}_-$.

Hint:

If $E \in \mathscr{N}_+ \cap \mathscr{N}_-$, then $0 \leq \mu(E) \leq 0 \Rightarrow \mu(E) = 0$, $\Rightarrow \mu^+(E) = \mu^-(E)$. Consequently, if $\mu(H) = 0$ for some $H \in \mathscr{A}$, then $\mu^+(H) = \mu^-(H)$ implying that $H \in \mathscr{N}_+ \cap \mathscr{N}_-$.

11. Let $(X, \mathscr{A}, \mu)$ be a finite measure space. If $\{E_n\}_1^\infty$ is a sequence from $\mathscr{A}$ such that $\mu(E_n) \geq \varepsilon$ for all $n = 1, 2, \cdots$, show that

$$\mu(\limsup E_n) \geq \varepsilon .$$

Hint:

Set $E = \limsup E_n$ and define

$$F_n = \bigcup_{k=n}^{\infty} E_k \quad \text{then clearly } F_n \downarrow E.$$

Since μ is finite, $\mu(F_n) \downarrow \mu(E)$. But $E_n \subset F_n \Rightarrow \mu(E_n) \leq \mu(F_n) \geq \varepsilon$. Hence,

$$\mu(E) = g.\ell.b. \ \mu(F_n) \geq \varepsilon.$$

34. THE RADON-NIKODYM DERIVATIVE

Let $(X, \mathscr{A}, \nu)$ be a σ-finite measure space. If μ is a σ-finite measure on $\mathscr{A}$, then according to the Lebesgue decomposition theorem, there exist measures μ_1 and μ_2 on $\mathscr{A}$ such that $\mu=\mu_1+\mu_2$ with $\mu_1<<\nu$ and $\mu_2 \perp \nu$. If $\mu<<\nu$ then from equation (4) of section 33, it follows that for every $A \in \mathscr{A}$

$$\mu(A) = \int_A f \, d\nu \tag{1}$$

This is the "Radon-Nikodym" theorem. The integrand $f(\cdot) \geq 0$ in (1) is called the "Radon-Nikodym derivative" of μ with respect to ν. It is usually denoted by

$$f = \frac{d\mu}{d\nu} \quad \text{or} \quad d\mu = f \, d\nu. \tag{2}$$

Remark 1

Let $h \in L_0(X, \mathscr{A}, \nu)$ be non-negative, then λ defined on $\mathscr{A}$ by

$$\lambda(H) = \int_H h \, d\nu \tag{3}$$

is obviously a measure. If $\nu(N)=0$ for some $N \in \mathscr{A}$ then, since $h \cdot I_N = 0$ ν–(a.e) it follows that $\lambda(N)=0$; thus, $\lambda<<\nu$. However, the converse is not necessarily true. In other words, if λ is a measure on $\mathscr{A}$ such that $\lambda<<\nu$, it does not follow from this that there exists a measurable map $h:X \rightarrow [0, \infty]$ such that (3) holds. For this to happen it is necessary that the measure ν is σ-finite, as the next counter example shows.

Example 1

Let X be an uncountable set and $\mathscr{A}=\{G \subset X;$ either G or G^c is countable$\}$, then $\mathscr{A}$ is a σ-algebra (problem 3 of section 18). Define

$$\mu: \mathscr{A} \rightarrow \{0, 1, 2, \ldots, \infty\}$$

by $\mu(G)=\text{Card}(G)$. Clearly then $(X, \mathscr{A}, \mu)$ is not a σ-finite measure space. Let $\lambda: \mathscr{A} \rightarrow \{0, 1\}$ be a measure defined by

$$\lambda(G) = \begin{cases} 0 & \text{if} \quad G \quad \text{is countable} \\ 1 & \text{if} \quad G^c \quad \text{is countable} \end{cases}.$$

Since $\lambda \leq \mu$ it is clear that $\lambda<<\mu$. However, there is not a positive $h \in L_1(X, \mathscr{K}, \mu)$ such that

$$\lambda(G) = \int_G h \, d\mu \quad \text{for all} \quad G \in \mathscr{K}$$

(notice that λ is finite), because $L_1(X, \mathscr{K}, \mu)=\{0\}$.

The result (1) can be sharpened by dropping the requirement that μ is σ-finite (the assumption that ν is σ-finite cannot be eliminated). This yields to a more "general form" of the Radon-Nikodym theorem which is formulated below.

Proposition 1

Let $(X, \mathcal{A}, \nu)$ be a σ-finite measure space. If μ is a measure on $\mathcal{A}$ such that $\mu \ll \nu$, there exists a non–negative function $h \in L_1(X, \mathcal{A}, \nu)$ unique up to equivalence, such that for any $H \in \mathcal{A}$

$$\mu(H) = \int_H h \, d\nu.$$

Proof:

We shall prove the proposition in the case when ν is finite. The extension to the σ-finite case is not difficult. Let

$$\mathcal{E} = \{E_t \in \mathcal{A}; t \in I\}$$

be a collection of disjoint sets such that

$$\nu(E_t) > 0 \quad \text{and} \quad \mu(E_t) < \infty \quad \text{for all } t \in I.$$

According to problem 5 of section 32, this class is at most countable. Let $\mathcal{E}_0$ be the maximal class with this property (see problem 9 of section 32). Clearly then

$$W = \bigcup_{E \in \mathcal{E}_0} E \in \mathcal{A}$$

and, since by assumption $\mu(E) < \infty$ for each $E \in \mathcal{E}_0$, the measure μ is σ-finite on W. Thus, there exists a measurable function $h \geq 0$ such that

$$\mu(A) = \int_A h \, d\nu \quad \text{for every} \quad A \in \mathcal{A} \cap W.$$

Let us show that for any $G \in \mathcal{A} \cap W^c$ $\nu(G)=0$. For if $\nu(G)>0$, then

$$\mathcal{E}_0 \subset \mathcal{E}_0 \cup \{G\},$$

which contradicts the maximality property of $\mathcal{E}_0$. Thus, $\nu(G)=0$, which completes the proof of the proposition when ν is finite. The extension to the case when ν is σ-finite is not difficult and is left to the reader (see the problem 1)

♥

Problems and complements

1. Prove the proposition 1 for σ-finite v.

   Hint:

   There exists a measurable partition $\{X_n\}_1^\infty$ of X such that $v(X_n)<\infty$ for every $n=1, 2, \cdots$.

According to proposition 1 there exists a measurable function

$$f_n : X_n \to [0, \infty]$$

such that for every $H \in \mathcal{A} \cap X_n$

$$\mu(H) = \int_H f_n \, dv.$$

The measurable function

$$f(\cdot) = \sum_{k=1}^\infty f_k(\cdot) I_{X_k}(\cdot)$$

is the Radon-Nikodym derivative $d\mu/dv$. For, given any $A \in \mathcal{A}$, we have

$$\int_A f \, dv = \lim_{n \to \infty} \int_A \left(\sum_{k=1}^n f_k I_{X_k} \right) dv = \sum_{k=1}^\infty \int_{A \cap X_n} f_k \, dv$$

$$= \sum_{k=1}^\infty \mu(A \cap X_n) = \mu(A).$$

2. Let $(X, \mathcal{A}, v)$ be a σ-finite measure space and $\{\mu_i\}_1^n$ measures on $\mathcal{A}$ such that $\mu_i << v$, $i=1, 2, \cdots, n$.

Show that

$$\frac{d}{dv}\left(\sum_{i=1}^n \mu_i \right) = \sum_{i=1}^n \frac{d\mu_i}{dv}.$$

   Hint:

   Clearly,

$$\sum_{i=1}^n \mu_i << v$$

so that for any $A \in \mathcal{A}$

$$\left(\sum_{i=1}^n \mu_i \right)(A) = \int_A f \, dv.$$

On the other hand,

$$\left(\sum_{i=1}^n \mu_i \right)(A) = \sum_{i=1}^n \mu_i(A) = \sum_{i=1}^n \int_A f_i \, dv = \int_A \left(\sum_{i=1}^n f_i \right) dv.$$

Thus, $f = \sum_{i=1}^n f_i$ v-(a.e). But, $f_i = d\mu_i/dv$.

3. Let $(X, \mathcal{A}, v)$ be a σ-finite measure space. If μ is a signed measure on $\mathcal{A}$ such that $\mu \ll v$, show that there exists $f \in L_0(X, \mathcal{A}, \mu)$ such that for all $A \in \mathcal{A}$

$$\mu(A) = \int_A f \, dv.$$

Hint:

Since $\mu^+ \ll v$ and $\mu^- \ll v$, the Radon-Nikodym theorem yields:

$$\mu^+(A) = \int_A f_1 \, dv, \qquad \mu^-(A) = \int_A f_2 \, dv.$$

Hence,

$$\mu(A) = \mu^+(A) - \mu^-(A) = \int_A (f_1 - f_2) \, dv.$$

The required function $f = f_1 - f_2$.

4. Let μ_0, μ_1, μ_2 be σ-finite measures on $(X, \mathcal{A})$ such that $\mu_0 \ll \mu_1$ and $\mu_1 \ll \mu_2$. Show that

$$\frac{d\mu_0}{d\mu_2} = \frac{d\mu_0}{d\mu_1} \cdot \frac{d\mu_1}{d\mu_2} \quad \mu_2 \ (a.e).$$

Hint:

Denote by

$$f = \frac{d\mu_0}{d\mu_1} \qquad g = \frac{d\mu_1}{d\mu_2}.$$

We must show that for every $A \in \mathcal{A}$

$$\mu_0(A) = \int_A f \cdot g \, d\mu_2.$$

Clearly, for any $E \in \mathcal{A}$

$$\int_A I_E \, d\mu_1 = \int_A I_E \, g \, d\mu_2.$$

Thus, if $\{h_n\}_1^\infty$ is a sequence of step functions on $(X, \mathcal{A})$ such that $0 \le h_n \uparrow f$, we have

$$\lim_{n \to \infty} \int_A h_n \, d\mu_1 = \int_A f \, d\mu_1 = \mu_0(A).$$

On the other hand,

$$\lim_{n \to \infty} \int_A h_n \, d\mu_1 = \lim_{n \to \infty} \int_A h_n g \, d\mu_2 = \int_A f g \, d\mu_2$$

which yields

$$\mu_0(A) = \int_A f g \, d\mu_2.$$

5. Let μ_1 and μ_2 be σ-finite measures on $(X, \mathcal{A})$ such that $\mu_1 \ll \mu_2$. Then for any non-negative measurable function h we have:

$$\int h \, d\mu_1 = \int h \, \frac{d\mu_1}{d\mu_2} \, d\mu_2 .$$

Hint:

If we set $\mu_0(A) = \int_A f g \, d\mu_1$, where $A \in \mathcal{A}$; then from the previous problem we have that

$$\int_A h \, d\mu_1 = \int_A h \cdot \frac{d\mu_1}{d\mu_2} \, d\mu_2 .$$

6. If μ_0 and μ_1 are two equivalent measures show that

$$\frac{d\mu_0}{d\mu_1} = \left(\frac{d\mu_1}{d\mu_0} \right)^{-1} .$$

Hint:

According to problem 6 of section 33, $\mathcal{M}_{\mu_0} = \mathcal{M}_{\mu_1}$. Now, since $\mu_0 \ll \mu_1 \ll \mu_2$ we have (see

problem 4) that

$$\frac{d\mu_2}{d\mu_0} = \frac{d\mu_2}{d\mu_1} \cdot \frac{d\mu_1}{d\mu_0} .$$

35. THE $\mathbf{L}_p$ SPACES

There is a vast literature dealing with the normed vector (linear) spaces. In this section we are concerned with the vector spaces of measurable numerical functions defined on a fixed measure space $(X, \mathcal{A}, \mu)$. We will stick to the (standard) notation introduced in section 31. Thus, $\mathbf{L}_0 = \mathbf{L}_0 (X, \mathcal{A}, \mu)$ is the collection of all (equivalence classes under the (a.e) equality) measurable functions $f: X \to R$. For $0 < p \le \infty$ $\mathbf{L}_p = \mathbf{L}_p (X, \mathcal{A}, \mu)$ is the subclass of $\mathbf{L}_0$ consisting of all $f \in \mathbf{L}_0$ for which

$$\|f\|_p = \left(\int |f|^p \, d\mu\right)^{1/p} < \infty \tag{1}$$

With the convention that two elements from $\mathbf{L}_0$ are identical if they are equal (a.e), the collection $\mathbf{L}_p$ is a "normed vector space" with the norm $\| \cdot \|_p$ if $1 \le p < \infty$. The space $\mathbf{L}_\infty \subset \mathbf{L}_0$ consists of all "essentially bounded "functions from $\mathbf{L}_0$ under the essential supremum norm $\| \cdot \|_\infty$ (see (7) of section 31).

A sequence $\{f_n\}_1^\infty$ of elements from $\mathbf{L}_p$ is said to converge to an element $f \in \mathbf{L}_p$ in "the mean of order p" if

$$\|f_n - f\|_p \to 0 \quad \text{as } n \to \infty \tag{2}$$

If $p=1$ we say that the sequence converges in "the mean" to f. From the Chebishev inequality (see proposition 1 of section 30) we have for any $\varepsilon > 0$ that

$$\mu\{|f_n - f| > \varepsilon\} \le \frac{1}{\varepsilon^p} \int |f_n - f|^p \, d\mu. \tag{3}$$

Thus, the convergence in the mean of order $p>0$ implies the convergence in measure.

A sequence $\{f_n\}_1^\infty$ from $\mathbf{L}_p$ is said to be "Cauchy" (or "fundamental") relative to the norm $\| \cdot \|_p$ if

$$\|f_m - f_n\|_p \to 0 \quad \text{as } m, n \to \infty.$$

Recall that a normed vector space is said to be "complete" if every Cauchy sequence in the space converges. Next, we shall prove that for every $1 \le p < \infty$ $\mathbf{L}_p$ is a complete, that is, for each Cauchy sequence $\{f_n\}_1^\infty$ from $\mathbf{L}_p$ there is an element $f \in \mathbf{L}_p$, unique up to equivalence, such that $\|f_n - f\|_p \to 0$ as $n \to \infty$.

Proposition 1 (Riesz – Fisher)

For any $1 \le p < \infty$ the space $\mathbf{L}_p$ is complete.

Proof:

Let $\{f_n\}_1^\infty$ be a Cauchy sequence in $\mathbf{L}_p$. It is obvious from (3) that $\{f_n\}_1^\infty$ is also Cauchy in measure. Then according to propositions 2 and 4 of section 26, there exists a subsequence $\{f_{n_k}\}_1^\infty$ of $\{f_n\}_1^\infty$ which converges (a.e) on X to a measurable function f. Consequently,

$$\lim_{i\to\infty} |f_{n_i} - f_{n_k}| = |f - f_{n_k}| \quad (a.e).$$

By Fatou's lemma,

$$\int \liminf_i |f_{n_i} - f_{n_k}|^p \, d\mu = \int |f - f_{n_k}|^p \, d\mu \leq \liminf_i \int |f_{n_i} - f_{n_k}|^p \, d\mu. \tag{4}$$

Since $\{f_n\}_1^\infty$ is Cauchy in L_p given any $\varepsilon > 0$ there is an integer $r_0 > 0$ such that $\left\| f_{n_i} - f_{n_k} \right\|_p < (\varepsilon/2)$ if

$i, k > r_0$, so that $\left\| f - f_{n_k} \right\|_p < \varepsilon/2$ and consequently, $f \in L_p$. Finally, by invoking the Minkowski inequality

we have:

$$\left\| f_n - f \right\|_p \leq \left\| f_n - f_{n_k} \right\|_p + \left\| f_{n_k} - f \right\|_p \leq \varepsilon.$$

Thus, $\left\| f_n - f \right\|_p \to 0$ as $n \to \infty$, which proves the proposition. ♥

Definition 1

A complete normed vector space is called a "Banach space."

Corollary 1

For every $p \geq 1$

$$\left\| f_n - f \right\|_p \to 0 \implies \left\| f_n \right\|_p \to \left\| f \right\|_p \quad \text{as } n \to \infty.$$

Indeed, by the Minkowski inequality

$$\left\| f \right\|_p \leq \left\| f - f_n \right\|_p + \left\| f_n \right\|_p$$

$$\left\| f_n \right\|_p \leq \left\| f_n - f \right\|_p + \left\| f \right\|_p.$$

Hence,

$$\left| \left\| f_n \right\|_p - \left\| f \right\|_p \right| \leq \left\| f_n - f \right\|_p.$$

The converse does not hold in general. However, we have the following:

Proposition 2

If $\{f_n\}_1^\infty$ and f are elements of $\mathbf{L}_p$ such that

$$\text{(i) } f_n \to f \ (a.e); \qquad \text{(ii) } \left\| f_n \right\|_p \to \left\| f \right\|_p,$$

then

$$\left\| f_n - f \right\|_p \to 0.$$

Proof:

Our method of proof requires the following elementary inequality:

$$|a + b|^p \leq \gamma_p (|a|^p + |b|^p) \tag{5}$$

where $\gamma_p = 1$ if $0 < p \leq 1$ and $\gamma_p = 2^{p-1}$ if $p \geq 1$ (for a proof see remark 1).

From (5), (i) and (ii) we deduce that

$$0 \leq 2^{p-1}(|f_n|^p + |f|^p) - |f_n - f|^p \to 2^p |f|^p \quad \text{(a.e.)}.$$

By Fatou's lemma we obtain from this as $n \to \infty$, that

$$2^p \int |f|^p \, d\mu \leq \liminf \int [2^{p-1}(|f_n|^p + |f|^p) - |f_n - f|^p] \, d\mu$$

$$= 2^{p-1} \|f\|_p^p + 2^{p-1} \lim \|f_n\|_p^p - \limsup \|f_n - f\|_p^p$$

$$= 2^p \|f\|_p^p - \limsup \|f_n - f\|_p^p .$$

Hence,

$$0 \geq \limsup \|f_n - f\|_p^p ,$$

which implies that $\lim \|f_n - f\|_p^p = 0$. This proves the proposition. ♥

Remark 1

Here is a sketch of a proof of inequality (5). For any $a, b \in R$ and $0 < p \leq 1$

$$|a + b|^p \leq (|a| + |b|)^p \leq (|a| + |b|)^{p-1}(|a| + |b|)$$

$$= \frac{|a|}{(|a| + |b|)^{1-p}} + \frac{|b|}{(|a| + |b|)^{1-p}} \leq \frac{|a|}{|a|^{1-p}} + \frac{|b|}{|b|^{1-p}} .$$

For $1 \leq p < \infty$ we have

$$|a + b|^p \leq 2^p \max(|a|, |b|)^p \leq 2^p (|a|^p + |b|^p).$$

A sharper inequality is obtained by observing that x^p is convex if $p > 1$. Thus, for $a, b > 0$ we have:

$$\left(\frac{a+b}{2}\right)^p \leq \frac{1}{2}(a^p + b^p) \Rightarrow (a + b)^p \leq 2^{p-1}(a^p + b^p).$$

As a matter of fact, various inequalities involving norm $\|\cdot\|_p$ are essentially convexity property and do not depend on the underlying measure space $(X, \mathcal{A}, \mu)$. Let us recall that **a real-valued function f defined on an interval $I \subset R$ is said to be "convex" if for all $x, y \in I$ and $0 \leq \lambda \leq 1$**

$$f(\lambda x + (1-\lambda)y) \leq \lambda f(x) + (1-\lambda) f(y). \tag{6}$$

Geometrically, this says that the chord joining any two points on the curve lies above the graph of $f(\cdot)$. A function $\psi: I \to R$ for which $-\psi$ is convex is called **"concave."** For example, e^x and x^p for $p \geq 1$ are convex. On the other hand, $\log x$ and $|x|^p$ where $0 < p < 1$ are concave. It is not difficult to show that a convex function is absolutely continuous on each closed subinterval of I. Thus f' exists (a.e) on I and it is increasing. If f is twice differentiable on I, we would have $f'' \geq 0$. As a matter of fact, a function $h: I \to R$ twice differentiable everywhere on I, is convex if and only if $f'' \geq 0$.

Example 1

As an application of convexity let us prove the Minkowski inequality. Given $f_1, f_2 \in L_p$, $1 \le p < \infty$, we have

$$\frac{|f_1 + f_2|}{\|f_1\|_p + \|f_2\|_p} \le \frac{|f_1| + |f_2|}{\|f_1\|_p + \|f_2\|_p} = \lambda \frac{|f_1|}{\|f_1\|_p} + (1-\lambda) \frac{|f_2|}{\|f_2\|_p}$$

where $\lambda = \|f_1\|_p / (\|f_1\|_p + \|f_2\|_p)$ and $1-\lambda = \|f_2\|_p / (\|f_1\|_p + \|f_2\|_p)$. Thus, since $|x|^p$ is convex,

$$\left(\frac{|f_1 + f_2|}{\|f_1\|_p + \|f_2\|_p}\right)^p \le \lambda \frac{|f_1|^p}{\|f_1\|_p^p} + (1-\lambda) \frac{|f_2|^p}{\|f_2\|_p^p} \,.$$

Integrating both sides leads to

$$\|f_1 + f_2\|_p^p / (\|f_1\|_p + \|f_2\|_p)^p \le 1.$$

Thus, $\|f_1 + f_2\|_p \le \|f_1\|_p + \|f_2\|_p$, which is the Minkowski inequality. Notice that for $0 < p < 1$ we have

$$\|f_1 + f_2\|_p \ge \|f_1\|_p + \|f_2\|_p$$

since $|x|^p$ is concave.

We are now in position to derive the classical inequality due **to Jensen**. Let $f : R \to R$ be a convex function. A straight line trough a point $(x_0, f(x_0))$ on the graph of $f(\cdot)$

$$y = m(x - x_0) + f(x_0)$$

is called a **"supporting line"** for the graph at x_0, if for all $x \in R$

$$m(x - x_0) + f(x_0) \le f(x).$$

In other words, if it lies below the graph of $f(\cdot)$.

Given $h \in L_1(X, A, \mu)$ denote by

$$\gamma = \int h \, d\mu$$

then

$$m(x - \gamma) + f(\gamma) \le f(x)$$

and consequently

$$m(h - \gamma) + f(\gamma) \le f(h).$$

(7)

Assume $\mu(X) < \infty$ then by integrating both sides of this inequality we obtain

$$m(\int h \, d\mu - \gamma \, \mu(X)) + \mu(X) f(\gamma) \le \int f(h) \, d\mu.$$

From this, for $\mu(X) = 1$ we have

$$f \left(\int h \, d\mu \right) \le \int f(h) \, d\mu, \tag{8}$$

which is the classical form of the Jensen inequality and which is of considerable interest in probability theory.

For an arbitrary measure μ we proceed as follows: Let $g \in L_1(X, \mathcal{A}, \mu)$ be non-negative with

$$\int g \, d\mu > 0 \text{ and } h g \in L_1,$$

and denote by

$$\gamma_0 = (\int h \, g \, d\mu) / \int g \, d\mu. \tag{9}$$

Multiplying both sides of (7) with $g(\cdot)$ and replacing γ with γ_0 we obtain, after integration, that

$$f(\gamma_0) \int g \, d\mu + m(\int h \, g \, d\mu - \gamma_0 \int g \, d\mu) \le \int f(h) \, g \, d\mu.$$

This and (9) yield

$$f\left(\frac{\int h \, g \, d\mu}{\int g \, d\mu}\right) \le \frac{\int f(h) \, g \, d\mu}{\int g \, d\mu}. \tag{10}$$

When $\mu(X)=1$ and $g=1$ (a.e) on X, (10) becomes (8). Notice also that the Jensen inequalities depend on the underling measure space.

In the rest of this section we shall describe briefly some basic features of linear functionals on $\mathbf{L}_p$ spaces. Recall that a map $\Phi:\mathbf{L}_p \to R$ is called a linear functional if for every $f_1, f_2 \in \mathbf{L}_p$ and $\alpha_1, \alpha_2 \in R$

$$\Phi(\alpha_1 f_1 + \alpha_2 f_2) = \alpha_1 \Phi(f_1) + \alpha_2 \Phi(f_2). \tag{11}$$

If there is a constant $C \ge 0$ such that

$$|\Phi(f)| \le C \|f\|_p,$$

the functional Φ is said to be "bounded." The smallest C for which this inequality holds is called the "norm" of Φ. Since

$$|\Phi\left(\frac{|f|}{\|f\|_p}\right)| \le C$$

the norm

$$\|\Phi\| = \sup_{f \in \mathbf{L}_p} \frac{|\Phi(f)|}{\|f\|_p}.$$

Example 2

Let $1<p<\infty$ and $q=p(1-p)$; given $g \in \mathbf{L}_q$, define the mapping $\Phi:\mathbf{L}_q \to R$ by

$$\Phi_g(f) = \int f \cdot g \, d\mu. \tag{12}$$

It is a consequence of the linearity of integral and of the Hölder inequality that Φ_g is a bounded linear functional, for

$$|\Phi_g(f)| \le \int |f \, g| \, d\mu \le \|f\|_p \|g\|_q.$$

Consequently

$$\|\Phi_g\| \le \|g\|_q. \tag{13}$$

As a matter of fact, $\left\|\Phi_g\right\| = \|g\|_q$. To prove this consider

$$h = |g|^{q-1}\,\mathrm{sgn}(g), \quad \text{where} \quad \mathrm{sgn}(x) = \begin{cases} 1 & \text{if} \quad x > 0 \\ 0 & \text{if} \quad x = 0 \\ -1 & \text{if} \quad x < 0 \end{cases}.$$

Clearly

$$|h|^p = |g|^{p(q-1)} = |g|^q,$$

which implies that $h \in L_p$. Since $hg = |g|^q$, it follows that

$$\Phi_g(h) = \int h\,g\,d\mu = \int |g|^2\,d\mu = (\int |g|^q\,d\mu)^{\frac{1}{q}} \, (\int |g|^q)^{\frac{1}{p}}\,d\mu$$

$$= (\int |g|^q\,d\mu)^{\frac{1}{q}} \, (\int |h|^p\,d\mu)^{\frac{1}{p}} = \|g\|_q\,\|h\|_p.$$

Thus,

$$\Phi_g\left(\frac{h}{\|h\|_p}\right) = \|g\|_q \implies \left\|\Phi_g\right\| \ge \|g\|_q.$$

This and (13) prove the assertion.

The next proposition, which represents a characterization of the bounded linear functionals for $1 \le p < \infty$, states that each such a functional on L_p must be in the form (12). This is so called the "Rietsz representation."

Proposition 3 (Rietsz)

Let $(X, \mathcal{A}, \mu)$ be a measure space and Φ a bounded linear functional on $L_p(X, \mathcal{A}, \mu)$, $1 < p < \infty$. Then there exists a $g \in L_q$, where $1/p + 1/q = 1$, such that

$$\Phi(f) = \int f\,g\,d\mu \quad \text{and} \quad \|\Phi\| = \|g\|_q.$$

Proof:

We shall assume that $\mu(X) < \infty$. Let μ_0 be a set function on $\mathcal{A}$ defined by

$$\mu_0(A) = \Phi(I_A).$$

Let $\{E_k\}_1^\infty$ be a sequence from $\mathcal{A}$ such that $E_n \uparrow E$, then $I_{E_n} \uparrow I_E$ pointwise. By the monotone convergence theorem we have that $\left\|I_{E_k}\right\|_p \to \|I_E\|_p$. Then according to proposition 2,

$$\left\|I_E - I_{E_k}\right\|_p \to 0 \quad \text{as} \quad k \to \infty.$$

Thus,

$$0 \le |\mu_0(E - E_k)| = |\Phi(I_E - I_{E_k})| \le \|\Phi\| \cdot \left\|I_E - I_{E_k}\right\|_p,$$

which implies (see problem 4 of section 17) that μ_0 is countably additive. From this inequality it also follows that $\mu_0 \ll \mu$. By the Radon-Nikodym theorem there exists $g \in L_1$ such that for all $A \in \mathcal{A}$

$$\Phi(I_A) = \mu_0(A) = \int_A g\, d\mu = \int g\, I_A\, d\mu.$$

By linearity of Φ it follows that

$$\Phi(h) = \int h\, g\, d\mu$$

for every step function h. Now set $A = \{g \geq 0\}$ and let $0 \leq f \in L_p$. Choose a sequence of step functions $\{h_n\}$ with $0 \leq h_n \uparrow f$. Then

$$h_n I_A\, g \uparrow f g^+ \quad \text{and} \quad \lim_{n \to \infty} \left\| (f - h_n) I_A \right\|_p = 0.$$

By the monotone convergence theorem

$$\lim_{n \to \infty} \int h_n g^+\, d\mu = \int f g^+\, d\mu = \Phi(f I_A) < \infty.$$

Similarly, $f g^- \in L_1$ and

$$\Phi(f I_{A^c}) = \int f g^-\, d\mu < \infty.$$

Thus, for every $f \in L_p$ we have

$$f g \in L_1 \quad \text{and} \quad \Phi(f) = \int f g\, d\mu.$$

It remains to show that $g \in L_q$. To this end set

$$E_n = \{|g| \leq n\}$$

and define $f_n = |g|^{q-1} I_{E_n} \operatorname{sgn}(g)$. Clearly, $f_n \in L_p$ and

$$f_n g = |f_n|^p = |g|^q I_{E_n} \uparrow |g|^q.|$$

Therefore,

$$\int |g|^q I_{E_n}\, d\mu = \int f_n g\, d\mu = \Phi(f_n) \leq \|\Phi\| \|f_n\|_p$$

$$= \|\Phi\| \left(\int |g|^q I_{E_n}\, d\mu \right)^{1/p}.$$

Hence,

$$\left(\int |g|^q I_{E_n}\, d\mu \right)^{1 - \frac{1}{p}} \leq \|\Phi\| < \infty,$$

which proves that $g \in L_q$ and consequently the proposition. ♥

Remark 2

The proof can be extended to the case of a σ-finite measure μ in a standard way.

Problems and complements

1. Given any $f_1, f_2 \in L_2$ show that $f_1 \cdot f_2 \in L_1$.

Hint:

By the Hölder inequality

$$\left| \int f_1 \cdot f_2 \, d\mu \right| \le \int |f_1 \cdot f_2| \, d\mu \le \|f_1\|_2 \cdot \|f_2\|_2.$$

This inequality is also known as the Cauchy-Schwartz inequality.

2. Given $0 \le f \in L_0$ show that $\{p; \|f\|_p < \infty\}$ is an interval.

Hint:

The idea is to establish that for any $s < t$ from $\{p; \|f\|_p < \infty\}$ and $r \in [s, t]$, $z \in \{p; \|f\|_p < \infty\}$. Set $A = \{f > 1\}$, then

$$\text{(i) } f^r I_A \le f^t I_A \; ; \qquad\qquad \text{(ii) } f^r I_{A^c} \le f^s I_A.$$

Hence,

$$0 \le \int f^r \, d\mu = \int_A f^r \, d\mu + \int_{A^c} f^r \, d\mu \le \int_A f^t \, d\mu + \int_{A^c} f^s \, d\mu < \infty.$$

3. Let $\{f_n\}_1^\infty$ be a Cauchy sequence from L_p, $p \ge 1$, which converges in the mean of order p to f. Show that f is unique up to equivalence.

Hint:

If there is another limit, say $f^* \in L_p$, then for any $n = 1, 2, \cdots$ we have (Minkowski's inequality)

$$\|f - f^*\|_p = \|(f - f_n) + (f_n - f^*)\|_p \le \|f - f_n\|_p + \|f_n - f^*\|_p \to 0$$

as $n \to 0$. Thus $\|f - f^*\|_p = 0$. On the other hand,

$$\mu\{|f{-}f^*| > 0\} \le \sum_{n=1}^\infty \mu\{|f - f| > \tfrac{1}{n}\} = 0,$$

since by the Chebishev inequality

$$\mu\{|f{-}f^*{}'| > \tfrac{1}{n}\} \le (\int |f - f^*|^p \, d\mu) \, n^p = 0.$$

Thus, $f = f^*$ (a.e).

4. Let $\{f_n\}_1^\infty$ be a sequence from L_0 such that for all $k = 1, 2, \cdots$, $0 \le f_k \le h$ (a.e) where $h \in L_1$. If

$f_n \xrightarrow{\ \mu\ } 0$ show that $\lim_{n \to \infty} \|f_n\| = 0$.

Hint:

Set $H = \{h > 0\}$ then $\{f_n > 0\} \subset H$ for each $n = 1, 2, \cdots$. If

$H_n = \{\frac{1}{n} \leq h < n\}$ then clearly $H_n \uparrow H$. Thus, $\int\limits_{H-H_n} h\, d\mu \downarrow 0$ as $n \to \infty$; hence, given $\varepsilon > 0$ there is

an integer n_0 such that $\int\limits_{H-H_n} h\, d\mu \leq \varepsilon$ if $n \geq n_0$. Consequently,

$$0 \leq \int f_k\, d\mu = \int\limits_{H} f_k\, d\mu = \int\limits_{H-H_n} f_k\, d\mu + \int\limits_{H_n} f_k\, d\mu \leq \varepsilon + \int\limits_{H_n} f_k\, d\mu.$$

Let us show that there exists an integer $k_0 > 0$ such that

$$\int\limits_{H_n} f_k\, d\mu \leq 2\varepsilon \quad \text{if } k \geq k_0.$$

Since $\mu(H_n) < \infty$ there exists $\delta > 0$ such that $\delta\mu(H_n) < \varepsilon$. Set $E_n = \{f_n > \delta\}$, then since $\mu(E_n) \to 0$ as $n \to \infty$,

$$\int\limits_{H_n} f_k\, d\mu = \int\limits_{H_n \cap E_k} f_k\, d\mu + \int\limits_{H_n \cap E_k^c} f_k\, d\mu \leq n\, \mu(E_k) + \delta\mu(H_n) \leq 2\varepsilon.$$

In other words, we have established that

$$0 \leq \int f_k\, d\mu \leq 2\varepsilon \quad \text{if } k \geq k_0.$$

5. Let $\{f_n\}_1^\infty$ be a sequence from $\mathbf{L}_0$ such that

$$|f_n| \leq |g| \quad (a.e) \quad \text{where } g \in \mathbf{L}_1.$$

If $f_n \xrightarrow{\mu} f$ show that $|f| \leq |g|$ (a.e) and that

$$\|f - f_n\| \to 0 \quad \text{as } n \to \infty.$$

Hint

It follows from remark 3 of section 26 that $|f| \leq |g|$ (a.e.). Now, from

$$|f_n - f| \leq |f_n| + |f| \leq 2|g|$$

and the fact that $|f_n - f| \xrightarrow{\mu} 0$ we deduce, taking into account the previous problem, that $\|f_n - f\| \to 0$.

This result is known as the "dominated converge theorem in measure".

6 Prove that the space $\mathbf{L}_\infty$ is complete.

Hint:

Two functions are considered identical if they are equal (a.e). Then

$$\|f\|_\infty = \text{ess sup } |f|$$

is a norm on L_∞. Observe that $|f| \leq \|f\|_\infty$ (a.e.). Let $\{f_n\}_1^\infty$ be a Cauchy sequence from $\mathbf{L}_\infty$. We must

show that there exists $f \in \mathbf{L}_\infty$ such that $\lim\limits_{n \to \infty} \|f - f_n\|_\infty = 0$. Indeed, since

$$|f_m - f_n| \leq \|f_m - f_n\|_\infty \quad (a.e),$$

given $\varepsilon > 0$ there is an index k_0 such that $\|f_m - f_n\|_\infty < \varepsilon$ if $m, n > k_0$. Consequently

$$|f_m - f_n| < \varepsilon \text{ (a.e).}$$

Thus $\lim\limits_{n \to \infty} f_n(x) = f(x)$ exists in R (a.e). Obviously, then

$$\|f\|_\infty \le \|f - f_n\|_\infty + \|f_n\|_\infty < \infty \Rightarrow f \in \mathbf{L}_\infty.$$

Finally, since for $n > k_0$,

$$|f\text{-}f_n| = \lim_{m \to \infty} |f_m - f_n| < \varepsilon \text{ (a.e)}$$

it is clear that $\text{essup}\,|f\text{-}f_n| = \|f - f_n\|_\infty < \varepsilon$, which implies that

$$\lim_{n \to \infty} \|f - f_n\|_\infty = 0.$$

7. Show that proposition 2 still holds if the convergence (a.e) of f_n to f is replaced by the convergence in measure provided that μ is finite

    Hint:

    Since μ is finite measure, (a.e) convergence implies the convergence in measure. Thus

$$0 \le 2^{p-1}(|f_n|^p + |f|^p) - |f_n\text{-}f|^p \to 2^p|f|^p$$

in measure. As Fatou's lemma is valid for convergence in measure, the argument used in proposition 2 shows that

$$2^p(\textstyle\int |f|^p\,d\mu) \le 2^p(\textstyle\int |f|^p\,d\mu) - \limsup_n (\textstyle\int |f_n\text{-}f|^p\,d\mu).$$

Consequently,

$$0 > \limsup_n \int |f_n\text{-}f|^p\,d\mu,$$

which proves the contention.

Chapter IV

INFINITE PRODUCT OF SPACES

36. PRODUCT OF MEASURE SPACES

Let $(X_1, \mathscr{A}_1, \mu_1)$ and $(X_2, \mathscr{A}_2, \mu_2)$ be two measure spaces. In order to formulate their product we shall first define the product of the measurable spaces $(X_1, \mathscr{A}_1)$ and $(X_2, \mathscr{A}_2)$. We start with $X_1 \times X_2$ and with the class of measurable rectangles

$$S^* = \{(A_1 \times A_2), A_1 \in \mathscr{A}_1, A_2 \in \mathscr{A}_2\}.$$

Set $\sigma\{S^*\} = \mathscr{A}_1 \otimes \mathscr{A}_2$, then

$$(X_1 \times X_2, \mathscr{A}_1 \otimes \mathscr{A}_2) \tag{1}$$

is the product of the measurable spaces.

The nontrivial part of the problem is the construction of a measure on this measurable space in terms of μ_1 and μ_2. We shall show that the set function $\mu = \mu_1 \times \mu_2$ defined on S^* by

$$\mu(A_1 \times A_2) = \mu_1(A_1) \cdot \mu_2(A_2) \tag{2}$$

is such a measure. To this end we need the following:

Proposition 1

The class S^* of measurable rectangles is a semialgebra (see problem 13 of section17).

Proof:

Clearly, $\theta, X_1 \times X_2 \in S^*$. On the other hand, using the identity

$$A_1 \times A_2 = (A_1 \times X_2) \cap (X_1 \times A_2) \tag{3}$$

we have that

$$(A_1 \times A_2) \cap (B_1 \times B_2) = (A_1 \times X_2) \cap (X_1 \times A_2) \cap (B_1 \times X_2) \cap (X_1 \times B_2)$$

$$= (A_1 \cap B_1 \times X_2) \cap (X_1 \times A_2 \cap B_2) = (A_1 \cap B_1 \times A_2 \cap B_2),$$

which implies that the class S^* is closed under finite intersection. Next,

$$(A \times B)^c = (A \times X_2)^c \cup (X_1 \times B)^c = (A^c \times X_2) \cup (X_1 \times B^c)$$

$$= (A^c \times B \cup B^c) \cup (A \cup A^c \times B^c) = (A^c \times B) \cup (A^c \times B^c) \cup (A \times B^c)$$

$$\cup (A \times B^c) \cup (A^c \times B^c) = (A^c \times B) \cup (A \times B^c) \cup (A^c \times B^c).$$

This completes the proof that S^* is a semialgebra. ♥

The next step is to show that the set function $\mu = \mu_1 \times \mu_2$ is a measure on S^*. In other words, given a sequence of disjoint rectangles from S^*, say $\{(A_i \times B_i)\}_1^\infty$ such that

$$\bigcup_{i=1}^{\infty} (A_i \times B_i) = (A \times B) \in S^*,$$

we must show that

$$\mu(A \times B) = \sum_{i=1}^{\infty} \mu(A_i \times B_i). \tag{4}$$

Here is a sketch of a proof. First, observe that

$$I_{A \times B} = I_A \cdot I_B,$$

then since

$$I_{A \times B} = \sum_{i=1}^{\infty} I_{A_i \times B_i}$$

we have that

$$I_A \cdot I_B = \sum_{i=1}^{\infty} I_{A_i} \cdot I_{B_i}.$$

Now (see problem 8 of section 29) by integrating both sides of this equation with respect to μ_2 we obtain:

$$I_A \, \mu_2(B) = \sum_{i=1}^{\infty} I_{A_i} \, \mu_2(B_i).$$

Integrating once more with respect to μ_1 we have that

$$\mu_1(A) \, \mu_2(B) = \sum_{i=1}^{\infty} \mu_1(A_i) \, \mu_2(B_i).$$

From this and (2) we deduce that

$$\mu(A \times B) = \sum_{i=1}^{\infty} \mu(A_i \times B_i).$$

This proves that μ is σ-additive on S^*.

Since S^* is also a semiring, we can use the results of section 19 to extend the measure μ on S^* to a measure on the σ-algebra

$$\sigma\{S^*\} = \mathscr{A}_1 \otimes \mathscr{A}_2.$$

The extended measure $\bar{\mu}$ is complete and is called the " product measure" of μ_1 and μ_2. The system

$$(X_1 \times X_2, \mathscr{A}_1 \otimes \mathscr{A}_2, \bar{\mu})$$

is called the "product measure space."

An element $C \in S^*$ is called a **"measurable cylinder" with base** $A \in \mathscr{A}_1$ if it is in the form $C = A \times X_2$. If $C = X_1 \times B$ it is a measurable cylinder with base $B \in \mathscr{A}_2$.

Proposition 2

Let $\hat{\mathcal{A}} \subset \mathscr{A}_1 \otimes \mathscr{A}_2$ be the class of measurable cylinders with basis in $\mathscr{A}_1$, then $\hat{\mathcal{A}}$ and $\mathscr{A}_1$ are isomorphic.

Proof.

We have to show that there exists a bijection $T: \mathscr{A}_1 \to \hat{\mathcal{A}}$ such that for any A, $B \in \mathscr{A}_1$, T(A-B)=

T(A)-T(B) and such that for any $\{A_i\}_1^\infty$ from $\mathscr{A}_1$

$$T(\bigcup_1^\infty A_i) = \bigcup_1^\infty T(A_i).$$

Indeed, define T on $\mathscr{A}_1$ by

$$T(A) = A \times X_2$$

then $T(A-B)=(A-B)\times X_2 =(A\times X_2) - (B\times X_2)=T(A) -T(B)$. Also

$$T(\bigcup_{i=1}^\infty A_i) = (\bigcup_{i=1}^\infty A_i) \times X_2 = \bigcup_{i=1}^\infty (A_i \times X_2) = \bigcup_{i=1}^\infty T(A_i).$$

Similar statement can be made for the class of measurable cylinders with basis in $\mathscr{A}_2$.

If $\{(X_i, \mathscr{A}_i, \mu_i)\}_1^n$ are measure spaces, their product space is the system

$$(X, \mathscr{A}, \mu)$$

where $X=X_1\times X_2 \times \cdots \times X_n$, $\mathscr{A}=\mathscr{A}_1\otimes \mathscr{A}_2\otimes \cdots \otimes \mathscr{A}_n$ and

$$\mu = \mu_1 \times \mu_2 \times \cdots \times \mu_n.$$

Of course,

$$\mathscr{A}_1\otimes\mathscr{A}_2\otimes \cdots \otimes\mathscr{A}_n = \sigma\{(A_1\times A_2\times \cdots \times A_n); A_i \in \mathscr{A}_i, \; i=1, 2, \cdots, n\}$$

and the product measure $\mu = \mu_1\times \mu_2 \times \cdots \times \mu_n$ is uniquely defined on the σ-algebra $\mathscr{A}$ by the requirement

that $\mu(A_1\times A_2 \times \cdots \times A_n)=\mu_1(A_1)\mu_2(A_2) \cdots \mu_n(A_n)$. The proof follows easily from (4) and an

induction on n. ♥

Problems and complements

1. Show that for any n=2, 3, $\cdots$

$$\bigcap_1^n (A_i \times B_i) = (\bigcap_1^n A_i) \times (\bigcap_1^n B_i)$$

Hint:

Use proposition 1 and induction on n.

2. Show that

$$(A_1 \times B_1) - (A_2 \times B_2) = [(A_1 - A_2) \times B_1] \cup [(A_1 \cap A_2) \times (B_1 - B_2)].$$

Hint:

See proposition 1.

3. Show that $A \times B \neq \theta$ if and only if $A \neq \theta$ and $B \neq \theta$.

Hint:

If $A \times B \neq \theta$ there exists $(u, v) \in A \times B$ and since $u \in A$ and $v \in B$, $A \neq \theta$ and $B \neq \theta$. If $A \neq \theta$ and $B \neq \theta$ there is an $u \in A$ and $v \in B$. Thus, $(u, v) \in A \times B \neq \theta$.

4. Let $M \in \mathcal{A}_1 \otimes \mathcal{A}_2$. Show that there exists a measurable rectangle $A \times B \supset M$.

Hint:

Let $\mathcal{A}_0$ be the algebra generated by the semialgebra $\mathbf{S}^*$. According to problem 13 of section 17, every element of $\mathcal{A}_0$ is a finite disjoint union of elements from $\mathbf{S}^*$. On the other hand, according to problem 11 of section 17, every member of $\sigma\{\mathcal{A}_0\} = \mathcal{A}_1 \otimes \mathcal{A}_2$ can be covered by a finite or countable union of elements from the algebra $\mathcal{A}_0$. Thus if $\{(A_i \times B_i)\}_1^\infty$ is a sequence from $\mathcal{A}_0$ such that

$$M \subset \bigcup_{i=1}^\infty (A_i \times B_1) \subset (\bigcup_1^\infty A_i \times \bigcup_1^\infty B_i).$$

5. Let $(X_1, \mathcal{A}_1, \mu_1)$ and $(X_2, \mathcal{A}_2, \mu_2)$ be σ-finite measure spaces. Show that the measure space $(X_1 \times X_2, \mathcal{A}_1 \otimes \mathcal{A}_2, \mu_1 \times \mu_2)$ is also σ-finite.

Hint:

There exists a measurable partition $\{E_i\}_1^\infty$ of X_1 such that $\mu_1(E_i) < \infty$ for all i=1, 2, $\cdots$. There is also a measurable partition $\{F_j\}_1^\infty$ of X_2 such that for all j=1, 2, $\cdots$, $\mu_2(F_j) < \infty$. But then, since

$$X_1 \times X_2 = (\bigcup_i E_i) \times (\bigcup_j F_j) = \bigcup_i \bigcup_j (E_i \times F_j)$$

and each $\mu_1 \times \mu_2 (E_i \times F_j) = \mu_1(E_i) \mu_2(F_j) < \infty$, the assertion follows.

6. Consider the measurable space $(R \times R, \mathcal{R} \otimes \mathcal{R})$ where $R = (-\infty, \infty)$ and $\mathcal{R}$ is the σ-algebra of Borel

subsets of R. Show that for every $E \in \mathscr{R}$ the set

$$D = \{(x, y) \in R \times R; x+y \in E\} \in \mathscr{R} \otimes \mathscr{R}.$$

Hint:

Clearly, $\{(x, y) \in R \times R; x+y \leq t\}$ is a closed set and thus measurable. If $E = (t_1, t_2]$, then

$$\{(x, y) \in R \times R; x+y \in (t_1, t_2]\} = \{(x, y) \in R \times R; x+y \leq t_2\}$$

$$- \{(x, y) \in R \times R; x+y \leq t_1\} \in \mathscr{R} \otimes \mathscr{R},$$

and so on.

7. Consider the measure space $(R, \mathscr{M}, \bar{\lambda})$ where $R = (-\infty, \infty)$, $\bar{\lambda}$ is the Lebesgue measure and $\mathscr{M}$ is the class of Lebesgue's measurable sets. This measure space is complete (see definition 1 of section 9). Show that the product space $(R \times R, \mathscr{M} \otimes \mathscr{M}, \bar{\lambda} \times \bar{\lambda})$ is not.

Hint:

Given a $\bar{\lambda}$-null set $N \in \mathscr{M}$, then $N \times R \in \mathscr{M} \otimes \mathscr{M}$ is clearly $\bar{\lambda} \times \bar{\lambda}$-null set. If $H \subset R$ is not measurable then $N \times H \subset N \times R$ but $N \times H \notin \mathscr{M} \otimes \mathscr{M}$.

8 Let $(X, \mathscr{A}_1)$ and $(Y, \mathscr{A}_2)$ be measurable spaces. If

$$\mathscr{E} \subset \mathscr{A}_1 \text{ is such that } \sigma\{\mathscr{E}\} = \mathscr{A}_1$$

$$\mathscr{F} \subset \mathscr{A}_2 \text{ is such that } \sigma\{\mathscr{F}\} = \mathscr{A}_2,$$

show that

$$\mathscr{E} \otimes \mathscr{F} = \mathscr{A}_1 \otimes \mathscr{A}_2.$$

Hint:

Denote by $\mathscr{C} = \{(A \times B); A \in \mathscr{E}, B \in \mathscr{F}\}$ then clearly

$$\mathscr{C} \subset \mathscr{A}_1 \otimes \mathscr{A}_2 \text{ and thus } \sigma\{\mathscr{C}\} = \mathscr{E} \otimes \mathscr{F} \subset \mathscr{A}_1 \otimes \mathscr{A}_2.$$

To prove the converse, fix $E \in \mathscr{E}$ and consider

$$\mathscr{H} = \{H \in Y; E \times H \in \mathscr{E} \otimes \mathscr{F}\}.$$

It is not difficult to verify that $\mathscr{H}$ is a σ-algebra such that $\mathscr{F} \subset \mathscr{H}$. Thus

$$\sigma\{\mathscr{F}\} \subset \mathscr{H} \Rightarrow \mathscr{A}_2 \subset \mathscr{H}.$$

Similarly, if we fix $F \in \mathscr{A}_1$ the class

$$\mathscr{D} = \{G \subset X; G \times F \in \mathscr{E} \otimes \mathscr{F}\}$$

is σ-algebra such that $\mathscr{E} \subset \mathscr{D}$ and $\sigma\{\mathscr{E}\} = \mathscr{A}_1 \subset \mathscr{D}$. Consequently, for any $A \in \mathscr{A}_1$ and $B \in \mathscr{A}_2$ $A \times B \subset \mathscr{E} \otimes \mathscr{F}$ implying that $\mathscr{A}_1 \otimes \mathscr{A}_2 \subset \mathscr{E} \otimes \mathscr{F}$.

37 SECTIONS

Let $(X \times Y, \mathscr{X} \otimes \mathscr{Y})$ be a **product measurable space. Given any subset** $E \subset X \times Y$ **and a fixed point** $x \in X$, **the subset** $E_X \subset Y$ **defined by**

$$E_X = \{y \in Y;\ (x, y) \in E\}$$

is called the "section" of E at x. If for some $x \in X$ **there is no** $y \in Y$ **such that** $(x, y) \in E$ **then obviously** $E_X = \emptyset$. **Similarly, for** $y \in Y$ **fixed, the subset** $E^y \subset X$ **defined by**

$$E^y = \{x \in X;\ (x, y) \in E\}$$

is the "section" of E at y.

For instance, if $A \subset X$ and $B \subset Y$ we have that

$$(A \times B)_X = \begin{cases} B & \text{if} \quad x \in A \\ \theta & \text{if} \quad x \notin A \end{cases} \qquad \text{and} \qquad (A \times B)^y = \begin{cases} A & \text{if} \quad x \in B \\ \theta & \text{if} \quad x \notin B \end{cases} \tag{1}$$

It is rather obvious that the operation of taking, say x section, is a mapping of $\mathscr{P}(X \times Y)$ onto $\mathscr{P}(Y)$. The mapping is a "homomorphism" for the operations of union, intersection and complement. In other words, if $\{E_t;\ t \in T\}$ is a subfamily of $\mathscr{P}(X \times Y)$, then

$$\left(\bigcup_T E_t\right)_X = \bigcup_T (E_t)_X, \qquad \left(\bigcap_T E_t\right)_X = \bigcap_T (E_t)_X, \tag{2}$$

and similarly for y-section. Of course, $(E^c)_X = (E_X)^c$.

The next proposition states that every $E \in \mathscr{X} \otimes \mathscr{Y}$ **has the property that** $E_X \in \mathscr{Y}$ **for all** $x \in X$ **and that** $E^y \in \mathscr{X}$ **for all** $y \in Y$.

Proposition 1

For every $M \in \mathscr{X} \otimes \mathscr{Y}$ and $x \in X$, $M_X \in \mathscr{Y}$ and for every $y \in Y$, $M^y \in \mathscr{X}$.

Proof:

For a fixed $x \in X$ define

$$\mathscr{M}_X = \{M \in \mathscr{X} \otimes \mathscr{Y};\ M_X \in \mathscr{Y}\}.$$

It is clear from (1) that the semi-algebra $S^* \subset \mathscr{M}_X$ and from (2) that $\mathscr{M}_X$ is a σ-algebra. From these two facts it follows at once that

$$\mathscr{M}_X = \mathscr{X} \otimes \mathscr{Y}.$$

In a similar way one proves that each $M^y \in \mathscr{X}$. ♥

Let $(X \times Y, \mathscr{X} \otimes \mathscr{Y}, \mu)$ be the product measure space where $\mu = \mu_1 \times \mu_2$. If $D = A \times B \in S^*$ then from (1) we have that

$$\mu_1(D^y) = \mu_1(A)\, I_B(y) \quad \text{and} \quad \mu_2(D_X) = \mu_2(B)\, I_A(x). \tag{3}$$

Thus, $\mu_1(D^y)$ is $\mathscr{Y}$-measurable and $\mu_2(D_x)$ is $\mathscr{X}$-measurable. From this it follows at once that

$$\int \mu_1(D^y)\, d\mu_2 = \int \mu_2(D_x)\, d\mu_1 = \mu_1(A)\, \mu_2(B) = \mu(A \times B).$$

The next proposition, which generalizes this observation, shows that for each $M \in \mathscr{X} \otimes \mathscr{Y}$, $\mu_1(M^y)$ and $\mu_2(M_x)$ are measurable functions. Recall, if μ_1 and μ_2 are σ-finite so is the product measure $\mu = \mu_1 \times \mu_2$ (see problem 5 of section 36).

Proposition 2

Assume that μ_1 and μ_2 are finite measures, then for any $M \in \mathscr{X} \otimes \mathscr{Y}$, $\mu_1(M^y)$ is $\mathscr{Y}$-measurable, $\mu_2(M_x)$ is $\mathscr{X}$-measurable and

$$\mu(M) = \int \mu_1(M^y)\, d\mu_2 = \int \mu_2(M_x)\, d\mu_1 . \qquad (4)$$

Here of course, $\mu(M) = \mu_1 \times \mu_2(M)$.

Proof:

Let $\mathscr{M} \subset \mathscr{X} \otimes \mathscr{Y}$ be the collection of all measurable sets for which the above assertions hold. Since $S^* \subset \mathscr{M}$ this class is not empty and obviously contains the algebra generated by S^*. In addition, $\mathscr{M}$ is a monotone class. Indeed, if $\{B_n\}_1^\infty$ is a monotone sequence from $\mathscr{M}$ such that $B_n \uparrow B$, then first, from the definition of $\mathscr{M}$ we have

$$\mu(B_n) = \int \mu_1(B_n^y)\, d\mu_2 = \int \mu_2((B_n)_x)\, d\mu_1 .$$

From this and the monotone convergence theorem it follows that

$$\mu(B) = \int \mu_1(B^y)\, d\mu_2 = \int \mu_2(B_x)\, d\mu_1,$$

which proves that $B \in \mathscr{M}$. Thus, $\mathscr{M}$ is a monotone class which contains the algebra generated by S^*. But then (see corollary 2 below), $\sigma\{S^*\} \subset \mathscr{M}$. The proof now follows from the inclusions

$$\sigma\{S^*\} \subset \mathscr{M} \subset \mathscr{X} \otimes \mathscr{Y} = \sigma\{S^*\}. \qquad \blacktriangledown$$

Corollary 1

If μ_1 and μ_2 are not finite, but σ-finite, there exists a sequence $\{A_n\}_1^\infty$ from $\mathscr{X}$ such that $A_n \uparrow X$ and $\mu_1(A_n) < \infty$. There is also a sequence $\{B_n\}_1^\infty$ from $\mathscr{Y}$ such that $\mu_2(B_n) < \infty$ and $B_n \uparrow Y$. Then the sets from $\mathscr{X} \otimes \mathscr{Y}$

$$M \cap (A_n \times B_n) \uparrow M \qquad \text{as } n \to \infty.$$

By applying the previous argument to this sequence and from the monotone convergence theorem we conclude that the proposition holds if μ is σ-finite. Note that for any $G \in \mathscr{X} \otimes \mathscr{Y}$ we have from (4) that

$$\mu(G) = 0 \iff \mu_2(G_x) = 0 \quad \text{(a.e)} \ \mu_1, \text{ and}$$

$$\mu(G) = 0 \iff \mu_1(G^y) = 0 \quad \text{(a.e)} \ \mu_2.$$

Remark 1

Recall that a monotone class $\mathcal{M}$ of sets is a class closed under formation of limits of monotone sequences of elements from $\mathcal{M}$. Every σ-algebra is a monotone class and every monotone algebra is a σ-algebra. The first assertion is obvious and the second follows from the fact that every countable union $\cup A_n$ and every countable intersection $\cap A_n$ is a monotone limit of sequences $\bigcup_{i=1}^{n} A_i$ and $\bigcap_{i=1}^{n} A_i$, respectively.

It is obvious that the intersection of monotone classes is a monotone class. Thus, if $\mathcal{C}$ is an arbitrary class of sets there exists the smallest monotone class, denoted by

$$m\{\,\mathcal{C}\,\},$$

which contains $\mathcal{C}$. Here we shall prove that for any algebra $\mathcal{A}_0$

$$m\{\mathcal{A}_0\} = \sigma\{\mathcal{A}_0\}.$$

This result was used to prove the proposition 2. It is obvious that $m\{\mathcal{A}_0\} \subset \sigma\{\mathcal{A}_0\}$. To prove the converse inclusion it suffices to show that $m\{\mathcal{A}_0\}$ is an algebra.

Proposition 3

The class $m\{\mathcal{A}_0\}$ is an algebra.

Proof:

(i) $m\{\mathcal{A}_0\}$ is closed under complementation, for if $\mathcal{L}$ is defined by

$$\mathcal{L} = \{A \in m\{\mathcal{A}_0\}; \ A^c \in m\{\mathcal{A}_0\}\}$$

then obviously $\mathcal{A}_0 \subset \mathcal{L}$. Since $\mathcal{L}$ is a monotone class (see problem 6) it follows that $m\{\mathcal{A}_0\} \subset \mathcal{L}$.

(ii) $m\{\mathcal{A}_0\}$ is closed under finite intersection. Indeed, for any $A \in m\{\mathcal{A}_0\}$ define the class $\mathcal{M}_A \subset m\{\mathcal{A}_0\}$ by

$$\mathcal{M}_A = \{B; \ A \cap B \in m\{\mathcal{A}_0\}\}.$$

It is not difficult to verify that $\mathcal{M}_A$ is a monotone class. If $A \in \mathcal{A}_0$ then obviously

$$\mathcal{A}_0 \subset \mathcal{M}_A \implies m\{\mathcal{A}_0\} \subset \mathcal{M}_A.$$

Cosequently,

$$m\{\mathcal{A}_0\} = \mathcal{M}_A \quad \text{for every } A \in \mathcal{A}_0.$$

Thus, if $B \in m\{\mathcal{A}_0\} \Rightarrow B \in \mathcal{M}_A \Rightarrow A \cap B \in \mathcal{M}_A \Rightarrow A \cap B \in m\{\mathcal{A}_0\} \Rightarrow A \in \mathcal{M}_B$. Consequently,

$\mathcal{A}_0 \subset \mathcal{M}_B$ and since $\mathcal{M}_B$ is monotone and contains $\mathcal{A}_0$, it follows that $m\{\mathcal{A}_0\} \subset \mathcal{M}_B$. Hence,

$$m\{\mathcal{A}_0\} = \mathcal{M}_A = \mathcal{M}_B,$$

which proves that $m\{\mathcal{A}_0\}$ is an algebra and thus the proposition. ♥

Corollary 2

If $\mathcal{M}$ is a monotone class which contains an algebra $\mathcal{A}_0$, then

$$\sigma\{\mathcal{A}_0\} \subset \mathcal{M},$$

for $\sigma\{\mathcal{A}_0\} = m\{\mathcal{A}_0\} \subset \mathcal{M}$.

Remark 2

Let $h : X \times Y \to R$ be $\mathcal{X} \otimes \mathcal{Y}$-measurable. Given $x \in X$, the x-section of h, denoted by h_x is the

mapping.

$$h_x : Y \to R \quad \text{defined by} \quad h_x(\cdot) = h(x, \cdot).$$

Similarly, for any $y \in Y$ fixed, the y-section of h, denoted by h^y, is the mapping

$$h^y : X \to R \quad \text{defined by} \quad h^y(\cdot) = h(\cdot, y).$$

The section h_x is $\mathcal{Y}$-measurable. Indeed, for any Borel subset $B \subset R$ we have $h^{-1}(B) \in \mathcal{X} \otimes \mathcal{Y}$.

Thus,

$$(h^{-1}(B))_x = \{(x, y) \in X \times Y; h(x, y) \in B\}_x$$

$$= \{y \in Y; h_x(y) \in B\} = h_x^{-1}(B) \in \mathcal{Y}.$$

In a similar fashion one can show that h^y is $\mathcal{X}$-measurable.

Problems and complements

1 Given measurable space $(R \times R, \mathcal{R} \otimes \mathcal{R})$, where $R=(-\infty, \infty)$ and $\mathcal{R}$ is the σ-algebra of Borel subsets of R, show that

$\quad$ (i) $\{(x, y); x+y=t\} \in \mathcal{R} \otimes \mathcal{R}$ for each $t \in R$.

$\quad$ (ii) $A = \{(x, y); x+y \in B\} \in \mathcal{R} \otimes \mathcal{R}$ for any $B \in \mathcal{R}$

$\quad$ (iii) Determine A_x and A^y.

$\quad$ Hint:

$\quad$ (i) Sets $\{(x, y); x+y \leq t\}$ and $\{(x, y); x+y \geq t\}$ are both closed subsets of $R \times R$ and thus measurable. Consequently, their intersection is also measurable.

$\quad$ (ii) Clearly,

$$A = \bigcup_{t \in B} \{(x, y); x+y=t\}.$$

Thus, if B is countable, A must be measurable. If $B=(\alpha, \beta)$, where $-\infty < \alpha < \beta < \infty$, then

$$A = \{(x, y); \alpha < x+y < \beta\} = \{(x, y); x+y < \beta\} - \{(x, y); x+y \leq \alpha\}$$

and is thus measurable. If B is an open set, and thus the union of at most countably many disjoint open intervals, A is again measurable, and so on.

$\quad$ (iii) $\quad A_x = \bigcup_{t \in B} \{(x, y); x+y=t\}_x = \bigcup_{t \in B} \{y; y=t-x\} = \{y; y \in B-x\} = B-x \in \mathcal{R}$

according to proposition 2 of section 22.

2 Let $(X \times Y, \mathcal{X} \otimes \mathcal{Y})$ be an arbitrary product measurable space. If $G \subset X \times Y$ is such that $G_x \in \mathcal{Y}$ and $G^y \in \mathcal{X}$ for all $x \in X$ and $y \in Y$, show by a counter example, that this does not necessarily imply that $G \in \mathcal{X} \otimes \mathcal{Y}$.

$\quad$ Hint:

$\quad$ Let X be uncountable and let $\mathcal{X}$ be generated by singletons $\{x\} \subset X$. Then each element of $\mathcal{X}$ is either countable or its complement is countable (see problem 12 of section 17). If $C \subset X$ is not measurable, then the "diagonal" set

$$\Delta(C) = \{(x, x); x \in C\}$$

is not countable, and since $(\Delta(C))^c = \Delta(C^c)$, it is not countable either. Thus,

$$\Delta(C) \notin \mathcal{X} \otimes \mathcal{X}.$$

$\quad$ On the other hand,

$$(\Delta(C))_x = \{x\} \text{ and } (\Delta(C))^y = \{y\}.$$

3 Let $(X \times R, \mathcal{X} \otimes \mathcal{R})$ be the product measure of space. If $f: X \to [0, \infty)$ is $\mathcal{X}$-measurable, show that

$$A(f) = \{(x, s) \in X \times R; 0 \leq s \leq f(x)\} \in \mathcal{X} \otimes \mathcal{R}.$$

Determine $A_X(f)$ and $A^S(f)$.

Hint:

If f is a step function, say

$$f(x) = \sum_{i=1}^{n} \alpha_i I_{C_i}(x)$$

then

$$A(f) = \{(x, s); 0 \le s \le f(x)\} =$$

$$= \{(x, s); 0 \le s \le f(x)\} \cap \bigcup_{i=1}^{n} C_i$$

$$= \bigcup_{i=1}^{n} \{(x, s); 0 \le s \le \alpha_i\} \cap C_i$$

$$= \bigcup_{i=1}^{n} \{(x, s); C_i \times [0, \alpha_i]\} \subset \mathcal{K} \otimes \mathcal{R}.$$

If $f \ge 0$ is an arbitrary $\mathcal{K}$-measurable function, there exists a sequence of step functions $\{f_k\}_1^{\infty}$ such that $f_k \uparrow f$ (a.e). Since

$$A(f_k) \subset A(f_{k+1}) \Rightarrow A(f_k) \uparrow A(f) \quad \text{where}$$

$$A(f) = \bigcup_{k=1}^{\infty} A(f_k) \in \mathcal{K} \otimes \mathcal{R}.$$

Next, for any fixed $x \in X$

$$A_X(f) = \{s \in R; 0 \le s \le f(x)\},$$

and for any $s \in R$

$$A^S(f) = \{x \in X; 0 \le s \le f(x)\}.$$

4 Let $(X \times Y, \mathcal{K} \otimes \mathcal{Y})$ be an arbitrary product measurable space and let

$$f : X \to R, \quad h : Y \to R$$

be measurable with respect to $\mathcal{K}$ and $\mathcal{Y}$, respectively. Show that the function $H : X \times Y \to R$ defined by

$$H(x, y) = f(x) + h(y)$$

is $\mathcal{K} \otimes \mathcal{Y}$-measurable.

Hint:

To establish if a mapping $\Phi : X \times Y \to R$ is measurable is not always easy. In many interesting cases the following approach can be useful. In analogy with the concept of a cylinder set we define the notion of a "cylinder function" on $X \times Y$. Let $f : X \to R$; a map $\Phi : X \times Y \to R$ is said to be a cylinder function on X if

$$\Phi(x, y) = f(x) \text{ for all } (x, y) \in X \times Y.$$

Clearly, Φ is $\mathcal{K} \otimes \mathcal{Y}$-measurable if and only if $f(\cdot)$ is $\mathcal{K}$-measurable, because given any $B \in \mathcal{R}$

$$\Phi^{-1}(B) = f^{-1}(B) \times Y \in \mathcal{K} \otimes \mathcal{Y}.$$

In a similar fashion, one defines a cylinder function on Y. Consequently,

$$H(x, y) = f(x) + h(y)$$

is measurable if and only if f is $\mathcal{K}$-measurable and h is $\mathcal{Y}$-measurable.

5 Let $h:R \to R$ be a Borel function. Show that the mapping $H:R \times R \to R$ defined by

$$H(x, y) = h(x + y)$$

Is $\mathcal{R} \times \mathcal{R}$-measurable.

    Hint:

    If $h = I_B$ with $B \in \mathcal{R}$, then

$$h(x+y) = I_B(x+y) = I_{\{(x,y); x+y \in B\}}$$

is $\mathcal{R} \times \mathcal{R}$-measurable (problem 1). The same holds if h is a step-function.

6 Show that $\mathcal{L}$ and $\mathcal{M}_A$ of proposition 3 are monotone classes.

    Hint:

    If $\{A_n\}_1^\infty$ is an increasing sequence from $\mathcal{L}$, then $\lim A_n \in m\{\mathcal{A}_0\}$. Since $\{A_n^c\}_1^\infty$ is

decreasing, $\lim A_n^c = (\lim A_n)^c \in m\{\mathcal{A}_0\}$. Consequently, $\mathcal{L} = m\{\mathcal{A}_0\}$.

    (ii) Let $\{B_n\}_1^\infty$ be an increasing sequence from $\mathcal{M}_A$ then $\bigcup_1^\infty B_n \in m\{\mathcal{A}_0\}$. Also, $(A \cap B_n\}_1^\infty$

is increasing so that

$$\bigcup_1^\infty (A \cap B_n) = A \cap (\bigcup_1^\infty B_n) \in m\{\mathcal{A}_0\}.$$

Let $(X, \mathscr{K}, \mu_1)$ and $(Y, \mathscr{Y}, \mu_2)$ be two σ-finite measure spaces. The product $(X \times Y, \mathscr{K} \otimes \mathscr{Y}, \mu)$, where $\mu = \mu_1 \times \mu_2$, is also σ-finite (see problem 5 of section 36). As we have established in the previous section, for every $A \in \mathscr{K} \otimes \mathscr{Y}$, the function $\mu_1(A^y)$ is $\mathscr{Y}$-measurable and $\mu_2(A_x)$ is $\mathscr{K}$-measurable. In addition, we have that

$$\mu(A) = \int_Y \mu_1(A^y)\, d\mu_2 = \int_X \mu_2(A_x)\, d\mu_1, \tag{1}$$

where, of course, for every $E_1 \in \mathscr{K}$, $E_2 \in \mathscr{Y}$

$$\mu(E_1 \times E_2) = \mu(E_1)\,\mu_2(E_2).$$

Let $h: X \times Y \to R$ be a $\mathscr{K} \otimes \mathscr{Y}$-measurable function and either μ-integrable or $h \geq 0$. In either case its integral with respect to μ exists, which we write as

$$\int h\, d\mu \quad \text{or} \quad \int_{X \times Y} h\, d(\mu_1 \times \mu_2).$$

In this section we are concerned with the relation between the integral of h and the integrals of the sections of h. The next result is due to Lebesgue and Fubini and is usually called the "Fubini theorem."

Proposition 1

Let $(X, \mathscr{K}, \mu_1)$ and $(Y, \mathscr{Y}, \mu_2)$ be σ-finite measure spaces and h an $\mathscr{K} \otimes \mathscr{Y}$-measurable function on $X \times Y$. If $h \geq 0$ or $\mu = \mu_1 \times \mu_2$-integrable then

$$\int h\, d\mu = \int_Y \left(\int_X h^y\, d\mu_1 \right) d\mu_2 = \int_X \left(\int_Y h_x\, d\mu_2 \right) \mu_1. \tag{2}$$

In addition, when h is μ-integrable the h^y is integrable for almost all $y \in Y$ and h_x is integrable for almost all $x \in X$. The last two integrals in (2) are called the **"iterated integrals"** of h.

Proof:

For $h = I_A$, where $A \in \mathscr{K} \otimes \mathscr{Y}$, we have that $h_x = I_{A_x}$ and $h^y = I_{A^y}$. Thus.

$$\int_Y I_{A_x}\, d\mu_2 = \mu_2(A_x), \qquad \int_X I_{A^y}\, d\mu_1 = \mu_1(A^y).$$

This and (1) prove the proposition when h is an indicator function. The proposition also holds when $h \geq 0$ is a step function, which is not difficult to verify. In the general case, there is a sequence of non-negative step functions $\{h_n\}_1^\infty$ such that $h_n \uparrow h$. Then, since $(h_n)_x \uparrow h_x$ and $(h_n)^y \uparrow h^y$, the assertion follows from the monotone convergence theorem.

If h is integrable then the functions

$$\int h_x\, d\mu_2 \quad \text{and} \quad \int h^y\, d\mu_1,$$

are also integrable and thus (a.e) finite, so that the functions h_X of y and h^y of x are almost all integrable.

Therefore, if $h = h^+ - h^-$ is integrable, that is h^+ and h^- are integrable, then

$$h_X = h_X^+ - h_X^- \quad \text{and} \quad h^y = (h^+)^y - (h^-)^y$$

are almost all integrable and almost everywhere finite. This proves the proposition. ♥

The next result is known as Tonelli's theorem. In a certain sense it is a converse to the Fubini theorem because it shows that the existence of one of the iterated integrals is sufficient condition for the integrability of the measurable function h.

Proposition 2

Assume that a measurable function

$$h : X \times Y \to R$$

is such that either

$$\int_X d\mu_1 \int_Y |h|\, d\mu_2 < \infty \quad \text{or} \quad \int_Y d\mu_2 \int_X |h|\, d\mu_1 < \infty,$$

then h is integrable and the Fubini theorem holds.

Proof:

Since the functions

$$\int_Y |h|\, d\mu_2 \geq 0 \quad \text{and} \quad \int_X |h|\, d\mu_1 \geq 0$$

are $\mathscr{X}$-measurable and $\mathscr{Y}$-measurable, respectively, the integrals

$$\int_X d\mu_1 \int_Y |h|\, d\mu_2 \quad \text{and} \quad \int_Y d\mu_2 \int |h|\, d\mu_1$$

are well defined. In addition, due to (2)

$$\int |h|\, d\mu = \int_X d\mu_1 \int_Y |h|\, d\mu_2 = \int_Y d\mu_2 \int_X |h|\, d\mu_1$$

so that $\int |h|\, d\mu < \infty$. This proves the proposition. ♥

Example 1

The next   example shows that the iterated integrals of a measurable function h may exist and have the same (finite) value, and yet h is not necessarily integrable. Let h be a function on the rectangle $[-1, 1] \times [-1, 1]$ defined as follows: $h(0, 0) = 0$ and

$$h(x, y) = \frac{xy}{(x^2 + y^2)^2} \quad \text{otherwise.}$$

It is not difficult to see that $h_X(\,\cdot\,)$ and $h^y(\,\cdot\,)$ are odd for every $x \in [-1, 1]$ and $y \in [-1, 1]$, respectively. Consequently,

$$\int_{-1}^{1} h^y(s)\, ds = \int_{-1}^{1} \frac{sy}{(s^2 + y^2)^2}\, ds = 0 \quad \text{and}$$

$$\int_{-1}^{1} h_x(t)\,dt = \int_{-1}^{1} \frac{xt}{(x^2+t^2)^2}\,dt = 0.$$

Thus,

$$\int_{-1}^{1} dy \int_{-1}^{1} h(x,y)dx = \int_{-1}^{1} dx \int_{-1}^{1} h(x,y)\,dy = 0.$$

On the other hand, the function h is not integrable over the rectangle, for if it were, it would be integrable over any measurable subset of this rectangle, say $[0,1]\times[0,1]$. However,

$$\int_{0}^{1} \frac{xy}{(x^2+y^2)^2}\,dy = \frac{1}{2x} - \frac{x}{2(x^2+1)}$$

and this function is not integrable over $[0, 1]$.

Example 2

Determine the integral

$$\int_{D} x^y\,dx\,dy$$

where $D=[0, 1]\times[a, b]$ and $0<a<b$, and use this result obtained to show that

$$\int_{0}^{1} \frac{x^b - x^a}{\ln x}\,dx = \ln\frac{1+b}{1+a}\,.$$

Since $x^y>0$ over D, the proposition 2 holds. Thus,

$$\int_{D} x^y\,dx\,dy = \int_{a}^{b} dy \int_{0}^{1} x^y\,dx = \ln\frac{1+b}{1+a}\,.$$

On the other hand,

$$\int_{D} x^y\,dx\,dy = \int_{0}^{1} dx \int_{a}^{b} x^y\,dy = \int_{0}^{1} \Big[\frac{x^y}{\ln x}\Big]_{a}^{b}\,dx = \int_{0}^{1} \frac{x^b - x^a}{\ln x}\,dx,$$

which proves our contention.

Problems and complements

1 Determine the integral

$$\int_0^\infty \int_0^\infty \frac{dx\, dy}{(1+y)(1+yx^2)}$$

and use the result obtained to evaluate

$$\int_0^\infty \frac{\ln x}{x^2-1}\, dx.$$

Hint:

Since the integrand is positive, the Fubini theorem is applicable. Thus,

$$\int_0^\infty \frac{dy}{1+y} \int_0^\infty \frac{dx}{1+yx^2} = \frac{\pi}{2} \int_0^\infty \frac{y^{-\frac{1}{2}}}{1+y}\, dy.$$

By means of the simple transformation $y=u^2$, we easily deduce that

$$\int_0^\infty \frac{y^{-\frac{1}{2}}}{1+y}\, dy = \pi.$$

Consequently, we have

$$\frac{\pi^2}{2} = \int_0^\infty dx \int_0^\infty \frac{dy}{(1+y)(1+x^2y)} = \int_0^\infty \frac{dx}{x^2-1} \int_0^\infty \left(\frac{dy}{1+x^2y} - \frac{1}{1+y}\right) dy$$

$$= \int_0^\infty \frac{dx}{x^2-1} [\ln \frac{1+x^2y}{1+y}]_0^\infty = 2 \int_0^\infty \frac{\ln x}{x^2-1}\, dx.$$

2 Let $B \subset R^2$ be measurable. Show that a function $h \geq 0$ with B as its domain of definition is measurable if the region under h, say $G \subset R^3$, is measurable.

Hint:

For any $t \in [0, \infty)$ fixed, we have that

$$\{(x, y) \in B;\ h(x, y) \geq t\} = \{(x, y) \in B;\ (x, y, t) \in G\}.$$

But the set on the right-hand side is the section of G at t. Since by assumption G is measurable, so is each its section for every $t \geq 0$.

3 Determine the integral

$$\int_0^\infty \int_0^\infty y \exp\{-(1+x^2)y^2\}\, dx\, dy.$$

Hint:

The integrand is clearly non-negative on $R_+ \times R_+$. In addition, it is continuous and thus a measurable function. Consequently, Fubini's theorem applies. Clearly,

$$\int_0^\infty dx \int_0^\infty y \exp\{-(1+x^2)y^2\}\,dy$$

$$= \frac{1}{2}\int_0^\infty \frac{dx}{1+x^2}\,\exp\{-(1+x^2)y^2\}\,d((1+x^2)y^2)$$

$$= \frac{1}{2}\int_0^\infty \frac{dx}{1+x^2} = \frac{\pi}{4}.$$

On the other hand,

$$\int_0^\infty dy \int_0^\infty y \exp\{-(1+x^2)y^2\}\,dx = \int_0^\infty y\,e^{-y^2}\,dy \int_0^\infty e^{-x^2y^2}\,dx$$

$$= \left(\int_0^\infty e^{-u^2}\,du\right)^2.$$

Consequently,

$$\int_0^\infty e^{-u^2}\,du = \sqrt{\pi}\,/2$$

and

$$\int_0^\infty \int_0^\infty y \exp\{-(1+x^2)y^2\}\,dx\,dy = \pi/4.$$

4   Let $D \subset R_+ \times R_+$ be the region bounded by the curves

$$y = x, \quad y = +\sqrt{1+x^2}, \quad x{\cdot}y = a \quad \text{and} \quad x{\cdot}y = b \qquad (0<a<b).$$

Determine the integral

$$\iint_D (y^2 - x^2)^{xy}(x^2 + y^2)\,dx\,dy.$$

Hint:

It is not difficult to verify that the integrand is non-negative on D. Changing variables x and y into u and v by means of the transformations

$$u^2 = y^2 - x^2 \quad \text{and} \quad v = x \cdot y,$$

the region D is transformed into rectangle

$$0 \le u \le 1,\; a \le v < b.$$

The corresponding Jacobian of the transformation is

$$\frac{D(u,v)}{D(x,y)} = -2\,(x^2 + y^2).$$

Thus,

$$\iint_D (y^2 - x^2)^{xy}(x^2 + y^2)\,dx\,dy = \frac{1}{2}\int_a^b dv \int_0^1 u^v = \frac{1}{2}\int_a^b \frac{dv}{1+v} = \frac{1}{2}\ln\frac{1+b}{1+a}.$$

5 Show that the integral

$$\int_0^\infty \int_0^\infty e^{-xy} \sin x \, dx \, dy$$

does not exist.

Hint:

Recall the well known theorem from analysis: If $I \subset R^n$ is a closed bounded n-dimensional rectangle and $h:I \to R$ bounded, then integrability of h over I implies integrability of $|h|$.

$$\int_0^\infty \int_0^\infty e^{-xy} \, |\sin x| \, dx = \int_0^\infty |\sin x| \, dx \int_0^\infty e^{-xy} \, dy = \int_0^\infty \frac{|\sin x|}{x} \, dx = +\infty.$$

On the other hand

$$\int_0^\infty dy \int_0^\infty e^{-xy} \sin x \, dx = \int_0^\infty \sin x \, dx \int_0^\infty e^{-xy} \, dy = \frac{\pi}{2}.$$

39. INFINITE PRODUCTS OF MEASURABLE SPACES

Let $\{(X_t, \mathcal{K}_t); t \in T\}$ be a family of measurable spaces, where T is an arbitrary index set. Recall (see section 8) that the (Cartesian) product

$$\Omega = \prod_T X_t$$

is the family of mappings

$$h : T \to \bigcup_T X_t \quad \text{such that} \quad h(t) \in X_t \quad \text{for each} \ t \in T.$$

The set X_t is called the t-th factor of the product Ω; given $s \in T$ the map

$$p_s : \prod_T X_t \to X_s \quad \text{defined by} \quad p_s(h) = h(s)$$

is termed the "projection" of Ω onto the s-th factor. According to the axiom of choice $\Omega \neq \theta$ if and only if $X_t \neq \theta$ for every $t \in T$ (proposition 2 of section 9). If $X_t = X$ for every $t \in T$, we write $\Omega = X^T$ (the collection of all maps $h : T \to X$).

A "rectangle" R is a subset of Ω of the form

$$R = \prod_T A_t \quad \text{where} \ A_t \subseteq X_t,$$

and $A_t \neq X_t$ for only finitely many $t \in T$. Each A_t is called a side of the rectangle R. A rectangle R is measurable if each side $A_t \in \mathcal{K}_t$.

Let $S \subset T$ be arbitrary and let $B \subset \prod_S X_t$; the subset $C \subset \Omega$ of the form

$$C = B \times \prod_{T-S} X_t$$

is called a "cylinder" (set) with the basis B. Clearly, every rectangle is a cylinder set.

Given $D \subset \Omega$ and a point $h(s) \in X_s$, the section of D at $h(s)$ is the subset $D(h(s)) \subset \prod_{T-\{s\}} X_t$ defined by

$$D(h(s)) = \{h \,|\, T -\{s\}; h \in D\}$$

where $h \,|\, T-\{s\}$ is the restriction of h to $T-\{s\}$. In general, if $\{t_1, \cdots, t_n\}; \subset T$, the section of D at the point

$$(h(t_1), \cdots, h(t_n)) \in X_{t_1} \times \cdots \times X_{t_2}$$

is the subset $D((h(t_1), \cdots, h(t_n)) \subset \prod_{T-\{t_1,\cdots,t_n\}} X_t$ defined by

$$D(h(t_1), \cdots, h(t_n)) = \{h \,|\, T -\{t_1, \cdots, t_n\}; h \in D\}.$$

Clearly, every section of a rectangle is a rectangle.

Denote by S^* the class of all measurable rectangles, then we have:

Proposition 1

The class S^* is a semialgebra.

Proof:

The proof is same as the proof of the proposition 1 of section 36, which it generalizes.

In analogy with the notation of section 36, we shall write

$$\sigma\{S^*\} = \underset{T}{\otimes}\, \mathscr{K}_t = \mathscr{B}. \tag{1}$$

The measurable space $(\Omega, \mathscr{B})$ is the product of measurable spaces $(X, \mathscr{K}_t)$, $t \in T$. To shade more light on the structure of the σ-algebra $\sigma\{S^*\}$ we shall outline another way of constructing $\sigma\{S^*\}$. ♥

Let $S \subset T$ be an arbitrary subset; by an abuse of notation which identifies the cylinder sets whose bases are in $\underset{S}{\prod} X_t$ with their basis, we shall denote by $\underset{S}{\otimes}\, \mathscr{K}_t$ the sub-σ-algebra of $\underset{T}{\otimes}\, \mathscr{K}_t$, consisting of measurable cylinders with basis in $\underset{S}{\prod} X_t$. Then we have:

Proposition 2

Let $\mathscr{U}$ be the union of all σ-algebras $\underset{S}{\otimes}\, \mathscr{K}_t$, as $S \subset T$ runs through all finite subsets of T, then $\mathscr{U}$ is an algebra.

Proof:

If $A \in \mathscr{U}$ then $A \in \underset{S}{\otimes}\, \mathscr{K}_t$, for some finite $S \subset T$ and obviously $A^c \in \underset{S}{\otimes}\, \mathscr{K}_t$, so that $A^c \in \mathscr{U}$. If $A_i \in \mathscr{U}$ for every $i \in I$ where I is a finite set, then

$$A_i \in \underset{S_i}{\otimes}\, \mathscr{K}_t \quad \text{for some finite subset } S_i \subset T.$$

Consequently, if $S = \underset{I}{\bigcup} S_i$, it follows that $\underset{I}{\bigcup} A_i \in \underset{S}{\otimes}\, \mathscr{K}_t => \underset{I}{\bigcup} A_i \in \mathscr{U}$. This proves the proposition. ♥

Corollary 1

Since

$$S^* \subset \mathscr{U} \subset \underset{T}{\otimes}\, \mathscr{K}_t => \sigma\{S^*\} = \sigma\{\mathscr{U}\}.$$

Proposition 3

Let $\mathscr{A}$ be the union of all σ-algebras $\underset{S'}{\otimes}\, \mathscr{K}_t$ where $S' \subset T$ runs through all countable subsets of T. Then $\mathscr{A}$ is a σ-algebra.

Proof:

Let $\{A_n\}_1^\infty$ be a sequence of sets from $\mathscr{A}$. Then each $A_i \in \underset{S'_i}{\otimes}\, \mathscr{K}_t$, $i = 1, 2, \cdots$. Consequently,

$$\overset{\infty}{\underset{1}{\bigcup}} A_i \in \underset{\overset{\infty}{\underset{1}{\bigcup}S'_i}}{\otimes} \mathscr{K}_t \in \mathscr{A} \quad \text{because} \quad \overset{\infty}{\underset{1}{\bigcup}} S'_i \subset T \quad \text{is countable.}$$

This proves the assertion. ♥

The σ-algebra $\mathscr{A}$ contains all measurable rectangles. Thus,

$$S^* \subset \mathscr{A} \subset \underset{T}{\otimes} \mathscr{K}_t,$$

which implies that

$$\mathscr{A} = \underset{T}{\otimes} \mathscr{K}_t.$$

Consequently, each measurable subset of $\underset{T}{\prod} X_t$ is an $S' \subset T$ cylinder, where S' is at most countable.

Remark 1

Consider the product measurable space

$$(\underset{T}{\prod} X_t, \underset{T}{\otimes} \mathscr{K}_t).$$

Given any $S \subset T$ we shall denote by $h_S = h \mid S$ (the restriction of h to S (section 3)). The mapping

$$p_S : \underset{T}{\prod} X_t \to \underset{S}{\prod} X_t$$

defined by $p_S(h) = h_S$, is called the "projection" of $\underset{T}{\prod} X_t$ onto $\underset{S}{\prod} X_t$. The term "coordinate mapping" is also used. For instance, if $S = \{t_1, \cdots, t_n\} \subset T$

$$p_S(h) = (h(t_1), \cdots, h(t_n)) = h \mid S.$$

Remark 2

For every countable subset $S \subset T$ the mapping p_S is $\underset{T}{\otimes} \mathscr{K}_t$-measurable since for any

$$G \in \underset{S}{\otimes} \mathscr{K}_t, \quad p_S^{-1}(G) = \{h; h_S \in G\} = G \times \underset{T-S}{\prod} X_t,$$

and this is obviously a measurable cylinder set in Ω. This implies that $\underset{T}{\otimes} \mathscr{K}_t$ is the minimal σ-algebra with

respect to which all p_S, for $S \subset T$ countable, are measurable.

40. INFINITE PRODUCT OF MEASURE SPACES

In section 36 it was established that the product of any finite collection of measure spaces, say $\{(X_i, \mathcal{K}_i, \mu_i)\}_1^n$, is a measure space

$$(\prod_1^n X_i, \otimes_1^n \mathcal{K}_i, \mu)$$

with $\mu = \mu_1 \times \cdots \times \mu_n$. Often we have to deal with the product of an infinite collection of measure spaces, say

$$\{(X_t, \mathcal{K}_t, \mu_t); t \in T\}.$$

Here it is not quite clear how to define the product

$$\mu = \prod_T \mu_t \tag{1}$$

on the product measurable space $(\prod_T X_t, \otimes_T \mathcal{K}_t)$. One of the difficulties is that the product (1) may not

converge. This is not the case if each μ_t is a "probability" measure, i.e., if $\mu_t(X_t)=1$ for all $t \in T$, which is

assumed in the rest of this section.

Recall that a measurable rectangle $M \in \otimes_T \mathcal{K}_t$ is a set of the form

$$\mathbf{M = \prod_T A_t} \text{ where } A_t \in \mathcal{K}_t \text{ for all } t \in T, \tag{2}$$

and where all but finitely many factors of the product $A_t = X_t$. According to this definition $\prod_T X_t$ may

be considered as a measurable rectangle. Recall that the class S^* of all measurable rectangles is a

semialgebra, and that

$$\otimes_T \mathcal{K}_t = \sigma\{S^*\}.$$

Each measurable subset of $\prod_T X_t$ is a measurable S-cylinder set, where $S \subset T$ is at most countable

(proposition 3 of section 39). Notice also that (see remark 1) the inclusions $S^* \subset \mathcal{A}_0(S^*) \subset \otimes_T \mathcal{K}_t$ imply

$$\sigma\{\mathcal{A}_0(S^*)\} = \otimes_T \mathcal{K}_t.$$

The set function μ given by (1) is now defined on the class S^* as follows: for every $M \in S^*$

$$\mu(M) = \prod_T \mu_t(A_t)$$

(see (2)). Since only finitely many $A_t \neq X_t$ we can write

$$\mu(M) = \prod_S \mu_t(A_t) \tag{3}$$

where $S \subset T$ consists of those $t \in T$ for which $A_t \neq X_t$. The principal problem here is to establish that the set function μ so defined is σ- additive.

Since the elements of the semialgebra $\boldsymbol{S}^*$ are rectangles and thus depend only on finitely many indices we may, without any loss of generality, assume that the index set $T = N_+ = \{1, 2, \cdots\}$. The argument below applies to this modification.

Proposition 1

The set function μ is well defined.

Proof:

Suppose that an element $M \in \boldsymbol{S}^*$ has two different representations, say

$$M = A \times \prod_{N_+ - S_1} X_t \quad \text{and} \quad M = B \times \prod_{N_+ - S_2} X_t \; ,$$

where $S_1 \subset S_2$ are two finite subsets of N_+. Since the first representation can be written as

$$M = A \times \prod_{S_2 - S_1} X_t \times \prod_{N_+ - S_2} X_t \; ,$$

it follows that

$$A \times \prod_{S_2 - S_1} X_t = B \implies \mu(A) = \mu(B),$$

proving the assertion.

The next proposition is an essential step toward our goal.

Proposition 2

The set function μ on $\boldsymbol{S}^*$ is finitely additive.

Proof:

Let $\{M_i\}_1^n$ be a sequence of disjoint elements from $\boldsymbol{S}^*$, such that

$$\bigcup_1^n M_i \in \boldsymbol{S}^*.$$

Since every M_i is a rectangle we have for each $i = 1, 2, \cdots, n$ that

$$\mu(M_i) = \prod_{j \in K_i} \mu_j(A_j^i) \; , \text{ where each } K_i \subset N \text{ is finite.}$$

Denote by

$$m = \max\{\bigcup_1^n K_i\},$$

then for every $i = 1, \cdots, n$

$$M_i = \prod_{j=1}^m A_j^i \times \prod_{j=m+1}^\infty X_j \; .$$

Consequently, (see (3)),

$$\mu\left(\bigcup_{i=1}^{n} M_i\right) = \mu\left(\bigcup_{i=1}^{n} (A_1^i \times \cdots \times A_m^i)\right)$$

$$= (\mu_1 \times \cdots \times \mu_m)\left(\bigcup_{i=1}^{n} (A_1^i \times \cdots \times A_m^i)\right)$$

$$= \sum_{i=1}^{n} \prod_{j=1}^{m} \mu_j(A_j^i) = \sum_{i=1}^{n} \mu(M_i),$$

because $\mu_1 \times \cdots \times \mu_m$ is a measure. This proves additivity of μ on S^*. ♥

Remark 1

Recall that the collection of all finite disjoint unions of elements from S^* is the algebra $\mathcal{A}_0(S^*)$ generated by S^* (section 17, problem 13). By proposition 5 of section 18, there is a unique extension of a measure μ on S^* to a measure ,denoted also by μ, on $\mathcal{A}_0(S^*)$.

We are now in a position to prove the principal result of this section.

Proposition 3

The set function μ on the algebra $\mathcal{A}_0(S^*)$ is σ-additive.

Proof:

Since μ is finitely additive on $\mathcal{A}_0(S^*)$, it suffices, on account of the problem 4 in section 17, to show that μ is continuous at θ from above . In other words, for any sequence $\{M_k\}_1^\infty$ in $\mathcal{A}_0(S^*)$ such that $M_n \downarrow \theta$ necessarily $\mu(M_n) \downarrow 0$. The standard way to prove this is by contradiction. If for some decreasing sequence $\{M_n\}_1^\infty$ in $\mathcal{A}_0(S^*)$ such that $M_n \downarrow \theta$ but $\mu(M_n) \nrightarrow 0$, then $\varepsilon \le \mu(M_n)$ for some $\varepsilon > 0$ and every $n=1, 2, \cdots$. This, however, implies that $\bigcap_1^\infty M_n \neq \theta$, contradicting the assumption that $M_n \downarrow \theta$.

To prove the proposition some new notation is required. Denote by $X^n = \prod_{i=n+1}^{\infty} X_i$ and by S_n^* the collection of all measurable rectangles from X^n. We also need the algebras $\mathcal{A}_0(S_n^*)$ and $\mu^n = \mu_{n+1} \times \mu_{n+2} \times \cdots$. Next, let $M_n(x_1)$, $M_n(x_1, x_2)$, $\cdots$ be the sections of M_n at $x_1 \in X_1, (x_1, x_2) \in X_1 \times X_2, \ldots$ and so on. Clearly $M_n(x_1, \cdots, x_n) \in \mathcal{A}_0(S_n^*)$ for every $n=1, 2, \cdots$. Denote by

$$E_1^n = \{x_1 \in X_1; \mu^1(M_n(x_1)) \ge \varepsilon/2\}$$

then, according to proposition 2 of section 37 and the assumption that $\varepsilon \le \mu(M_n)$ for all $n=1, 2, \cdots$, we have that

$$\varepsilon \le \mu(M_n) = \int \mu^1(M_n(x_1))\, d\mu_1 = \int_{E_1^n} \mu^1(M_n(x_1))\, d\mu_1$$

$$+ \int_{(E_1^n)^c} \mu^1(M_n(x_1))\, d\mu_1 \le \mu_1(E_1^n) + \varepsilon/2.$$

Thus, $\mu_1(E_1^n) \ge \varepsilon/2$ for every $n=1, 2, \cdots$. Since $\{E_1^n\}_1^\infty$ is decreasing and μ_1 is a measure and thus

continuous from above at θ, $\bigcap_1^\infty E_1^n \neq \theta$. Thus, there exists a point $y_1 \in X_1$, such that for every $n=1, 2, \cdots$,

$\mu^1(M_n(y_1)) \ge \varepsilon/2$. Next, consider the sequence $\{M_n(y_1)\}_1^\infty$ which obviously decreases. The above

argument applied to this sequence leads to a point $y_2 \in X_2$ such that

$$\mu^2(M_n(y_1, y_2)) \ge \varepsilon/2^2 \quad \text{for all} \quad n=1, 2, \cdots,$$

and so on. By an induction argument we obtain a point $(y_1, y_2, \cdots)$ such that $\mu^k(M_n(y_1, \ldots, y_k)) \ge \varepsilon/2^k$.

Since each M_n is a finite dimensional cylinder set, there exists at least one point $(x_1, x_2, \ldots) \in M_n(y_1,$

$\ldots, y_k)$ such that $x_i = y_i$, $i=1, \ldots, k$. Let us show that $(y_1, y_2, \ldots) \in \bigcap_1^\infty M_n$. To this end assume, for the

sake of definiteness, that M_n is $\{1,\ldots,k\}$-cylinder set then, according to what we have just said, it follows

that $(y_1, y_2, \ldots) \in M_n$ for all $n=1,2,\ldots$, and thus

$$\bigcap_1^\infty M_n \neq \theta$$

which is the desired contradiction. This proves the proposition. ♥

Corollary 1

One of the principal results of section 19 is that a measure μ on a semiring S can be extended to the

σ-algebra $\mathcal{M}$ of μ^*-measurable sets, which contain S. From this and the last proposition we conclude that

there is a unique measure μ on the σ-algebra $\otimes_i \mathcal{K}_i$, with the property that for any measurable cylinder set

of the form

$$C = B \times \prod_{k=n+1}^\infty X_k, \text{ we have that } \mu(C) = (\mu_1 \times \cdots \times \mu_k)(B).$$

41. COIN TOSSING PROBABILITY SPACE

Events associated with an experiment or trial are classified into conditions under which the experiment is conducted and outcomes. Conditions are events which are known or made to occur. Outcomes are events which may or may not occur in a performance of the experiment.

An experiment is called "random" if its repetition under the same conditions does not necessarily produce the same result. Note that the possibility of "repeated trials" is one of the fundamental premises in science. A mathematical description or a "model" of a random experiment is a probability space $(X, \mathcal{X}, \mu)$. Here X is the set of all possible outcomes of the experiment, $\mathcal{X}$ is a class (a σ-algebra) of random events associated with this experiment and μ is a probability measure.

Consider the random (thought) experiment which consists of an infinite sequence of tosses of a coin. Each outcome ω of this experiment is an infinite sequence of T's (tails) and H's (heads). Or, if we replace T by 0 and H by 1, an outcome ω is an infinite sequence of the form

$$\omega : (\varepsilon_1, \varepsilon_2, \cdots)$$

where $\varepsilon_i = 0$ or 1 for each i=1, 2, $\cdots$. The set Ω of all such sequences is the set of all possible outcomes of the random experiment. Clearly,

$$\Omega = \{0, 1\}^{N_+} \quad \text{and} \quad \mathrm{Card}\,(\Omega) = 2^{\aleph_0} = \mathbf{c},$$

(see (2) of section 7). If $\{0, 1\}$ is endowed with the discrete topology (i.e., every subset of $\{0, 1\}$ is an open set), then since $\{0, 1\}$ is compact, by the Tyhonoff theorem the set Ω, as the Cartesian product of countably many copies of $\{0, 1\}$, is also compact.

To obtain a suitable class of random events let $\{\eta_i\}_1^\infty$ be a sequence of coordinate mappings

$$\eta_i : \Omega \to \{0, 1\} \quad \text{defined by} \quad \eta_i(\omega) = \varepsilon_i \quad i=1, 2, \cdots \tag{1}$$

Let $\mathcal{B} \subset \mathcal{P}(\Omega)$ be the smallest σ-algebra with respect to which all the mappings η_i are measurable, i.e.,

$$\mathcal{B} = \sigma\{\eta_1, \eta_2, \cdots\} = \sigma\{\bigcup_1^\infty \eta_i^{-1}(\mathcal{U})\} \tag{2}$$

where $\mathcal{U} = \mathcal{P}(\{0, 1\})$. For instance, for every n=1, 2, $\cdots$

$$\{\omega; \varepsilon_1 = x_1, \cdots, \varepsilon_n = x_n\} \in \mathcal{B}, \quad \text{where } x_i = 0, 1, \quad i=1, 2, \cdots.$$

It is clear from this that

$$\{\sum_{i=1}^n \varepsilon_i = k\} \in \mathcal{B} \quad \text{for any } k=0, 1, \cdots, n.$$

To obtain a probability measure on $(\Omega, \mathcal{B})$ we proceed as follow: Assume that the coin, which is tossed repeatedly, is well balanced (fair) so that each side has the same chance to occur during each individual toss. Denote by $I^1 = [0, 1]$ and let $\mathcal{R}_{I^1}$ be the Borel algebra of subsets of I^1. As we know (see (4) of section 22) $\mathrm{Card}(\mathcal{R}_{I^1}) = \mathbf{c}$. Since every $t \in I^1$ has a unique binary expansion

$$t = \sum_{i=1}^{\infty} \varepsilon_i / 2^i$$

(problem 1 of section 7), it follows that $I^1 \sim \Omega$. Consequently, there is a bijection

$$\varphi : I^1 \to \Omega \qquad \text{where} \quad \varphi(t) = (\varepsilon_1, \varepsilon_2, \cdots). \tag{3}$$

Clearly, for each $i=1, 2, \cdots$, $\varepsilon_i = \varepsilon_i(t)$.

We have (see problem 2 of section 7) that

$$\varepsilon_n(t) = \sum_{i=1}^{2^{n-1}} I_{[\frac{2i-1}{2^n}, \frac{2i}{2^n})}(t), \tag{4}$$

which is obviously $\mathcal{R}_{I^1}$ -measurable for any $n=1, 2, \cdots$. On the other hand (see problem 1 of section 7),

$$\Delta x_1, \cdots, x_n = [\sum_{i=1}^{n} x_i 2^{-i}, \sum_{i=1}^{n} x_i 2^{-i} + 2^{-n}) \tag{5}$$

$$= \varphi^{-1}(\{\varepsilon_1 = x_1, \cdots, \varepsilon_n = x_n\})$$

where $x_i = 0, 1$ for all $i=1, 2, \cdots$. This shows that the map φ in (3) is $\mathcal{R}_{I^1}$ -measurable. In other words,

$$\varphi^{-1}(\mathcal{B}) \subset \mathcal{R}_{I^1}.$$

Consider now the probability space $(I^1, \mathcal{R}_{I^1}, \overline{\lambda})$ where $\overline{\lambda}$ is the Lebesgue measure on R (the uniform distribution on I^1). The map φ and $\overline{\lambda}$ induce the probability measure P defined on $\mathcal{B}$ as follows: given any $A \in \mathcal{B}$

$$P(A) = \overline{\lambda}(\varphi^{-1}(A)). \tag{6}$$

For instance,

$$\varphi^{-1}(\{\varepsilon_1 = x_1\}) = \Delta x_1 = [\frac{x_1}{2}, \frac{x_1}{2} + \frac{1}{2})$$

so that

$$P\{\varepsilon_1 = x_1\} = \overline{\lambda}(\Delta x_1) = \frac{1}{2}.$$

Similarly,

$$P\{\varepsilon_1 = x_1, \cdots, \varepsilon_n = x_n\} = \overline{\lambda}(\Delta x_1, \cdots, x_n) = \frac{1}{2^n}.$$

On the other hand, since

$$P\{\varepsilon_2 = x_2\} = P\{\varepsilon_1 = 0, \varepsilon_2 = x_2\} + P\{\varepsilon_1 = 1, \varepsilon_2 = x_2\} = \frac{1}{2}$$

and in general

$$P\{\varepsilon_n = x_n\} = \frac{1}{2},$$

we have that

$$P\{\varepsilon_1=1, \cdots, \varepsilon_n=x_n\} = \prod_{i=1}^{n} P\{\varepsilon_i=x_i\}, \qquad (7)$$

which means that $\{\varepsilon_i\}_1^{\infty}$ is a sequence of "independent" identically distributed random variables.

The probability space $(\Omega, \mathcal{B}, P)$ is a mathematical model of the coin tossing random experiment. Probabilities of various random events can now be calculated. For instance, to obtain the probability that in the first n-tosses the "head" will appear k times ($0 \leq k \leq n$) we have to determine probability of the event that $\sum_{i=1}^{n} \varepsilon_i = k$. A simple combinatorial argument shows that

$$P\{\sum_{i=1}^{n} \varepsilon_i = k\} = \binom{n}{k} / 2^n .$$

A non-trivial problem is to show that

$$\frac{1}{n} \sum_{i=1}^{n} \varepsilon_i \to \frac{1}{2} \quad (a.e),$$

(the Kolmogorov strong law of large numbers). To simplify notation somewhat, write

$$\zeta_i = \varepsilon_i - \frac{1}{2}, \quad \text{then} \quad \zeta_i = -\frac{1}{2}, \frac{1}{2}, \quad i=1, 2, \cdots .$$

Next, given $\varepsilon > 0$, we have:

$$\int (\frac{1}{n} \sum_{i=1}^{n} \zeta_i)^4 \, dP \geq \int_{\{\frac{1}{n}|\sum_{i=1}^{n} \zeta_i| > \varepsilon\}} \varepsilon^4 \, dP = \varepsilon^4 \, P\{\frac{1}{n} | \sum_{i=1}^{n} \zeta_i | > \varepsilon\}.$$

Thus

$$P\{\frac{1}{n} | \sum_{i=1}^{n} \zeta_i | > \varepsilon\} \leq \frac{1}{(\varepsilon n)^4} \int (\sum_{i=1}^{n} \zeta_i)^4 \, dP, \qquad (8)$$

which is a version of the Chebishev inequality. Bearing in mind that $\zeta_1, \zeta_2, \cdots$ are independent identically distributed random variables such that for all $i=1, 2, \cdots$

$$\int \zeta_i \, dP = \int_{\{\zeta_i = -\frac{1}{2}\}} (-\frac{1}{2}) \, dP + \int_{\{\zeta_i = \frac{1}{2}\}} (\frac{1}{2}) \, dP = 0,$$

then for every $i \neq j \neq k \neq \ell$

$$\int \zeta_i \zeta_j \zeta_k \zeta_\ell \, dP = (\int \zeta_i \, dP) (\int \zeta_j \, dP) (\int \zeta_k \, dP) (\int \zeta_\ell \, dP) = 0.$$

From this it follows that the only non-null members of the sum

$$\int (\sum_{i=1}^{n} \zeta_i)^4 \, dP = \sum \int (\zeta_i \cdot \zeta_j \cdot \zeta_k \cdot \zeta_\ell) \, dP$$

are

$$\int \zeta_i^4 \; dP = \frac{1}{2^4} \qquad \text{and} \qquad \int \zeta_i^2 \, \zeta_j^2 \; dP = \frac{1}{2^4}.$$

Thus,

$$\int \Big(\sum_{i=1}^{n} \zeta_i \Big)^4 dP = \sum_{i=1}^{n} \int \zeta_i^4 \; dP + \sum_{1 \le i < j < n} \int \zeta_i^2 \, \zeta_j^2 \; dP$$

$$= \frac{n}{2^4} + \frac{3n(n-1)}{2^4} = \frac{n}{2^4}(3n-2).$$

Consequently,

$$\sum_{n=1}^{\infty} P\{ \frac{1}{n} \; | \sum_{i=1}^{n} \zeta_i \; | > \varepsilon \} \le \frac{1}{(2\,\varepsilon)^4} \sum_{n=1}^{\infty} \frac{3n-2}{n^3} < \infty$$

which, taking into account corollary 1 of section 25, proves that

$$\frac{1}{n} \sum_{i=1}^{n} \zeta_i \to 0 \quad (a.e).$$

Remark 1

There is an interesting interpretation of this result: Since $\sum_{i=1}^{n} \varepsilon_i$ represents the number of 1's in the first n-tosses of a coin, the proportion of 1's and 0's in "almost all ω" is $\frac{1}{2}$. This famous result is due to E. Borel which states that, except for a subset of the Lebesgue measure 0, every number from [0, 1] is "normal", in the sense that in its binary expansion, 0's and 1's occur with equal limiting frequency.

Remark 2

The method (equation 6) used to calculate the probability measure P on the coin tossing measurable space $(\Omega, \mathcal{B})$ is not limited to this particular problem. As a matter of fact, if $(\Omega, \mathcal{B})$ is an arbitrary measurable space and $h: I^1 \to \Omega$ a measurable map, i. e., $h^{-1}(\mathcal{B}) \subset \mathcal{R}_{I^1}$, then

$$P = \bar{\lambda} \circ h^{-1} = \bar{\lambda}(h^{-1}(\cdot))$$

is a probability measure on Ω. It is important to point out that nearly all probability measures used in statistics can be obtained in this fashion.

A probability space is a measure space $(\Omega, \mathscr{B}, P)$ where Ω is the set of all possible outcomes of a random experiment. The elements of the σ-algebra $\mathscr{B}$ are called random events or just "events" associated with this experiment, and the measure P is a probability, i.e., $P(\Omega)=1$. The set Ω is the "sure" or "certain" event, while the empty set θ is the "impossible" event. If $A \in \mathscr{B}$ and $P(A)=1$, it is called "almost sure" (abbreviated (a.s.)) event relative to the probability P. On the other hand, if $N \in \mathscr{B}$ satisfies $P(N)=0$, it is called a "null event" relative to P. Notice that a null event is not impossible event ($P(\theta)=0$ for any probability measure P). Sometimes one uses abbreviation $A \overset{a.s}{=\!=} \Omega$ and $N \overset{a.s}{=\!=} \theta$.

Consider the measurable space $(R, \mathscr{R})$ where $R=(-\infty, \infty)$ and $\mathscr{R}$ is the σ-algebra of Borel subsets R. A $\mathscr{B}$-measurable map $\zeta:\Omega \to R$ (i.e., $\zeta^{-1}(\mathscr{R}) \subset \mathscr{B}$) is called a (real-valued) "random variable." The" image measure" on R, $\mu=P \circ \zeta^{-1}$ is a probability on $(R, \mathscr{R})$. We call

$$F_\zeta(x) = P\{\omega; \zeta(\omega) \le x\}, \quad \text{where} \quad x \in R,$$

the "distribution function" of the random variable ζ. If $\zeta^1: \Omega \to R$ is another random variable such that

$$P\{\omega; \zeta(\omega)=\zeta^1(\omega)\} = 1, \quad \text{then} \quad F_{\zeta^1}(x) = F_\zeta(x)$$

for all $x \in R$.

Let $(U, \mathscr{U})$ be a measurable space. A map $\xi: \Omega \to U$ such that $\xi^{-1}(\mathscr{U}) \subset \mathscr{B}$ is usually called a "**random transformation**" with the "**state space**" $(U, \mathscr{U})$. In applications the state space $(U, \mathscr{U})$ is often the Euclidean measurable space $(R^n, \mathscr{R}^n)$ where

$$R^n = \underbrace{R \times \cdots \times R}_{n-\text{times}} \quad \text{and} \quad \mathscr{R}^n = \underbrace{\mathscr{R} \otimes \cdots \otimes \mathscr{R}}_{n-\text{times}}$$

As we have seen (problem 10 of section 24) a mapping $\xi:\Omega \to R^n$ is a vector

$$\xi(\omega) = (X_1(\omega), \cdots, X_n(\omega))$$

where each $X_i:\Omega \to R$, $i=1, \cdots, n$. A version of the next proposition was discussed in section 24, problem 10.

Proposition 1

$$\xi^{-1}(\mathscr{R}^n) \subset \mathscr{B} <=> X_i^{-1}(\mathscr{R}) \subset \mathscr{B} \quad \text{for all} \quad i=1, 2, \cdots, n.$$

Proof:

Assume that ξ is $\mathscr{B}$-measure and that $\{B_i\}_1^n$ is a sequence of Borel subsets of R. Since

$$X_1^{-1}(B_1) = \xi^{-1}(B_1 \times R^{n-1}) \in \mathscr{B}$$

$$X_i^{-1}(B_i) = \xi^{-1}(R^{i-1} \times B_i \times R^{n-i}) \in \mathscr{B} \qquad i=2, \cdots, n,$$

the necessity is proved. Now, assume that each X_i is $\mathscr{B}$-measurable, then since

$$\bigcap_{i=1}^{\infty} X_i^{-1}(B_i) = \xi^{-1}\left(\bigcap_{i=1}^{\infty} R^{i-1} \cap B_i \cap R^{n-i}\right) = \xi^{-1}(B_1 \times \cdots \times B_n)$$

$$\in \mathscr{B} => \xi^{-1}(S^*) \subset \mathscr{B},$$

where S^* is the semialgebra of measurable rectangles of R^n. Consequently, from the inclusion $\sigma\{\xi^{-1}(S^*)\}$ $\subset \mathscr{B}$ and proposition 1 of section 17, we have

$$\sigma\{\xi^{-1}(S^*)\} = \xi^{-1}(\sigma\{S^*\}) = \xi^{-1}(\mathscr{R}^n) \subset \mathscr{B},$$

which completes the proof. ♥

The probability

$$P\{X_1 \in B_1, \cdots, X_n \in B_n\}$$

Is called the "**joint distribution**" of the system of random variables $\{X_i\}_1^n$, while

$$P\{X_{i_1} \in B_{i_1}, \cdots, X_{i_k} \in B_{i_k}\} \qquad k<n$$

is called a "**marginal distribution**" of $\{X_i\}_1^n$. A particular form of the joint distribution is the function on R^n defined by

$$F(x_1, \cdots, x_n) = P\{X_1 \leq x_1, \cdots, X_n \leq x_n\},$$

which is called the "**distribution function**" of the system $\{X_i\}_1^n$.

A random variable ζ on $(\Omega, \mathscr{B}, P)$ induces the image probability measure

$$\mu_\zeta = P \circ \zeta^{-1} \qquad \text{on the state space } (R, \mathscr{R}).$$

If μ_ζ is absolutely continuous with respect to $\overline{\lambda}$ (the Lebesgue measure on R), then according to the Radon-Nikodym theorem (section 34) there exists a Borel function $f_\zeta(\cdot) \geq 0$ defined on R such that for any $B \in \mathscr{R}$

$$\mu_\zeta(B) = \int_B f_\zeta \, d\overline{\lambda} \;.$$

The function $f_\zeta(\cdot)$ is called the "**probability density**" of the random variable ζ. Clearly,

$$F_\zeta(x) = \int_{-\infty}^{x} f_\zeta(s) \, ds.$$

The **(mathematical)"expectation"** or the "**mean value $E(\zeta)$**" of the random variable ζ is defined by

$$E(\zeta) = \int \zeta \, dP = \int_{-\infty}^{\infty} x \, dF_\zeta(x).$$

The following result is often useful.

Proposition 2

Let X and Y be (two real) random variables on a probability space $(\Omega, \mathscr{B}, P)$. If $X^{-1}(\mathscr{R}) \supset Y^{-1}(\mathscr{R})$, there exists a Borel function $h: R \to R$ such that

$$Y = h(X) \tag{1}$$

Proof:

Actually, the condition is necessary and sufficient for (1) to hold, since

$$Y^{-1}(\mathscr{R}) = (h \circ X)^{-1}(\mathscr{R}) = X^{-1}(h^{-1}(\mathscr{R})) \subset X^{-1}(\mathscr{R}),$$

which proves necessity. Next, assume that $Y^{-1}(\mathscr{R}) \subset X^{-1}(\mathscr{R})$. If

$$Y = \sum_{i=1}^{n} y_i I_{B_i}, \text{ where } \{B_i\}_1^n \text{ is a partition of } \Omega,$$

then by the last assumption, $B_i \in X^{-1}(\mathscr{R})$. Consequently, there exists $\Gamma_i \in \mathscr{R}$ such that

$$B_i = X^{-1}(\Gamma_i), \quad i = 1, \cdots, n.$$

If $\{\Gamma_i\}_1^n$ are disjoint, then

$$h(\,\cdot\,) = \sum_{i=1}^{n} y_i I_{\Gamma_i}(\,\cdot\,),$$

yielding

$$h(X) = \sum_{i=1}^{n} y_i I_{\Gamma_i}(X) = \sum_{i=1}^{n} y_i I_{X^{-1}(\Gamma_i)} = \sum_{i=1}^{n} y_i I_{B_i} = Y.$$

This proves the assertion when Y is a step function and $\Gamma_i \cap \Gamma_j = \theta$. If $\{\Gamma_i\}_1^n$ are not disjoint, the sets

$$H_1 = \Gamma_1, \quad H_k = \Gamma_k - \bigcup_{i=1}^{k-1} \Gamma_i, \quad k = 2, \cdots, n \quad \text{are}$$

and since

$$X^{-1}(\Gamma_i \cap \Gamma_j) = X^{-1}(\Gamma_i) X^{-1}(\Gamma_j) = B_i \cap B_j = \theta$$

we have:

$$X^{-1}(H_k) = X^{-1}(\Gamma_k - \bigcup_1^{k-1} \Gamma_i) = X^{-1}(\Gamma_k) - \bigcup_{i=1}^{k-1} X^{-1}(\Gamma_k \cap \Gamma_i)$$

$$= X^{-1}(\Gamma_k) = B_k.$$

Consequently, for

$$h(\cdot) = \sum_{i=1}^{n} I_{H_i}(\cdot) \quad \text{we have that } h(X) = Y.$$

This proves the assertion when Y is a step function. In general, for any Y which is $X^{-1}(\mathscr{R})$-measurable there exists a sequence of $X^{-1}(\mathscr{R})$-measurable step functions which converges pointwise to Y. This completes our proof. ♥

The next problem is often encountered in various applications of probability theory. Given a probability space $(\Omega, \mathscr{B}, P)$ and a sequence of events $\{A_k\}_1^\infty$ from $\mathscr{B}$, it is of some interest to find out how many of these events can occur in a single performance of the random experiment? The question of special interests is if the number is finite or not. Recall (see section 2) that

$$A^* = \lim \sup A_n$$

is the event that infinitely many A_k will occur in a performance a the random experiment. This, of course, does not preclude the possibility that infinitely many may not occur. We often write $A^*=\{A_k \ i.o\}$, where i.o stands for "infinitely often." Denote by η the number of A_k which occur in a performance of the random experiment, then clearly

$$\eta(\omega) = \sum_{k=1}^{\infty} I_{A_k}(\omega).$$

Obviously, η is a non-negative, integer valued random variable. The mean value of η is

$$E(\eta) = \int \eta \, dP = \sum_{k=1}^{\infty} P(A_k).$$

The next result is known as the Borel-Cantelly lemma.

Proposition 3

$$\sum_{k=1}^{\infty} P(A_k) < \infty \implies P\{\eta = +\infty\} = 0.$$

Proof:

By the Chebishev inequality

$$P\{\eta \geq n\} \leq \frac{1}{n} \int \eta \, dP = \frac{1}{n} \sum_{i=1}^{\infty} (A_i) \to 0$$

as $n \to \infty$. Since

$$P\{\eta = +\infty\} = P(\bigcap_{k=0}^{\infty} \{\eta \geq k\}) \leq P\{\eta \geq m\}, \quad m = 1, 2, \cdots,$$

the assertion follows. ♥

Problems and complements

1. Let Y and $\xi_n = (X_1, \cdots, X_n)$ be a random variable and a random vector, respectively, on a probability space $(\Omega, \mathscr{B}, P)$. If $Y^{-1}(\mathscr{R}) \subset \xi_n^{-1}(\mathscr{R}^n)$ show that there exists a Borel function

$$f: R^n \to R \quad \text{such that} \quad Y = f(\xi_n).$$

Hint:

First (see proposition 2) we consider the case when Y is a step function, say

$$Y = \sum_{i=1}^{N} y_i I_{A_i}, \qquad (A_i)_1^N \text{ being partition of } \Omega.$$

Since by assumption each $A_i \in \xi_n^{-1}(\mathscr{R}^n)$ there exists $L_i \in \mathscr{R}^n$ such that

$$A_i = \xi_n^{-1}(L_i) \quad i=1, \cdots, N.$$

Consequently,

$$Y = \sum_{i=1}^{N} y_i I_{\xi_n^{-1}(L_i)} = \sum_{i=1}^{N} y_i I_{L_i}(\xi_n).$$

From this we conclude that

$$Y = f(\xi_n) \quad \text{where} \quad f(\cdot) = \sum_{i=1}^{N} y_i I_{L_i}(\cdot).$$

Let Y be an arbitrary $\xi_n^{-1}(\mathscr{R}^n)$-measurable random variable; there exists a sequence $\{Y_k\}_1^\infty$ of

$\xi_n^{-1}(\mathscr{R}^n)$-measurable step functions converging pointwise to Y. For each Y_k there exists a Borel

measurable function $f_k : R^n \to R$ for which

$$Y_k = f_k(\xi_n).$$

If $C \subset R^n$ is the set on which the sequence $\{f_n\}_1^\infty$ converges, we define

$$f = \lim_{k \to \infty} f_k \quad \text{and} \quad f = 0 \quad \text{on} \quad C^c.$$

Then $f(\cdot)$ is the required Borel function for which
$$Y = f(\xi_n).$$

2. Let $Z_m = (Y_1, \cdots, Y_m)$ and $\xi_n = (X_1, \cdots, X_n)$ be random vectors on $(\Omega, \mathscr{B}, P)$ such that

$$Z_m^{-1}(\mathscr{R}^m) \subset \xi_n^{-1}(\mathscr{R}^n).$$

Show that there is a Borel function $h: R^n \to R^m$ for which
$$Z_m = f(\xi_n),$$

Hint:

Apply the previous result to each component Y_i of Z_m to obtain

$$(Y_1, \cdots, Y_m) = (h_1(X_1, \cdots, X_n), \cdots h_m(X_1, \cdots, X_n)).$$

3. Let $h{:}R^n \to R^n$ be defined as follows:

$$h((x_1, \cdots, x_n)) = (x_n, x_{n-1}, \cdots, x_1).$$

Show that $h(\cdot)$ is a Borel function.

Hint:

If $(x_1, \cdots, x_n)$ and $(y_1, \cdots, y_n)$ are two arbitrary points of R^n we have

$$h((x_1, \cdots, x_n)) + h((y_1, \cdots, y_n)] = (x_n + y_n, \cdots, x_1 + y_1) =$$

$$= h((x_1 + y_1, \cdots, x_n + y_n)).$$

Thus, h is a linear function and therefore Borel measurable.

4. Let X be a random variable on $(\Omega, \mathcal{B}, P)$ with a distribution on function $F(x) = P\{\omega; X(\omega) \leq x\}$. Show that

(i) $F(x+0) = F(x)$ for each $x \in R$, and that $\lim\limits_{x \to -\infty} F(x)=0$, $\lim\limits_{x \to +\infty} F(x)=1$.

(ii) $F(x)$ is continuous at x if and only if $P\{X=x\}=0$;

(iii) The set of discontinuity points of F is at most countable.

Hint:

(i) Consider $\mu = P \circ X^{-1}$, then for any $\varepsilon > 0$ and $x \in R$

$$F(x + \varepsilon) - F(x) = \mu((x, x + \varepsilon]).$$

Consequently,

$$F(x+0) = \lim\limits_{\varepsilon \to 0} F(x+\varepsilon) = F(x) + \lim\limits_{\varepsilon \to 0} \mu((x, x+\varepsilon]) = F(x) + \mu(\theta) = F(x).$$

Finally,

$$\lim\limits_{x \to -\infty} F(x) = \lim\limits_{x \to -\infty} \mu((-\infty, x]) = 0$$

$$\lim\limits_{x \to \infty} F(x) = \lim\limits_{x \to \infty} \mu((-\infty, x]) = 1.$$

(ii) $P\{X=x\} = \mu((-\infty, x]) - \mu((-\infty, x)) = F(x) + F(x-0)$.

(iii) The set of discontinuity points of any monotone function is at most countable. Here is a simple proof for F. Let $x \in R$ be a discontinuity point of F, then

$$d(x) = F(x) - F(x-0) > 0$$

is the "jump" of F at x. Set

$$D_n = \{x \in R; \ \frac{1}{n+1} < d(x) \leq \frac{1}{n}\}.$$

It is clear that D_n is finite set for each $n=1, 2, \cdots$. Thus

$$D = \bigcup_{n=1}^{\infty} D_n,$$

the set of all discontinuity points of F is at most countable.

5. Let X be a random variable with strictly increasing distribution function F(·). Show that

   (i) F(X) is a random variable;

   (ii) P{F(X)≤t} = t.

   Hint:

   (i) F(·) is obviously a Borel function so that the composition F o X is measurable.

   (ii) Denote by

$$F^{-1}(t) = \inf\{x;\ F(x)\geq t\}.$$

Since F(·) is right continuous,

$$\{x;\ F(x)\geq t\} = [F^{-1}(t),\ \infty).$$

Hence,

$$F^{-1}(t) > x \iff F(x) > t.$$

Thus,

$$P\{F(X) > t\} = P\{F^{-1}(t) < X\}$$

which implies that

$$P\{F(X) \leq t\} = P\{X \leq F^{-1}(t)\} = F(F^{-1}(t)) = t.$$

In other words, F(X) is a random variable with the uniform distribution.

6. Let X be a measurable map on $(\Omega, \mathscr{B}, P)$ with a state space $(U, \mathscr{U})$. If $h: U \to R$ is $\mathscr{U}$-measurable, i.e.

$h^{-1}(\mathscr{R}) \subset \mathscr{U}$, show that

$$\int_{\Omega} (h\ o\ X)\ dP = \int_{U} h\ d\mu \qquad\qquad (*)$$

where $\mu = F\ o\ X^{-1}$.

   Hint:

   For $h = I_M$, where $M \in \mathscr{U}$, we have:

$$h(X) = I_M(X) = I_{X^{-1}(M)} \qquad \text{yielding}$$

$$\int h(X)\ dP = P(X^{-1}(M)) = (F\ o\ X^{-1})(M) = \mu(M) = \int h\ d\mu.$$

From this we conclude that $(*)$ holds for every non-negative step function h on U. If $\{h_n\}_1^{\infty}$ is a sequence of non-negative step functions on U such that $h_n \uparrow h_0$ then $h_n(X) \uparrow h_0(X)$. The assertion now follows by taking the limit.

7. Let X be a random variable on a probability space $(\Omega, \mathscr{B}, P)$. If $A \in \mathscr{B}$ is an atom, show that X=constant (a.e) on A.

   Proof:

   Denote by

$$F_A(x) = P(\{X \le x\} \cap A),$$

then $\lim_{x \to -\infty} F_A(x) = 0$ and $\lim_{x \to \infty} F_A(x) = P(A)$. Now denote by

$$x_0 = \inf\{x; F_A(x) = P(A)\},$$

then clearly

$$F_A(x) = \begin{cases} P(A) & \text{if } x \ge x_0 \\ 0 & \text{if } x < x_0 \end{cases}.$$

Consequently, for any $\alpha < x_0 \le \beta$

$$P(\{\alpha < X < \beta\} \cap A) = P(A).$$

Since

$$\{x_0\} = \bigcap_{n=1}^{\infty} (x_0 - \frac{1}{n}, x_0 + \frac{1}{n}),$$

it follows that

$$P(\{X = x_0\} \cap A) = \lim_{n \to \infty} P(\{x_0 - \frac{1}{n} < X \le x_0 + \frac{1}{n}\} \cap A) = P(A).$$

In other words, $\{\omega \in A; X(\omega) = x_0\} \overset{a.e}{=\!=} A$.

8. Let X be a random variable on $(\Omega, \mathscr{B}, P)$ such that $E|X| < \infty$. Show that

$$\sum_{n=1}^{\infty} P\{|X| \ge n\} \le E|X| \le \sum_{n=1}^{\infty} P\{|X| \ge n\} + 1.$$

Hint:

Set $B_k = \{k \le |X| < k+1\}$, $k = 0, 1, \cdots$, then clearly

$$k P(B_k) \le \int_{B_k} |X|\, dP \le (k+1) P(B_k).$$

Hence

$$\sum_{k=1}^{\infty} k P(B_k) \le \int |X|\, dP \le 1 + \sum_{k=1}^{\infty} k P(B_k).$$

Since

$$\sum_{k=1}^{\infty} k P(B_k) = \sum_{k=1}^{\infty} \sum_{n=1}^{\infty} P(B_k) = \sum_{n=1}^{\infty} \sum_{k=1}^{\infty} P(B_k) = \sum_{n=1}^{\infty} P\{|X| \ge n\},$$

our contention is proved.

9. Let $D = \{x_k\}_1^{\infty}$ be the set of discontinuity points of a distribution function $F(x)$. Define by

$$M(x) = \sum_{x_i \le x} d(x_i) \qquad \text{where} \quad d(x_i) = F(x_i) - F(x_i - 0).$$

Show that $H(x) = F(x) - M(x)$ is nondecreasing and continuous.

Hint:

It is clear that $M(\cdot)$ is a nondecreasing step function continuous from the right. Thus, H is continuous from the right. Consequently, for any $\varepsilon>0$ and $x>R$

$$H(x) - H(x-\varepsilon) = (F(x)-F(x-\varepsilon)) - (M(x)-M(x-\varepsilon)) \geq 0.$$

Thus, H is nondecreasing and since

$$\lim_{\varepsilon \to 0} (H(x)-F(x-\varepsilon)) = 0 \text{ the assertion follows.}$$

10. Let $\{X_n\}_1^\infty$ be a sequence of random variables on a probability space $(\Omega, \mathcal{B}, P)$. If for every $\varepsilon>0$

$$\sum_{n=1}^\infty P\{|X_n|>\varepsilon\} < \infty, \quad \text{show that } X\_\to 0 \text{ (a.e).}$$

Show also that the converse does not hold in general.

Hint:

See corollary 1 of section 26. To prove the second part take $\Omega=[0, 1)$ and $P=\bar{\lambda}$ (the Lebesgue measure). Define

$$X_n(x) = \begin{cases} 0 & \text{if} \quad 0 \leq x \leq 1-\dfrac{1}{n} \\ 1 & \text{if} \quad 1-\dfrac{1}{n} < x \leq 1 \end{cases}.$$

Clearly, $X_n(x) \to 0$ as $n \to \infty$ for every $x \in [0, 1]$. However, for any $\varepsilon>0$ sufficiently small

$$P\{X_n>\varepsilon\} = \frac{1}{n} \Rightarrow \sum_{n=1}^\infty P\{X_n>\varepsilon\} = +\infty.$$

11 Let X be a random variable on $(\Omega, \mathcal{B}, P)$ such that for some $t>0$, $E(|X|^t)<\infty$. Show that for any $0<s<t$,

$$(E(|X|^s))^{1/s} \leq (E(|X|^t))^{1/t} \qquad \text{(Liapunov).}$$

Hint:

Set $u=t/s$ and $Z=|X|^s$. Since $\varphi(x)=|x|^u$ is a convex function (see (6) of section 36), an application of the Jensen inequality (see (8) of section 35) yields

$$(E|Z|)^u \leq E(|Z|^u).$$

43. THE KOLMOGOROV CONSISTENCY THEOREM

Let $(\Omega, \mathscr{B})$ be the product of measurable spaces $\{(X_t, \mathscr{K}_t); t \in T\}$. Recall that any $x \in \Omega$ is a map

$$x : T \to \bigcup_T X_t \quad \text{such that} \quad x(t) \in X_t$$

for every $t \in T$. Let $\mathscr{T}$ be the collection of all finite subsets of T. Given any $S \in \mathscr{T}$ we shall, by an abuse of notation, identify the measurable cylinder sets with their basis in $\prod_S X_t$, with their basis. Thus

$$\underset{S}{\otimes} \mathscr{K}_t \subset \mathscr{B}$$

is the sub-σ-algebra of measurable cylinder sets with basis in $\prod_S X_t$.

Let P be a probability measure on $(\Omega, \mathscr{B})$; we denote by P_S the restriction of P to $\underset{S}{\otimes} \mathscr{K}_t$, i.e.,

$$P_S = P \mid \underset{S}{\otimes} \mathscr{K}_t. \tag{1}$$

Clearly, P_S is a probability measure on the measurable space $(\prod_S X_t, \underset{S}{\otimes} \mathscr{K}_t)$, where for any $B \in \underset{S}{\otimes} \mathscr{K}_t$

$$P_S(B) = P(B \times \prod_{T-S} X_t).$$

If $S_1, S_2 \in \mathscr{T}$ and $S_1 \subset S_2$, it is clear that

$$P_{S_1} = P_{S_2} \mid \underset{S_1}{\otimes} \mathscr{K}_t. \tag{2}$$

Thus the probability measure P on $(\Omega, \mathscr{B})$ generates a unique family $\{P_S ; S \in \mathscr{T}\}$ of probability measures (the finite dimensional distributions of P) **satisfying the consistency condition (2).**

Is the converse true? In other words, given a family of finite dimensional distributions $\{P_S ; S \in \mathscr{T}\}$ on the measurable spaces $(\prod_S X_t, \underset{S}{\otimes} \mathscr{K}_t)$, satisfying the consistency condition (2), is there a probability measure P on $(\prod_T X_t, \underset{T}{\otimes} \mathscr{K}_t)$, with $\{P_S ; S \in \mathscr{T}\}$ as the marginal distributions? In some special cases (see proposition 3 of section 40) the answer is affirmative. In general, however, the best what we can do is the following:

Proposition 1

Given a family $\{P_S ; S \in \mathscr{T}\}$ of probability measures on measurable spaces $(\prod_S X_t, \underset{S}{\otimes} \mathscr{K}_t)$ satisfying the consistency condition (2). There exists an additive set function $\bar{P}$ on $(\Omega, \mathscr{U})$ (here $\mathscr{U}$ is the algebra defined by

$$\mathscr{U} = \underset{S \in \mathscr{T}}{\bigcup} \underset{S}{\otimes} \mathscr{K}_t),$$

such that

$$\overline{P}\,|\,\underset{S}{\otimes}\,\mathscr{K} = P_S\,.$$

Proof:

Since $\overline{P}$ is an extension of each P_S, it is defined on the semialgebra $\mathscr{S}^*$ of measurable rectangles as follows: Since any $M \in \mathscr{S}^*$ is in the form

$$M = A \times \prod_{T\text{-}S} X_t \quad \text{for some } S \in \mathscr{T}, \quad \overline{P}(M) = P_S(A). \tag{3}$$

The set function $\overline{P}$ is well defined (see proposition 1 of section 40) and (finitely) additive on $\mathscr{S}^*$. To prove this let $\{M_i\}_1^n$ be a disjoint sequence form $\mathscr{S}^*$ such that $\bigcup_1^n M_i \in \mathscr{S}^*$. Since each M_i is a measurable rectangle, it can be written in the form

$$M_i = A_i \times \prod_{T\text{-}S_i} X_t = B_i \times \prod_{T-\bigcup_1^n S_i} X_t\,.$$

From this we have than that

$$\overline{P}(\bigcup_1^n M_i) = \overline{P}(\bigcup_1^n (B_i \times \prod_{T-\bigcup_1^n S_i} X_t)$$

$$= \overline{P}((\bigcup_1^n B_i) \times \prod_{T-\bigcup_1^n S_i} X_t) = P_{\bigcup_1^n S_i}(\bigcup_1^n B_i)$$

$$= \sum_1^n P_{\bigcup_1^n S_i}(B_i) = \sum_1^n P_{S_i}(A_i) = \sum_1^n \overline{P}(M_i).$$

This proves the proposition. ♥

To summarize, we have established that the set function $\overline{P}$ defined by (3) is finitely additive on $\mathscr{S}^*$. The consistency condition (2) does not imply that $\overline{P}$ is σ-additive on $\mathscr{S}^*$. However, if $\Omega = R^T$, where $R = (-\infty, \infty)$, the set function $\overline{P}$ is a measure on $\mathscr{S}^*$. The Kolmogorov proof of this fact requires an auxiliary result which we will discuss first.

Let $(R^n, \mathscr{R}^n)$ be a n-dimensional Euclidian measurable space. Recall (see problem 12 of section 29) that every probability measure P_n on it is "regular", i.e., given any $B \in \mathscr{R}^n$ and $\varepsilon > 0$, there exists a closed set F_ε and an open set O_ε such that

$$F_\varepsilon \subset B \subset O_\varepsilon \quad \text{and} \quad P_n(O_\varepsilon - F_\varepsilon) \le \varepsilon.$$

Proposition 2

Given $B \in \mathscr{R}^n$ and $\varepsilon > 0$, there exists a compact subset $K_\varepsilon \subset B$ such that

$$P_n(B - K_\varepsilon) \le \varepsilon.$$

Proof:

Recall that every probability measure on $(R^n, \mathscr{R}^n)$ is "tight", i.e., given $\varepsilon > 0$ there exists a

compact subset $K_\varepsilon \subset R^n$ such that $P_n(R^n - K_\varepsilon) \le \varepsilon$. For instance, if

$$K_m = \{x \in R^n; \ \|x\| \le m\}, \quad m = 1, 2, \cdots,$$

then since $K_m \uparrow R^n$ as $m \to \infty$, given any $\varepsilon > 0$ there exists an integer $i > 0$ such that $P_n(R^n - K_i) \le \varepsilon/2$.

Since P_n is regular, given any $B \in \mathscr{R}^n$ there exists a closed set $F_\varepsilon \subset B$ such that $P_n(B - F_\varepsilon) \le \varepsilon/2$. Since

$\overline{K_i} = K_i \cap F_\varepsilon$ is also compact and

$$P_n(B - \overline{K_i}) \le P_n(B - K_i) + P_n(B - F_\varepsilon) \le \varepsilon,$$

the assertion follows. ♥

Proposition 3 (Kolmogorov)

Let $\{P_S; \ S \in \mathscr{T}\}$ be a consistent family of finite dimensional distributions on the measurable spaces

$(R^S, \mathscr{R}^S)$. There exist a unique probability measure $\overline{P}$ (which is an extension of each P_S) on $(\Omega, \mathscr{B})$

such that

$$\overline{P} \mid \mathscr{R}^S = P_S \quad \text{for every } \ S \in \mathscr{T}.$$

Proof:

Here, of course, $\Omega = R^T$ and $\mathscr{B} = \sigma\{S^*\}$. Let $\overline{P}$ be the set function defined on S^* as in proposition

1. It is clearly well defined and additive on S^*. We contend that $\overline{P}$ is σ-additive, i.e., a probability measure

on S^*. To prove this we will have to show that $\overline{P}$ is continuous at θ.

Our proof is by contradiction. Assume that there exists a sequence of sets $\{M_n\}_1^\infty$ from S^* such

that $M_n \downarrow \theta$ and $\overline{P}(M_n) \ge \delta$ for some fixed $\delta > 0$ and all $n = 1, 2, \cdots$. By demonstrating that this leads to a

contradiction will prove the proposition. To this end consider a sequence of sets $\{T_k\}_1^\infty$ from $\mathscr{T}$ such that

$$T_1 \subset T_2 \subset \cdots.$$

Let $A_j \in \mathscr{R}^{T_j}$ be a rectangle and let set $M_j \in \mathcal{S}^*$ be defined by

$$M_j = \{x \in \Omega;\ x|T_j \in A_j\} = A_j \times R^{T-T_j}.$$

According to proposition 3, there exists compact subset $K_j \subset A_j$ such that

$$P_{T_j}(A_j - K_j) \le \delta/2^{j+1}.$$

Set

$$F_j = K_j \times R^{T-T_j},$$

then we have that

$$\overline{P}(M_j - F_j) = P_{T_j}(A_j - K_j) \le \delta/2^{j+1}.$$

Let $\{\overline{F}_k\}_1^\infty$ be a sequence defined by

$$\overline{F}_k = \bigcap_1^k F_j, \qquad (\overline{F}_1 \supset \overline{F}_2 \supset \cdots)$$

then, since

$$\overline{F}_n \subset F_n \subset M_n \qquad \text{and} \qquad \overline{P}(M_n) \ge \delta,\ n=1, 2, \cdots,$$

we have that

$$\overline{P}(\overline{F}_n) = \overline{P}(\bigcap_1^n (M_j - (M_j - F_j))) = \overline{P}(\bigcap_1^n M_j - \bigcup_1^n (M_j - F_j))$$

$$\ge \overline{P}(\bigcap_1^n M_j) - \overline{P}(\bigcup_1^n (M_j - F_j)) \ge \overline{P}(M_n) - \frac{\delta}{2} \sum_1^\infty \frac{1}{2^j} \ge \frac{\delta}{2},$$

because $\overline{P}(M_n) \ge \delta$ for all $n=1, 2, \cdots$. Hence,

$$\overline{P}(\overline{F}_n) \ge \frac{\delta}{2} \implies \overline{F}_n \ne \theta, \quad n=1, 2, \cdots.$$

Let us show that this contradicts assumption that $\overline{F}_n \downarrow \theta$.

Since the elements of the semialgebra $\mathcal{S}^*$ are rectangles and thus depend only on finitely many indices we may, without any loss of generality, assume that $T=N_+ = \{1, 2, \cdots\}$, and that each

$$\overline{F}_j = \{x, (x_1, \cdots x_j) \in K_j\},$$

where $K_j \subset R^j$ is compact.

Take a point $x^{(n)} \in \bar{F}_n$. Since $\bar{F}_1 \supset \bar{F}_2 \supset \cdots$ it is clear that $x^{(n)} \in \bar{F}_j$ for all $j \leq n$. Thus for any $p = 0, 1, \cdots$, $x^{(n+p)} \in \bar{F}_n$. Clearly then

$$(x_1^{(n+p)}, \cdots, x_n^{(n+p)}) \in K_n.$$

Consider the sequence $\{x^{(n)}\}_1^\infty$, where $x^{(n)} = (x_1^{(n)}, x_2^{(n)}, \cdots)$. Since each $K_n \subset R^n$ is compact there is a subsequence $\{x^{(n_{1k})}\}_1^\infty$ of $\{x^{(n)}\}_1^\infty$ such that

$$x_1^{(n_{1k})} \to \bar{x}_1 \qquad \text{as} \ \ k \to \infty.$$

From the sequence $\{x^{(n_{1k})}\}_1^\infty$ we can choose a subsequence $\{x^{(n_{2k})}\}$ such that $x_2^{(n_{2k})} \to \bar{x}_2$ as $k \to \infty$, and so on. The main diagonal of the matrix $[x^{n_{ij}}]$ is the sequence $\{x^{(n_{ii})}\}_1^\infty$, where

$$x^{(n_{ii})} = (x_1^{(n_{ii})}, x_2^{(n_{ii})}, \cdots).$$

Clearly,

$$x^{(n_{ii})} \to \bar{x} = (\bar{x}_1, \bar{x}_2, \cdots) \qquad \text{as} \ i \to \infty.$$

But $(x_1^{(n_{ii})}, \cdots, x_k^{(n_{ii})}) \to (\bar{x}_1, \cdots, \bar{x}_k) \in K_k$ for every $k = 1, 2, \cdots$. Therefore,

$$\bar{x} \in \bar{F}_n \subset F_n \qquad \text{for all} \ n = 1, 2, \cdots,$$

which implies that

$$\bar{x} \in \bigcap_1^\infty F_n,$$

and that establishes the desired contradiction. ♥

To summarize, starting with a consistent family of finite dimensional distributions

$$\{P_{T_0}; T_0 \in \mathcal{T}\},$$

on the measurable spaces $(R^S, \mathcal{R}^S)$, we have constructed a probability measure P on the function space $\Omega = R^T$. One of the problems with this construction is that for T uncountable, the space Ω is too big while the σ-algebra $\mathcal{B}$ is too small. As a result, many interesting subsets of Ω may not be measurable. For instance, if $\{A_t; t \in T\}$ is a family from $\mathcal{B}$, the set

$$\sup_T A_t = \bigcup_T A_t$$

is not necessarily measurable. If $T=[0, 1]$ the set $C(T) \subset R^T$ of all continuous functions defined on T is not measurable either, i,e., $C(T) \notin \mathscr{B}$, because

$$C(T) = \bigcap_{n=1}^{\infty} \bigcup_{k=1}^{\infty} \{\omega;\ \sup_{|t-s| \leq \frac{1}{k}} |\omega(t) - \omega(s)| \leq \frac{1}{n} \}.$$

Since each $\omega \in C(T)$ is uniformly continuous on $T=[0, 1]$, the space $C(T)$ with the supremum norm

$$\|\omega\| = \sup_{0 \leq t \leq 1} |\omega(t)|,$$

is a complete separable metric space. The map $\|\cdot\| : C(T) \to [0, \infty)$ is continuous since for any $\omega_1, \omega_2 \in C(T)$

$$|\ \|\omega_1\| - \|\omega_2\|\ | \leq \|\omega_1 - \omega_2\|.$$

The projection map $p_t : C(T) \to R$ defined by

$$p_t(\omega) = \omega(t), \quad t \in [0, 1]$$

is also continuous. Let $\mathscr{O}_{C(T)}$ be the class of all open subsets of $C(T)$, then

$$\mathscr{B}_{C(T)} = \sigma\{\mathscr{O}_{C(T)}\}$$

is the Borel algebra of subsets of $C(T)$. Denote by

$$\mathscr{A}_T = \sigma\{\bigcup_{t \in T} p_t^{-1}(\mathcal{R})\},$$

where $\mathcal{R}$ is the Borel algebra of subsets of R, then we have:

Proposition 4

$$\mathscr{B}_{C(T)} = \mathscr{A}_T.$$

Proof:

Let $\mathscr{O}$ be the class of all open subsets of R then, since $p_t(\cdot)$ is continuous for every $t \in T$, it is clear that $p_t^{-1}(\mathscr{O}) \subset \mathscr{O}_{C(T)}$ for every t. Consequently, $\mathscr{A}_T \subset \mathscr{B}_{C(T)}$. To prove the converse inclusion it suffices to show that $\mathscr{O}_{C(T)} \subset \mathscr{A}_T$. Since $C(T)$ is separable, every open set is a union of countably many closed spheres. Thus, it is enough to show that $\mathscr{A}_T$ contains all closed spheres. Let $r_1, r_2, \cdots$ be an enumeration of rationales in $[0, 1]$. Let $\omega_0 \in C(T)$ be fixed, then for any $x > 0$

$$\{\omega;\ \|\omega - \omega_0\| \leq x\} = \{\omega;\ \sup_{t \in T} |\omega(t) - \omega_0(t)| \leq x\}$$

$$= \{\omega;\ \sup_{1 \leq i < \infty} |\omega(r_i) - \omega_0(r_i)| \leq x\}$$

$$= \bigcap_{i=1}^{\infty} \{\omega;\ |\omega(r_i) - \omega_0(r_i)| \leq x\} \in \mathscr{A}_T,$$

which implies that $\mathscr{O}_{C(T)} \subset \mathscr{A}_T$. This proves the proposition.   ♥

Proposition 5

Let P_0 and P_1 be two measures on C(T). A necessary and sufficient condition that $P_0 = P_1$ is that for all $T_0 \in \mathscr{T}$

$$P_0 \mid \mathscr{R}^{T_0} = P_1 \mid \mathscr{R}^{T_0}.$$

Proof:

Denote by $S_{T_0} = p_{T_0}^{-1}(\mathscr{R}^{T_0})$ then the class $\mathscr{L} = \cup\{S_{T_0}; T_0 \in \mathscr{T}\}$ is an algebra such that $\sigma\{\mathscr{L}\} = \mathscr{A}_T$. Clearly,

$$P_0 \mid \mathscr{R}^{T_0} = P_1 \mid \mathscr{R}^{T_0} \quad \text{for all } T_0 \in \mathscr{T} \Rightarrow P_0(E) = P_1(E)$$

for all $E \in \mathscr{L}$, and consequently $P_0 = P_1$ on $\mathscr{A}_T$.

BIBLIOGRAPHY

1. Berberian, S. K. (1965). Measure and Integration, The Macmilan, New York.

2. Carathéodory, C. (1927). Varlesungen über reelle Functionen, 2 Auflage. Leipzing: Teubner

3. Dugundji, J. (1966). Topology, Allyn and Bacon, Boston.

4. Dunford, N. and Schwartz, J. T. (1958). Linear Operators, Part 1. General Theory. Inter-science, New York.

5. Halmos, P. R. (1950). Measure Theory. Van Nostrand, New York

6. Hausdorff, F. (1962). Set Theory, Chelsa, New York.

7. Kelly, J. L. (1955).General Topology. Van Nostrand.

8. Lebesgue, H. (1902). Intégral, longuer, aire. Annali di Mat. Pura Appl., Ser. 3, 7; 231 – 359.

9. Kolmogorov, A. N. (1956). Fundations of the Theory of Probability. Chelsa, New York.

10. Loéve, M. (1953). Probability Theory, 3rd ed., Van Nostrand, Princeton, N. J.

11. Munroe, M. E. (1953). Introduction to Measure and Integration. Addison-Wesley, Cambridge, Mass.

12. Natanson, I. P. (1964). Theory of Function of Real Variable, Ungar, New York.

13. Neveu, J. (1965) Mathematical Foundation of the Calculus of Probability. Holden-Day. San Francisco.

14, Riesz, F. and Sz.-Nagu, B.(1956) Functional Analysis (English ed.) New York, Ungar.

15. Royden, H. L. (1968). Real Analysis (second ed.), Macmillan, New York.

16. Saks, S. (1964). Theory of the Integral. 2nd ed. Dover, New York

17. Shiryayev, A. N. (1984). Probability, Springer-Verlag, New York.

18. Suppes, P. (1972). Axiomatic Set Theory. Dover Publications.

Index